César Camacho

Alcides Lins Neto

Geometric Theory of Foliations

Translated by Sue E. Goodman

BIRKHÄUSER
Boston · Basel · Stuttgart

Library of Congress Cataloging in Publication Data

Camacho, Cesar, 1943-
Geometric theory of foliations.

Translation of: Teoria geometrica das folheacoes.
Bibliography: p.
Includes index.
1. Foliations (Mathematics) 2. Geometry, Differential.
I. Lins Neto, Alcides, 1947- . II. Title.
QA613.62.C3613 1985 514′.73 84-10978
ISBN 0-8176-3139-9

CIP-Kurztitelaufnahme der Deutschen Bibliothek

Camacho, César:
Geometric theory of foliations/César Camacho;
Alcides Lins Neto. Transl. by Sue E. Goodman –
Boston; Basel; Stuttgart: Birkhäuser, 1985.

Einheitssacht.: Theoria geométrica das
folheações ⟨dt.⟩
ISBN 3-7643-3139-9

NE: Lins Neto, Alcides: 27

ISBN 0-8176-3139-9
ISBN 3-7643-3139-9

CONTENTS

INTRODUCTION

Intuitively, a foliation corresponds to a decomposition of a manifold into a union of connected, disjoint submanifolds of the same dimension, called leaves, which pile up locally like pages of a book.

The theory of foliations, as it is known, began with the work of C. Ehresmann and G. Reeb, in the 1940's; however, as Reeb has himself observed, already in the last century P. Painlevé saw the necessity of creating a geometric theory (of foliations) in order to better understand the problems in the study of solutions of holomorphic differential equations in the complex field.

The development of the theory of foliations was however provoked by the following question about the topology of manifolds proposed by H. Hopf in the 1930's: "Does there exist on the Euclidean sphere S^3 a completely integrable vector field, that is, a field X such that $X \cdot \operatorname{curl} X \equiv 0$?" By Frobenius' theorem, this question is equivalent to the following: "Does there exist on the sphere S^3 a two-dimensional foliation?"

This question was answered affirmatively by Reeb in his thesis, where he presents an example of a foliation of S^3 with the following characteristics: There exists one compact leaf homeomorphic to the two-dimensional torus, while the other leaves are homeomorphic to two-dimensional planes which accumulate asymptotically on the compact leaf. Further, the foliation is C^∞. Also in the work are proved the stability theorems, one of which, valid for any dimension, states that if a leaf is compact and has finite fundamental group then it has a neighborhood consisting of compact leaves with finite fundamental group. Reeb's thesis motivated the research of other mathematicians, among whom was A. Haefliger, who proved in his thesis in 1958, that there exist no analytic two-dimensional foliations on S^3. In fact Haefliger's theorem is true in higher dimensions.

The example of Reeb and others, which were constructed later, posed the following question, folkloric in the midst of mathematics: "Is it true that every foliation of dimension two on S^3 has a compact leaf?" This question was answered affirmatively by S. P. Novikov in 1965, using in part the methods introduced by Haefliger in his thesis. In fact Novikov's theorem is much stronger. It says that on any three-dimensional, compact simply connected manifold, there exists a compact leaf homeomorphic to the two-dimensional torus, bounding a solid torus, where the leaves are homeomorphic to two-dimensional planes which accumulate on the compact leaf, in the same way as in the Reeb foliation of S^3.

One presumes that the question initially proposed by Hopf, was motivated by the intuition that there must exist nonhomotopic invariants which would serve to classify three-dimensional manifolds. In fact this question did not succeed in this objective, since any three-dimensional manifold does admit a two-dimensional foliation. However, a refinement proposed by J. Milnor with the same motivation, had better results. In effect, Milnor defined the rank of a manifold as the maximum number of pairwise commutative vector fields, linearly independent at each point, which it is possible to construct on the manifold. This concept translates naturally in terms of foliations associated to actions of the group $\mathbb{R}^n$. The problem proposed by Milnor was to calculate the rank of S^3. This problem was solved by E. Lima in 1963 by showing that the rank of a compact, simply connected, three-dimensional manifold is one. Later H. Rosenberg, R. Roussarie and D. Weil classified the compact three-dimensional manifolds of rank two.

In this book we intend to present to the reader, in a systematic manner, the sequence of results mentioned above. The later development of the theory of foliations, has accelerated, especially in the last ten years. We hope that this book motivates the reading of works not treated here. Some of these are listed in the bibliography.

We wish to express here our appreciation to Airton Medeiros and Roberto Mendes for various suggestions and especially to Paulo Sad for his collaboration in the reading and criticism of the text.

Rio de Janeiro, May 1979

César Camacho
Alcides Lins Neto

Addendum to the English edition

This book is a translation of TEORIA GEOMÉTRICA DAS FOLHEAÇÕES, *no. 12* of the Series Projeto Euclides, published by IMPA – CNPq (BRAZIL). In this translation the arguments of some theorems were improved and an appendix about elementary properties of the fundamental group was suppressed. We wish to acknowledge Sue Goodman for the excellent work of translation.

Rio de Janeiro, March 1984

César Camacho
Alcides Lins Neto

I. DIFFERENTIABLE MANIFOLDS

In this chapter, we state the basics of the theory of differentiable manifolds and maps with the intention of establishing the principal theorems and notation which will be used in the book.

§1. Differentiable manifolds

Just as topological spaces form the natural domain of continuous functions, differentiable manifolds are the natural domain of differentiable maps. In order to better understand the definition of manifold, we begin by recalling some aspects of differential calculus.

A map $f: U \longrightarrow \mathbb{R}^n$ from an open set $U \subset \mathbb{R}^m$ to $\mathbb{R}^n$ is differentiable at $x \in U$ if there is a linear transformation $T: \mathbb{R}^m \longrightarrow \mathbb{R}^n$ which approximates f in a neighborhood of x in the following sense:

$$f(x + v) = f(x) + T \cdot v + R(v)$$

and

$$\lim_{v \to 0} \frac{R(v)}{|v|} = 0$$

for all sufficiently small $v \in \mathbb{R}^m$.

The map T, when it exists, is unique. It is called the derivative of f at x and is denoted $Df(x)$.

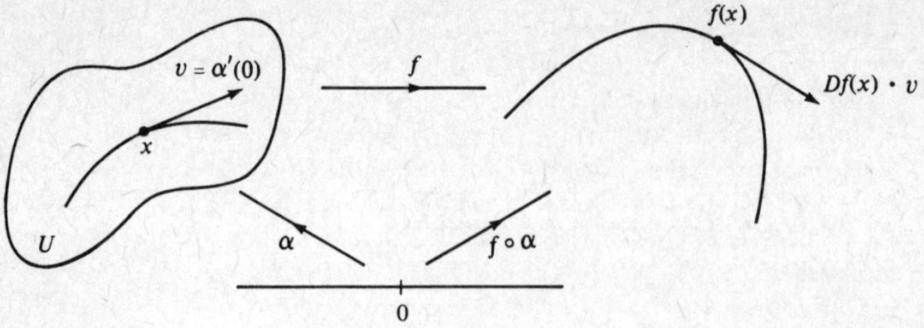

Figure 1

The derivative has the following geometric interpretation. Given $v \in \mathbb{R}^m$ we take a differentiable curve $\alpha : I \longrightarrow U$ defined on an open interval $I \subseteq \mathbb{R}$ containing $0 \in \mathbb{R}$ such that $\alpha(0) = x$ and $\alpha'(0) = v$. Then

$$Df(x) \cdot v = \lim_{t \to 0} \frac{f(\alpha(t)) - f(x)}{t} = \frac{d}{dt} f(\alpha(t)) \Big|_{t=0} .$$

Fixing the canonical basis $\{e_1, \ldots, e_m\}$ of $\mathbb{R}^m$, we define the partial derivatives of f at x by $(\partial f/\partial x_i)(x) = Df(x) \cdot e_i$. In this manner we have that for each vector $v = \sum_{i=1}^m \alpha_i e_i$ in $\mathbb{R}^m$

$$Df(x) \cdot v = \sum_{i=1}^{m} \alpha_i Df(x) \cdot e_i = \sum_{i=1}^{m} \frac{\partial f}{\partial x_i}(x) \cdot \alpha_i .$$

We say that f is of class C^1 on U when all the partial derivatives $(\partial f/\partial x_i)(x)$ are continuous as functions of $x \in U$. Proceeding inductively on r, f is of class C^r when all the partial derivatives of f are of class C^{r-1} on U. When f is of class C^r for all $r \in \mathbb{N}$, we say f is of class C^∞.

The differentiability of the composition of two maps is determined by the chain rule which says that if $f : U \longrightarrow \mathbb{R}^n$ and $g : V \longrightarrow \mathbb{R}^k$ are of class C^r and $f(U) \subset V$ then $g \circ f : U \longrightarrow \mathbb{R}^k$ is of class C^r and $D(g \circ f)(x) = Dg(f(x)) \cdot Df(x)$.

A map of class C^r, $r \geq 1$, $f : U \longrightarrow V = f(U)$ between open sets U, V in $\mathbb{R}^m$ is called a C^r diffeomorphism if f possesses an inverse $f^{-1} : V \longrightarrow U$ of class C^r. In particular, a diffeomorphism is a homeomorphism. Moreover, for each $x \in U$, $Df(x) : \mathbb{R}^m \longrightarrow \mathbb{R}^m$ is an isomorphism and $(Df(x))^{-1} = Df^{-1}(f(x))$.

Let $f : W \longrightarrow \mathbb{R}^m$, $W \subseteq \mathbb{R}^m$ be an open set. We say that f is a local diffeomorphism when for each $p \in W$ there exists a neighborhood $U \subset W$ of p such that $f \mid U : U \longrightarrow f(U) \subset \mathbb{R}^m$ is a diffeomorphism.

The notion of differentiability, which until now was associated with maps defined on open sets of Euclidean spaces, will be extended next to maps defined on certain topological spaces locally homeomorphic to $\mathbb{R}^m$.

With this objective we define a *local chart* or a *system of coordinates* on a topological space M as a pair (U,φ) where U is an open set in M and $\varphi : U \longrightarrow \mathbb{R}^m$ is a homeomorphism from U to an open set $\varphi(U)$ in $\mathbb{R}^m$. An atlas $\mathcal{A}$ of dimension m is a collection of local charts whose domains cover M and such that if (U,φ), $(\tilde{U},\tilde{\varphi}) \in \mathcal{A}$ and $U \cap \tilde{U} \neq \emptyset$ then the map $\tilde{\varphi} \circ \varphi^{-1} : \varphi(U \cap \tilde{U}) \longrightarrow \tilde{\varphi}(U \cap \tilde{U})$ is a C^r diffeomorphism between open sets of $\mathbb{R}^m$. The diffeomorphisms $\tilde{\varphi} \circ \varphi^{-1}$ as above are called *changes of coordinates*.

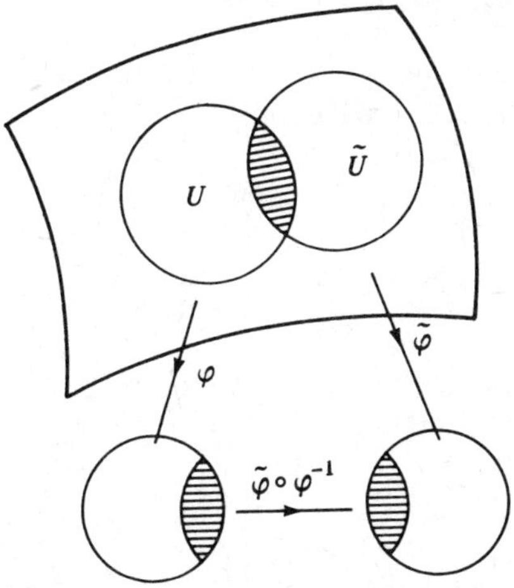

Figure 2

The concept of differentiability can now be extended to maps between topological spaces which possess an atlas of class C^r, $r \geq 1$.

Indeed, let M and N be topological spaces and suppose that $\mathcal{A}$ and $\mathcal{B}$ are atlases of class C^r on M and N respectively. We say that $f : M \longrightarrow N$ is differentiable of class C^k, $k \leq r$, if f is continuous and for each $x \in M$ there exist local charts $(U,\varphi) \in \mathcal{A}$ and $(V,\psi) \in \mathcal{B}$ with $x \in U$ and $f(x) \in V$ such that

$$\psi \circ f \circ \varphi^{-1} : \varphi(U \cap f^{-1}(V)) \subset \mathbb{R}^m \longrightarrow \psi(V) \subset \mathbb{R}^m$$

is of class C^k.

Since the changes of variable are C^r diffeomorphisms, $r \geq k$, this definition is independent of the local charts chosen. It is clear that in the Euclidean spaces $\mathbb{R}^m$, if we consider the single local chart ($\mathbb{R}^m$, identity map), the notion of differentiability given above coincides with the usual one.

An atlas $\mathcal{A}$ of class C^r on M is called maximal when it contains all the local charts $(\tilde{U},\tilde{\varphi})$ whose changes of coordinates with elements $(U,\varphi) \in \mathcal{A}$

$$\tilde{\varphi} \circ \varphi^{-1} : \varphi(U \cap \tilde{U}) \longrightarrow \tilde{\varphi}(U \cap \tilde{U}) \qquad (*)$$

are C^r diffeomorphisms. The advantage of considering a maximal atlas $\mathcal{A}$ is that in this case the domains of local charts of $\mathcal{A}$ form a topological basis of M. On the other hand, each atlas $\mathcal{A}$ is contained in a unique maximal atlas $\overline{\mathcal{A}}$. Indeed, $\overline{\mathcal{A}}$ is defined as the union of all the local charts $(\tilde{U}, \tilde{\varphi})$ such that if $(U,\varphi) \in \mathcal{A}$ and $U \cap \tilde{U} \neq \varnothing$ then the changes of coordinates $(*)$ are of class C^r.

A maximal atlas of dimension m and class C^r on M is also called a *differentiable structure of dimension m and class C^r on M*. It is now clear that a continuous map $f : M \longrightarrow N$ between spaces M and N with C^r differentiable structures is of class C^k, $k \leq r$, if and only if for each $x \in M$ there exist local charts (U,φ), (V,ψ) on M and N such that $x \in U$, $f(U) \subset V$ and $\psi \circ f \circ \varphi^{-1}$: $\varphi(U) \longrightarrow \psi(V) \subset \mathbb{R}^n$ is of class C^k.

A *differentiable manifold of class C^r and dimension m* is a Hausdorff topological space M, with a countable basis, provided with a differentiable structure of dimension m and class C^r.

To denote a manifold M of dimension m we will at times use the notation M^m.

Unless otherwise specified, all manifolds considered hereafter will be of class C^∞.

A map $f : M \longrightarrow N$ of class C^r, $r \geq 1$, is a diffeomorphism when it possesses an inverse $f^{-1} : N \longrightarrow M$ of class C^r. In this case we say that the manifolds are diffeomorphic and we write $M \simeq N$.

Example 1. The sphere S^m.

The sphere S^m defined by

$$S^m = \{x = (x_1, \dots, x_{m+1}) \in \mathbb{R}^{m+1} \mid x_1^2 + \dots + x_{m+1}^2 = 1\} ,$$

with the topology induced from $\mathbb{R}^{m+1}$ possesses a C^∞ manifold structure of dimension m.

Indeed, let $U_i^\sigma \subset S^m$, $\sigma = \pm 1$, $1 \leq i \leq m+1$, be the set $U_i^\sigma = \{x \in S^m \mid \sigma x_i > 0\}$. The local charts $\varphi_i^\sigma : U_i^\sigma \longrightarrow \mathbb{R}^m$ defined by $\varphi_i^\sigma(x_1, \dots, x_{i-1}, x_i, x_{i+1}, \dots, x_{m+1}) = (x_1, \dots, x_{i-1}, x_{i+1}, \dots, x_{m+1})$ cover S^m and the changes of coordinates are C^∞.

Example 2. Real projective spaces.

Let $\mathbb{P}^n$, $n \geq 1$, be the set of lines of $\mathbb{R}^{n+1}$ which pass through the origin. $\mathbb{P}^n$ can be considered as the quotient space of $\mathbb{R}^{n+1} - \{0\}$ with the equivalence relation which identifies two vectors $u, v \neq 0$ if $u = tv$, where $t \in \mathbb{R} - \{0\}$. We consider $\mathbb{P}^n$ with the quotient topology, that is, $U \subset \mathbb{P}^n$ is open if $\pi^{-1}(U)$ is open in $\mathbb{R}^{n+1} - \{0\}$, where $\pi : \mathbb{R}^{n+1} - \{0\} \longrightarrow \mathbb{P}^n$ is the quotient map which associates to each element $u \in \mathbb{R}^{n+1} - \{0\}$ its equivalence class $\pi[u]$. The space $\mathbb{P}^n$ with this topology is compact and Hausdorff and it is called the real projective space of dimension n.

Let us see how one constructs a C^∞ atlas on $\mathbb{P}^n$. The equivalence class $[x_1, \dots, x_{n+1}]$ of the vector $(x_1, \dots, x_{n+1})$ can be thought of as being the line in $\mathbb{R}^{n+1}$ which passes through the origin and through the point $(x_1, \dots, x_{n+1}) \neq 0$. For each $i \in \{1, \dots, n+1\}$, let $U_i = \{[x_1, \dots, x_{n+1}] \in \mathbb{P}^n \mid x_i \neq 0\}$. We define $\varphi_i : U_i \longrightarrow \mathbb{R}^n$ by $\varphi_i[x_1, \dots, x_i, \dots, x_{n+1}] = (1/x_i)(x_1, \dots, x_{i-1}, x_{i+1}, \dots, x_{n+1})$.

It is easily shown that U_i is an open subset of $\mathbb{P}^n$, that φ_i is a homeomorphism and that $\bigcup_{i=1}^{n+1} U_i = \mathbb{P}^n$. Moreover, if $i, j \in \{1, \dots, n+1\}$ with $i < j$, we have that $U_i \cap U_j \neq \varnothing$ and

$$\varphi_i \circ \varphi_j^{-1} : \varphi_j(U_i \cap U_j) \longrightarrow \varphi_i(U_i \cap U_j)$$

is given by the expression

$$\varphi_i \circ \varphi_j^{-1}(y_1, \dots, y_n) = \frac{1}{y_i}(y_1, \dots, y_{i-1}, y_{i+1}, \dots, y_{j-1}, 1, y_j, \dots, y_n).$$

Then $\{(U_i, \varphi_i) \mid i = 1, \dots, n+1\}$ is a C^∞ atlas on $\mathbb{P}^n$.

Example 3. Classification of manifolds.

A natural problem in the theory of differentiable manifolds is the problem of classification. We will say that two manifolds M and N are equivalent if they are homeomorphic. The problem then consists of determining explicitly a representative of each equivalence class.

For example it is very easy to show that if M has dimension 1 and is connected, then M is homeomorphic to S^1 or to $\mathbb{R}$ (see exercise 4). In the case of dimension 2 the problem is already complicated, since its solution implies in particular a classification, modulo homeomorphisms, of all open subsets of $\mathbb{R}^2$. However if we restrict to compact connected manifolds of dimension 2 a classification is possible. One proves ([32] and [61]) that every compact connected surface is homeomorphic to the quotient of a polygon by an equivalence relation that pairwise identifies the sides of the polygon according to the following rules.

ORIENTABLE SURFACES

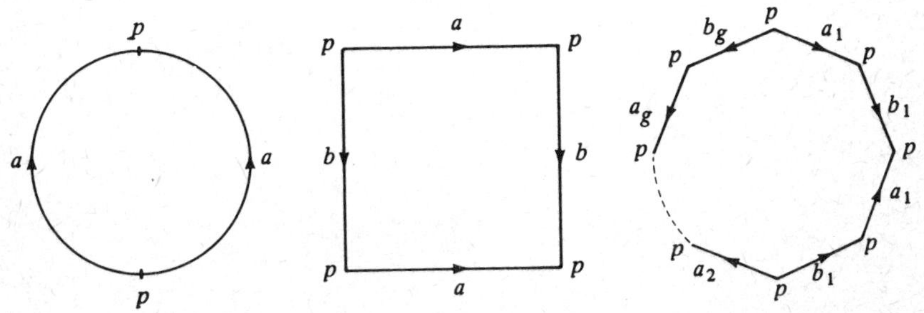

sphere

torus

surface of genus g or g-torus

Figure 3

NONORIENTABLE SURFACES

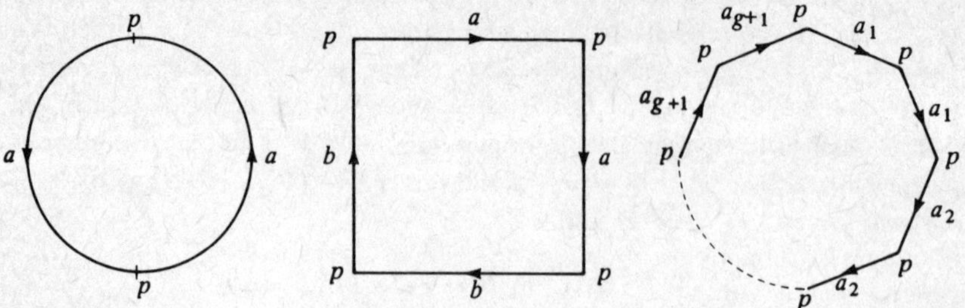

projective plane | Klein bottle | nonorientable surface of genus g

Figure 4

In figures 3 and 4 the sides with the same letter are identified to preserve the sense of the arrows.

In the sequence of figures below we illustrate how a surface of genus 2 can be obtained by identifications from an octagon.

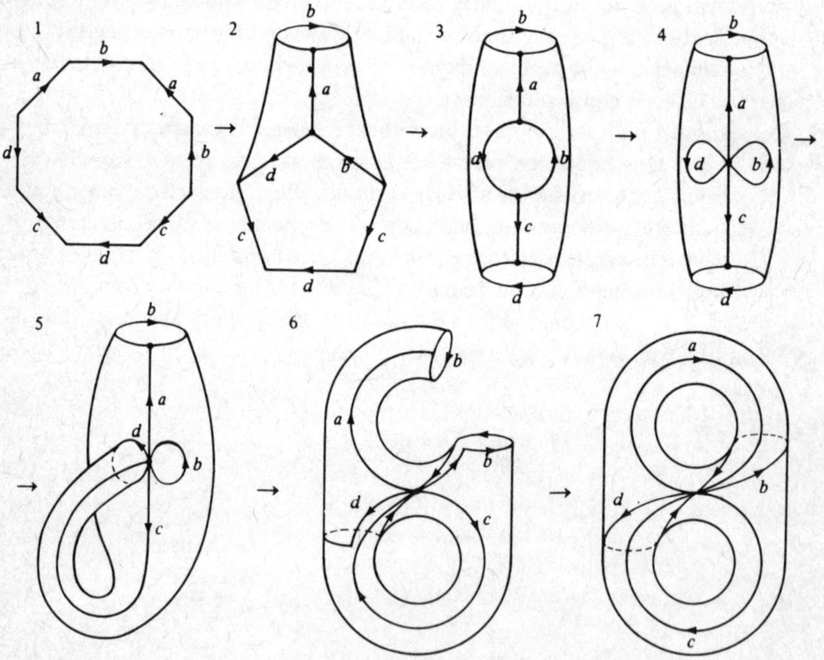

Figure 5

In figure 5, we first identify sides a, followed by sides c. Observing that the vertices of the octagon are all identified to the same point, in the fourth figure we identify the common endpoints of sides b and d that in the third figure form the boundary of a hole. In the fifth figure we identify the sides d. In the sixth figure we change the position of the figure, deforming it a little. Finally in the seventh figure we see the final position of the curves a, b, c, d on the surface of genus two.

Example 4. Manifolds of dimension 3.

The problem of classification of compact manifolds of dimension 3 is still open. Even a classification of those that are simply connected is not known. However some useful properties of manifolds of dimension 3 can be obtained in an elementary manner. For example one can show that any compact connected manifold of dimension 3 can be decomposed into the union of two handlebodies bounded by a surface of genus g in $\mathbb{R}^3$.

This can be shown using the fact that such manifolds can be triangulated, i.e., decomposed into a union of tetrahedra such that any two of them intersect along faces, edges or vertices [51]. The union of all the edges and vertices form a connected set of dimension one which will be the "skeleton" of one of the two handlebodies T_1. The complement of T_1 is a handlebody T_2 whose skeleton is formed by the edges and vertices of a "triangulation" dual to the first. The dual triangulation is obtained in the following manner. In the center of each tetrahedron put a vertex. The vertices that are in adjacent tetrahedra are joined by an edge which cuts the common face in one point. The faces are taken so that they intercept each edge of the first triangulation in only one point. In this manner we obtain a decomposition of the manifold into polyhedra (not necessarily tetrahedra). For more details see [51].

Example 5. Non-Hausdorff manifolds.

In the definition of manifold it was required that every differentiable manifold is a Hausdorff space. Nevertheless there exist spaces that satisfy all the axioms of the definition of manifold except that of Hausdorff. Such spaces are called non-Hausdorff manifolds and they occur naturally in certain discussions. Let us look at an example.

Let $f : \mathbb{R}^2 \longrightarrow \mathbb{R}$ be defined by $f(x,y) = \alpha(x^2)e^y$ where $\alpha : \mathbb{R} \longrightarrow \mathbb{R}$ is a function class C^∞ such that $\alpha(t) = 1$ if $t \in (-\epsilon,\epsilon)$, $\alpha(1) = 0$ and $\alpha'(t) < 0$ for $t > \epsilon$.

It is easy to see that the level curves of $f, f^{-1}(t)$, t fixed, have the following form:

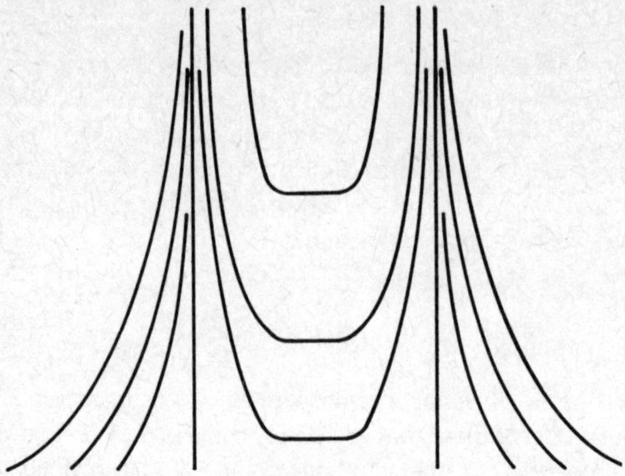

Figure 6

Each connected component of a level curve of f is called a leaf of f. Let M be the quotient space of $\mathbb{R}^2$ for the equivalence relation which identifies points on the same leaf of f. Let $\pi : \mathbb{R}^2 \longrightarrow M$ be the projection map for the quotient. The space M with the quotient topology is not Hausdorff since the points $a = \pi(1,t)$ and $b = \pi(-1,t)$, $t \in \mathbb{R}$, are points of ramification, that is, they do not admit disjoint neighborhoods of M. However if it is possible to define a C^∞ atlas on M of dimension 1.

In fact, the sets $U_1 = \{\pi(x,y) \in M \mid x < 1\}$ and $U_2 = \{\pi(x,y) \in M \mid x > -1\}$ are open in $M = U_1 \cup U_2$. Define $\varphi_i : U_i \longrightarrow \mathbb{R}$ by $\varphi_i(\pi(x,y)) = f(x,y)$, $i = 1,2$. It is easy to show that φ_1 and φ_2 are homeomorphisms, that $\varphi_i(U_1 \cap U_2) = (0, +\infty)$ and that $\varphi_1 \circ \varphi_2^{-1}$ is the identity on $(0, +\infty)$. Hence $\{(U_1,\varphi_1),(U_2,\varphi_2)\}$ is a C^∞ atlas for M.

Observe that M is homeomorphic to the quotient space of two disjoint copies of $\mathbb{R}$ for the equivalence relation which identifies points with the same negative coordinate.

Example 6. A manifold structure for covering spaces.

Let X and Y be locally path connected topological spaces, and Y be connected. We say that a continuous, surjective map $\pi : X \longrightarrow Y$ is a *covering map* if it satisfies the following properties:

(a) For each $y \in Y$ there exists a connected neighborhood U of y such that $\pi^{-1}(U) = \bigcup_{x \in \pi^{-1}(y)} V_x$ where for each $x \in \pi^{-1}(y)$ the restriction $\pi \mid V_x : V_x \longrightarrow U$ is a homeomorphism.

(b) If $x_1, x_2 \in \pi^{-1}(y)$ and $x_1 \neq x_2$ then $V_{x_1} \cap V_{x_2} = \varnothing$.

In particular the set $\pi^{-1}(y)$ is discrete and has no accumulation points. The neighborhood U is called a distinguished neighborhood of y and the set $\pi^{-1}(y)$ the fiber over y.

We will consider the following two situations which occur at various times throughout the text:

(A) Suppose that Y is a differentiable manifold. In this case it is possible to define a differentiable manifold structure on X so that $\dim(X) = \dim(Y)$ and π is a local C^∞ diffeomorphism.
Indeed let $x \in X$ and let $y = \pi(x)$. Consider neighborhoods V of x and U of y such that $\pi \mid V : V \longrightarrow U$ is a homeomorphism. Since Y is a manifold, let $\varphi : W \longrightarrow \mathbb{R}^n$ be a local chart such that $y \in W \subset U$. Take $V_1 = \pi^{-1}(W) \cap U$, which is open in X. Then the composition $\psi = \varphi \circ (\pi \mid V_1) : V_1 \longrightarrow \varphi(W)$ is a homeomorphism. Also the collection of charts (V_1, ψ), constructed as above, is a C^∞ atlas of dimension n on X, as can be easily shown. This structure is called the structure co-induced by π. It is the unique structure which makes π a local C^∞ diffeomorphism.

(B) In the case that X is a differentiable manifold (but not necessarily Y), with one more natural hypothesis, we can induce on Y a C^r differentiable manifold structure by means of π. With this structure π will be a C^r local diffeomorphism.

Hypothesis. *Suppose that given $y \in Y$ there exists a distinguished neighborhood U of y with the following property: if $x_1, x_2 \in \pi^{-1}(y)$ and V_1 and V_2 are neighborhoods of x_1 and x_2 such that $\pi \mid V_i : V_i \longrightarrow U$ $(i = 1,2)$ is a homeomorphism on U then $(\pi \mid V_2)^{-1} \circ (\pi \mid V_1) : V_1 \longrightarrow V_2$ is a C^r diffeomorphism $(r \geq 1)$.*

Construction of the atlas on Y. Given $y \in Y$, we fix U, a distinguished neighborhood as in the hypothesis. Let $x \in \pi^{-1}(y)$ and take V a neighborhood of x such that $\pi \mid V : V \longrightarrow U$ is a homeomorphism. Since X is a differentiable manifold, there exists a local chart $\varphi : W \longrightarrow \mathbb{R}^n$, with $x \in W \subset V$. Let $\psi = \varphi \circ (\pi \mid W)^{-1} : \pi(W) \longrightarrow \mathbb{R}^n$. Then ψ is a homeomorphism. By using the hypothesis it is possible to prove that the collection of all the maps (W, ψ) constructed as above, constitutes a C^r atlas of Y. We leave the details to the reader.

A specific example is the following. Let M be the quotient space of $\mathbb{R}^2$ for the equivalence relation $\sim$ where $(x,y) \sim (x',y')$ if and only if $(x - x', y - y') \in \mathbb{Z}^2$. Let $\pi : \mathbb{R}^2 \longrightarrow M$ be the quotient projection. It is easy to verify that π is a covering map and satisfies the hypothesis of B. Hence M, with the structure induced by π, is a C^∞ differentiable manifold of dimension 2. It is possible to show that M is diffeomorphic to the torus $T^2 = S^1 \times S^1$.

§2. The derivative

We introduce now the notion of derivative of a differentiable map between two manifolds, taking as a model the geometric interpretation of the derivative of a differentiable map in $\mathbb{R}^m$ given in the beginning.

Fixing $x \in M$, we consider the set $C_x(M)$ of all C^∞ curves, $\alpha : (-\epsilon, \epsilon) \longrightarrow M$ with $\epsilon > 0$ and $\alpha(0) = x$. Given a local chart $\varphi : U \longrightarrow \mathbb{R}^m$, $x \in U$, the curves $\alpha_u(t) = \varphi^{-1}(\varphi(x) + tu)$, $u \in \mathbb{R}^m$, are in $C_x(M)$. So $C_x(M) \neq \neq \emptyset$. We introduce the following equivalence relation on $C_x(M) : \alpha \sim \beta$ if for any local chart (U, φ), $x \in U$, we have $(d(\varphi \circ \alpha)/dt)(t)\big|_{t=0} = =(d(\varphi \circ \beta)/dt)(t)\big|_{t=0}$. Observe that if (V, ψ) is another local chart with $x \in V$, then $\psi \circ \alpha = (\psi \circ \varphi^{-1}) \circ (\varphi \circ \alpha)$, so by the chain rule, the relation $\sim$ is independent of the chart chosen.

The quotient of $C_x(M)$ by $\sim$ is called the *tangent space to* M *at* x and is denoted T_xM. The space T_xM has a natural structure of a real vector space of dimension m. Indeed, if $[\alpha]$ denotes the equivalence class of a curve $\alpha \in C_x(M)$, for $u = [\alpha]$, $v = [\beta]$ and $\lambda \in \mathbb{R}$, we define $u + v = = [\varphi^{-1}(\varphi \circ \alpha + \varphi \circ \beta)]$ and $\lambda u = [\varphi^{-1}(\lambda \cdot \varphi \circ \alpha)]$, where $\varphi : U \longrightarrow \mathbb{R}^m$ is a local chart such that $\varphi(x) = 0$. It is easy to show that these definitions are independent of the chart chosen and satisfy the axioms for a vector space. It is also true that if $\varphi : U \longrightarrow \mathbb{R}^m$ is a local chart with $\varphi(x) = 0$ and $\{e_1, \ldots, e_m\}$ is a basis for $\mathbb{R}^m$, then the set $\{[\varphi^{-1}(te_1)], \ldots, [\varphi^{-1}(te_m)]\}$ is a basis for T_xM, te_i being the curve $t \longmapsto te_i$, $t \in \mathbb{R}$. Therefore T_xM is a vector space of dimension m. When $\{e_1, \ldots, e_m\}$ is the canonical basis for $\mathbb{R}^m$ we also use the notation $[\varphi^{-1}(te_i)] = \partial/\partial x_i$, $i = 1, \ldots, m$.

Every differentiable map $f : M \longrightarrow N$ induces at each point $x \in M$ a linear map $Df(x) : T_x(M) \longrightarrow T_y(N)$, $y = f(x)$, in the following manner. Given $u = [\alpha] \in T_xM$ we take $Df(x) \cdot u = [f \circ \alpha]$, the equivalence class of the curve $f \circ \alpha$ in T_yN. It is easy to see that $Df(x)$ is well-defined and linear. It is called the *derivative of* f *at* x.

A differentiable manifold can be thought of as being a topological space that is locally Euclidean. Taking into account this pictorial image, a tangent vector to M at x is, essentially, the linear approximation of a differentiable curve which goes through x. The derivative of a differentiable function $f : M \longrightarrow N$ at $x \in M$ is a linear approximation of the function in a neighborhood of x. In practice, in order to work with differentiable functions, one should always have this intuitive idea in mind.

§3. Immersions and submersions

Given a C^r map $(r \geq 1)$, $f : M \longrightarrow N$, we say f is an *immersion* if for every $x \in M$, $Df(x) : T_xM \longrightarrow T_yN$, $y = f(x)$, is injective. We say f is a *submersion* if for every $x \in M$, $Df(x) : T_xM \longrightarrow T_yN$ is surjective.

An immersion $f : M \longrightarrow N$ is called an *imbedding* if $f : M \longrightarrow f(M) \subset N$ is a homeomorphism, when one gives $f(M)$ the topology induced by N. In particular, if $f : M \longrightarrow N$ is an imbedding, then f is a one-to-one immersion; however the converse is false, that is, there are one-to-one immersions which are not imbeddings, as is illustrated by the immersion in figure 7.

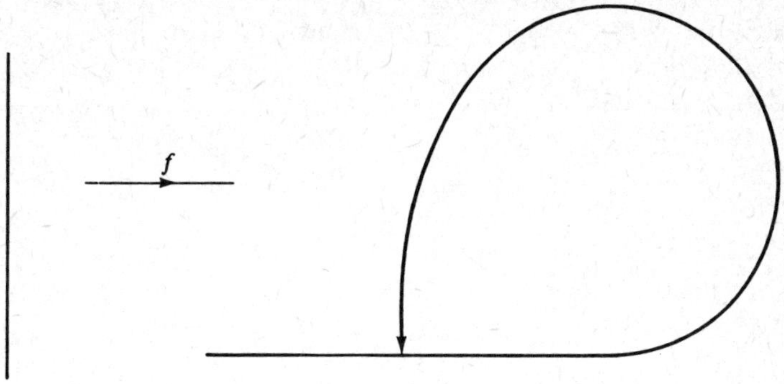

Figure 7

The following theorems follow from the inverse function theorem in $\mathbb{R}^m$.

Theorem *(Local form of immersions). Let $f: M^m \longrightarrow N^n$ be of class C^r $(r \geq 1)$. Suppose $Df(p) : T_pM \longrightarrow T_qN$, $q = f(p)$, is injective. Then there exist local charts $\varphi : U \longrightarrow \mathbb{R}^m$, $p \in U$, and $\psi : V \longrightarrow \mathbb{R}^n$, $q \in V$ and a decomposition $\mathbb{R}^n = \mathbb{R}^m \times \mathbb{R}^{n-m}$ such that $f(U) \subset V$ and f is expressed in the charts (U,φ), (V,ψ) as $\psi \circ f \circ \varphi^{-1}(x) = (x,0)$ if $x \in \varphi(U)$. In other words f is locally equivalent to the linear immersion $x \mapsto (x,0)$.*

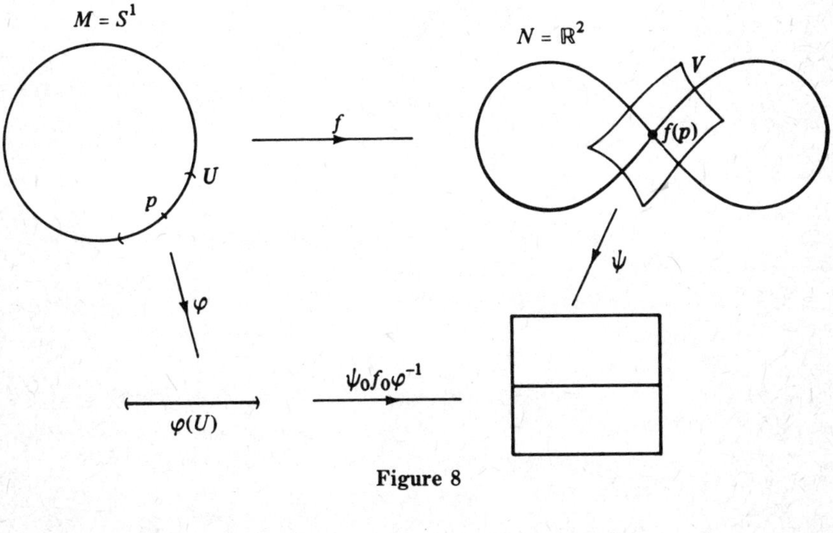

Figure 8

Theorem *(Local form of submersions). Let $f : M^m \to N^n$ be of class C^r $(r \geq 1)$. Suppose $Df(p) : T_pM \to T_qN$, $q = f(p)$, is surjective. Then there exist local charts $\varphi : U \to \mathbb{R}^m$, $p \in U$, $\psi : V \to \mathbb{R}^n$, $q \in V$ and a decomposition $\mathbb{R}^m = \mathbb{R}^n \times \mathbb{R}^{m-n}$ such that $f(U) \subset V$ and f is expressed in the charts (U,φ), (V,ψ) as $\psi \circ f \circ \varphi^{-1}(x,y) = x$. In other words, f is locally equivalent to the projection $(x,y) \mapsto x$.*

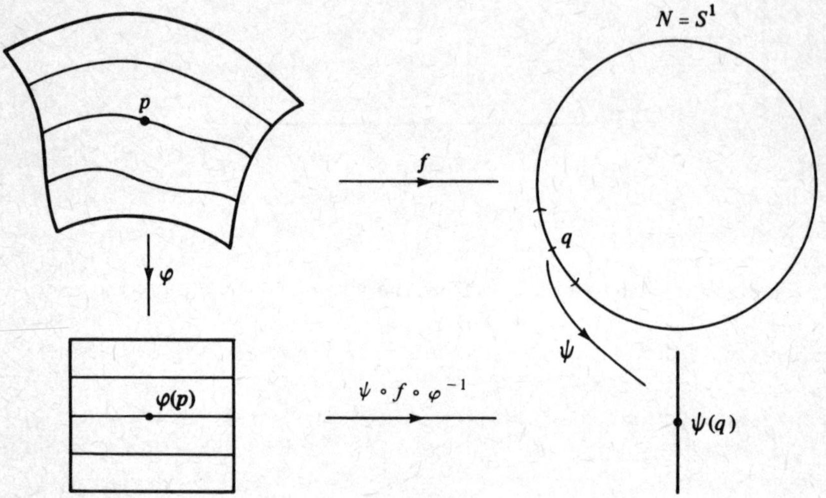

Figure 9

§4. Submanifolds

A subset $N \subset M^m$ is called a submanifold of M of dimension n and class C^r $(r \geq 1)$ if for every $p \in N$ there exists a C^r local chart, (U,φ), with $\varphi(U) = V \times W$ where $0 \in V \subset \mathbb{R}^n$, $0 \in W \subset \mathbb{R}^{m-n}$ and V, W are Euclidean balls such that $\varphi(N \cap U) = V \times 0$. In this situation we also say that the codimension of N is $m - n = \dim(M) - \dim(N)$.

It follows that a submanifold N of M is, in particular, a C^r manifold. For example S^n is a C^∞ submanifold of dimension n of $\mathbb{R}^{n+1}$.

From the theorem giving the local form of immersions one can obtain the following consequence.

Theorem. *Let $f : M \to N$ be a C^r map $(r \geq 1)$ such that $Df(p) : T_pM \to T_qN$, $q = f(p)$, is injective. Then there is a neighborhood U of p in M such that $f(U)$ is a C^r submanifold of N of dimension equal to* $\dim(M)$.

Corollary. *If $f : M \longrightarrow N$ is a C^r imbedding then $f(M)$ is a C^r submanifold of N of dimension* $\dim(M)$.

When $f : M \longrightarrow N$ is an immersion we say that $f(M)$ is an *immersed submanifold in N*. The example in figure 7 shows that an immersed submanifold may not be a submanifold, even when f is one-to-one.

§5. Regular values

Let $f : M \longrightarrow N$ be a C^r map $(r \geq 1)$. When $p \in M$ is such that $Df(p) : T_pM \longrightarrow T_qN$, $q = f(p)$, is surjective, we say p is a *regular point of f*. When $q \in N$ is such that $f^{-1}(q) = \varnothing$ or $f^{-1}(q)$ consists entirely of regular points, we say that q is a *regular value of f*. A point $q \in N$ which is not a regular value of f is called a *critical value of f*.

The following result is an immediate consequence of the local form of submersions.

Theorem. *Let $f : M \longrightarrow N$ be of class C^r $(r \geq 1)$. If $q \in N$ is a regular value of f and $f^{-1}(q) \neq \varnothing$, then $f^{-1}(q)$ is a C^r submanifold of M of codimension equal to* $\dim(N)$.

An example of the above situation is $S^n = f^{-1}(1)$ where $f(x_1, \ldots, x_{n+1}) = \sum_{i=1}^{n+1} x_i^2$.

An important theorem in the theory of differentiable functions is the following ([35]).

Theorem *(Sard). Let $f : M \longrightarrow N$ be a C^∞ map. Then the set of critical values of f has Lebesgue measure* 0. *In particular the set of regular values of f is dense in N.*

§6. Transversality

Let $f : M \longrightarrow N$ be a C^r map $(r \geq 1)$ and $S \subset N$ a submanifold of N. We say f is *transverse to S at $x \in M$* if $y = f(x) \notin S$ or if $y = f(x) \in S$ and the following condition is satisfied:

$$T_yN = T_yS + Df(x) \cdot (T_xM).$$

When f is transverse to S at every point of M we say *f is transverse to S*.

A local characterization of transversality is the following.

Theorem. *Let $f : M \longrightarrow N^n$ be a C^r map $(r \geq 1)$ and $S^s \subset N$ a C^r submanifold of N. Let $p \in M$ and $q = f(p) \in S$. A necessary and sufficient condi-*

tion for f to be transverse to S at p is that there exist a local chart $\psi : V \longrightarrow \mathbb{R}^n$, $q \in V$, a neighborhood U of p in M and a decomposition $\mathbb{R}^n = \mathbb{R}^s \times \mathbb{R}^{n-s}$ such that $f(U) \subset V$, $\psi(S \cap V) \subset \mathbb{R}^s \times \{0\}$ and the map $\pi_2 \circ \psi \circ f : U \longrightarrow \mathbb{R}^{n-s}$ is a submersion, where $\pi_2 : \mathbb{R}^s \times \mathbb{R}^{n-s} \longrightarrow \mathbb{R}^{n-s}$ is the projection on the second factor.

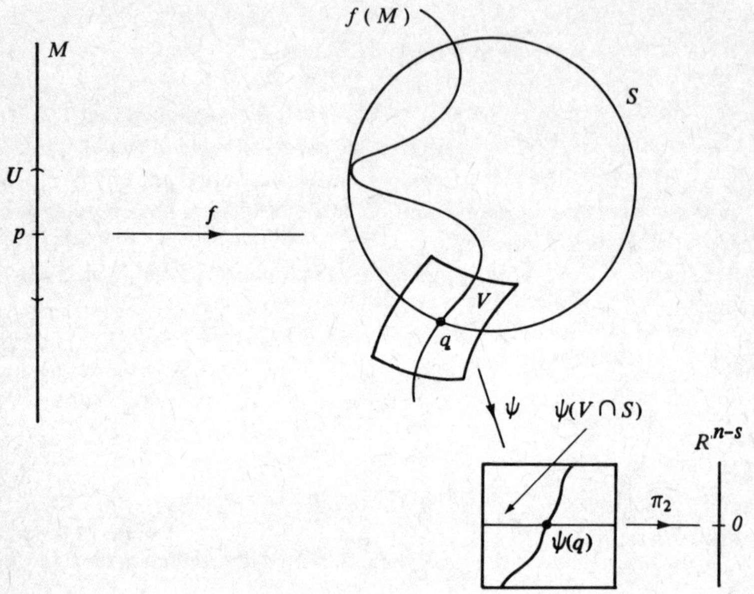

Figure 10

Theorem. *Let $f : M \longrightarrow N$ be of class C^r $(r \geq 1)$ and S^s a C^r submanifold of N. If f is transverse to S and $f^{-1}(S) \neq \emptyset$ then $f^{-1}(S)$ is a C^r submanifold of M, with the codimension of $f^{-1}(S)$ equal to the codimension of S.*

When $S^s, M^m \subset N^n$ are two C^r submanifolds of N, we say that S meets M transversally at $x \in S$ if the imbedding $i : S \longrightarrow N$, $i(y) = y$, is transverse to M at x. This condition is equivalent to $T_xN = T_xM + T_xS$ in case $x \in S \cap M$.

For more information about the theory of transversality we refer the reader to [1] or [26].

§7. Partitions of unity

Consider a countable open cover (U_n) of a manifold M. We say this cover is locally finite if every point of M has a neighborhood which meets only a finite number of U_n's.

Definition. A partition of unity subordinate to the cover (U_n) is a collection of nonnegative functions (φ_n) such that

(1) The support of φ_n is contained in U_n. Recall that the support of a function is the closure of the set of points where the function is nonzero.

(2) $\sum_n \varphi_n(p) = 1$ for every $p \in M$.

Theorem. *Every countable, locally finite, open cover of a manifold admits a partition of unity subordinate to it.*

The proof can be found in [27].

A Riemannian metric on a manifold M is a map that associates to each point $p \in M$ an inner product $\langle\ ,\ \rangle$ defined on the tangent space to M at p. The metric is said to be C^r if for each point $p \in M$ there is a system of coordinates $\varphi = (x_1, \ldots, x_m) : U \longrightarrow \mathbb{R}^m$ with $p \in U$ such that the maps $g_{ij}(q) = \langle \frac{\partial}{\partial x_i}(q), \frac{\partial}{\partial x_j}(q) \rangle$ are of class C^r for all $i,j = 1, \ldots, m$.

It is easy to see that on any manifold there exists a C^∞ Riemannian metric. In fact, take a countable, locally finite cover (U_n) of open coordinate neighborhoods of a manifold M. On each U_n, fix a system of coordinates $(x_1, \ldots, x_m) : U_n \longrightarrow \mathbb{R}^m$ and define the following metric $\langle\ ,\ \rangle(q)$, $q \in U_n$, by the values on the basis elements of $T_q(M)$: $\langle \frac{\partial}{\partial x_i}(q), \frac{\partial}{\partial x_j}(q) \rangle = \delta_{ij}$ where $\delta_{ij} = 0$ if $i \neq j$ and $\delta_{ii} = 1$. Now let (φ_n) be a partition of unity subordinate to (U_n). The metric on M is defined at each point $p \in M$ by the expression $\langle\ ,\ \rangle(p) = \sum_n \varphi_n(p) \langle\ ,\ \rangle$.

II. FOLIATIONS

We introduce in this chapter the notion of a foliation and the more elementary properties which will be used in the rest of the book. We will also see various examples illustrating the concept.

§1. Foliations

A foliation of dimension n of a differentiable manifold M^m is, roughly speaking, a decomposition of M into connected submanifolds of dimension n called leaves, which locally stack up like the subsets of $\mathbb{R}^m = \mathbb{R}^n \times \mathbb{R}^{m-n}$ with the second coordinate constant.

The simplest example of a foliation of dimension n is the foliation of $\mathbb{R}^m = \mathbb{R}^n \times \mathbb{R}^{m-n}$ where the leaves are n-planes of the form $\mathbb{R}^n \times \{c\}$ with $c \in \mathbb{R}^{m-n}$.

The diffeomorphisms $h : U \subset \mathbb{R}^m \longrightarrow V \subset \mathbb{R}^m$ which preserve the leaves of this foliation locally have the following form

$$(*) \qquad h(x,y) = (h_1(x,y), h_2(y)), \qquad (x,y) \in \mathbb{R}^n \times \mathbb{R}^{m-n} .$$

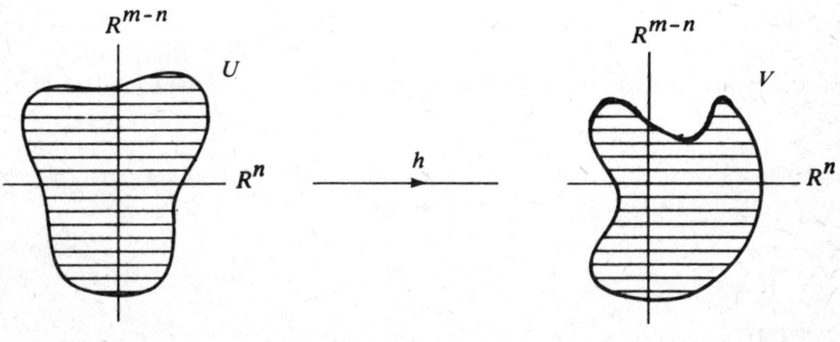

Figure 1

Definition. Let M be a C^∞ manifold of dimension m. A C^r foliation of dimension n of M is a C^r atlas $\mathfrak{F}$ on M which is maximal with the following properties:

(a) If $(U,\varphi) \in \mathfrak{F}$ then $\varphi(U) = U_1 \times U_2 \subset \mathbb{R}^n \times \mathbb{R}^{m-n}$ where U_1 and U_2 are open disks in $\mathbb{R}^n$ and $\mathbb{R}^{m-n}$ respectively.

(b) If (U,φ) and $(V,\psi) \in \mathfrak{F}$ are such that $U \cap V \neq \emptyset$ then the change of coordinates map $\psi \circ \varphi^{-1} : \varphi(U \cap V) \longrightarrow \psi(U \cap V)$ is of the form $(*)$, that is, $\psi \circ \varphi^{-1}(x,y) = (h_1(x,y), h_2(y))$. We say that M is foliated by $\mathfrak{F}$, or that $\mathfrak{F}$ is a foliated structure of dimension n and class C^r on M.

In figure 2, we illustrate the local aspect of a two-dimensional manifold foliated by a one-dimensional foliation.

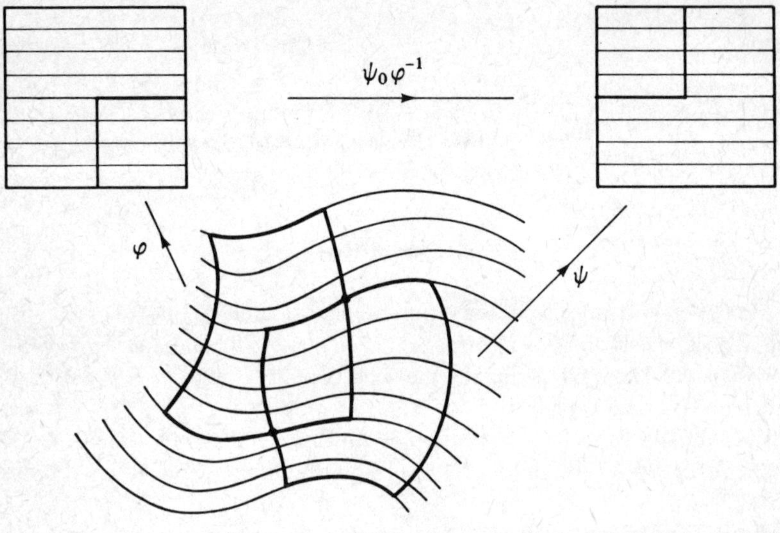

Figure 2

Remark 1. When we say M is a C^∞ manifold which has an atlas $\mathfrak{F}$ as above, we are implicitly saying that M has an atlas $\mathfrak{A}$ whose changes of coordinates are C^∞; however, if $(U,\varphi) \in \mathfrak{A}$ and $(V,\psi) \in \mathfrak{F}$ and $U \cap V \neq \emptyset$ then $\varphi \circ \psi^{-1}$ and $\psi \circ \varphi^{-1}$ are C^r. The only relation between $\mathfrak{F}$ and $\mathfrak{A}$ is that the mixed change of variables, as above, are C^r. As is clear from the definition, the atlas $\mathfrak{F}$ is very special (due to condition $(*)$) and, as we will see further in §4, not all manifolds of dimension m possess a foliation of dimension n, where

$0 < n < m$.

We observe that if $A = \{(U_i,\varphi_i) \mid i \in I\}$ is an atlas, not maximal, whose local charts satisfy (a) and (b), then there is a unique maximal atlas which contains A and satisfies (a) and (b). This atlas is defined as the set of all charts $\varphi : U \longrightarrow \mathbb{R}^n$ such that if $U \cap U_i \neq \emptyset$ for some $i \in I$, then $h = \varphi \circ \varphi_i^{-1} : \varphi_i(U \cap U_i) \longrightarrow \varphi(U \cap U_i)$ is a C^r diffeomorphism of the form $(*)$. Although the condition of the atlas being maximal is not necessary in the definition, it is convenient since in that case the set of all domains of local charts form a basis for the topology on M.

From now on we consider only foliations of class C^r, $r \geq 1$. The charts $(U,\varphi) \in \mathfrak{F}$ will be called *foliation charts.*

Let $\mathfrak{F}$ be a C^r foliation of dimension n, $0 < n < m$, of a manifold M^m. Consider a local chart (U,φ) of $\mathfrak{F}$ such that $\varphi(U) = U_1 \times U_2 \subset \mathbb{R}^n \times \mathbb{R}^{m-n}$. The sets of the form $\varphi^{-1}(U_1 \times \{c\})$, $c \in U_2$ are called *plaques of* U, or else *plaques of* $\mathfrak{F}$. Fixing $c \in U_2$, the map $f = \varphi^{-1}/U_1 \times \{c\} : U_1 \times \{c\} \longrightarrow U$ is a C^r imbedding, so the plaques are connected n-dimensional C^r submanifolds of M. Further if α and β are plaques of U then $\alpha \cap \beta = \emptyset$ or $\alpha = \beta$.

A *path of plaques of* $\mathfrak{F}$ is a sequence $\alpha_1 ,\ldots, \alpha_k$ of plaques of $\mathfrak{F}$ such that $\alpha_j \cap \alpha_{j+1} \neq \emptyset$ for all $j \in \{1 ,\ldots, k-1\}$. Since M is covered by plaques of $\mathfrak{F}$, we can define on M the following equivalence relation: "pRq if there exists a path of plaques $\alpha_1 ,\ldots, \alpha_k$ with $p \in \alpha_1$, $q \in \alpha_k$". The equivalence classes of the relation R are called *leaves of* $\mathfrak{F}$.

From the definition it follows that a leaf of $\mathfrak{F}$ is a subset of M connected by paths. Indeed, if F is a leaf of $\mathfrak{F}$ and $p,q \in F$, then there is a path of plaques $\alpha_1 ,\ldots, \alpha_k$ such that $p \in \alpha_1$ and $q \in \alpha_k$. Since the plaques α_j are connected by paths and $\alpha_j \cap \alpha_{j+1} \neq \emptyset$, it is immediate that $\alpha_1 \cup \ldots \cup \alpha_k \subset F$ is connected by paths, so there is a continuous path in F connecting p to q.

Another important fact is that every leaf F of $\mathfrak{F}$ has the structure of a C^r manifold of dimension n naturally induced by charts of $\mathfrak{F}$. Before proving this, we will look at some examples of foliations.

Example 1. Foliations defined by submersions.

Let $f : M^m \longrightarrow N^n$ be a C^r submersion. From the local form of submersions, we have that given $p \in M$ and $q = f(p) \in N$ there exist local charts (U,φ) on M, (V,ψ) on N such that $p \in U$, $q \in V$, $\varphi(U) = U_1 \times U_2 \subset \mathbb{R}^{m-n} \times \mathbb{R}^n$ and $\psi(V) = V_2 \supset U_2$ and $\psi \circ f \circ \varphi^{-1} : U_1 \times U_2 \longrightarrow U_2$ coincides with the projection $(x,y) \mapsto y$.

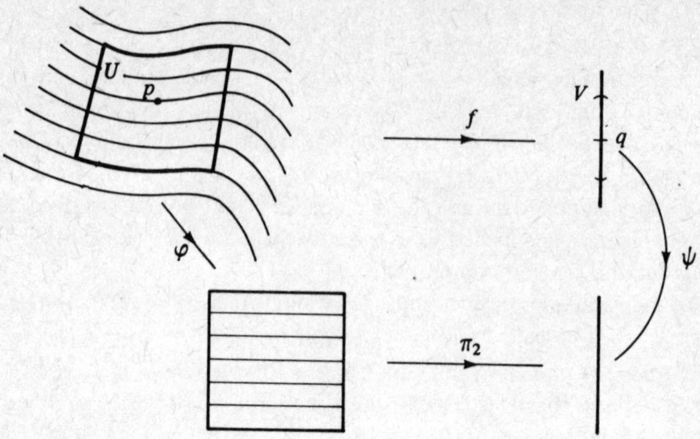

Figure 3

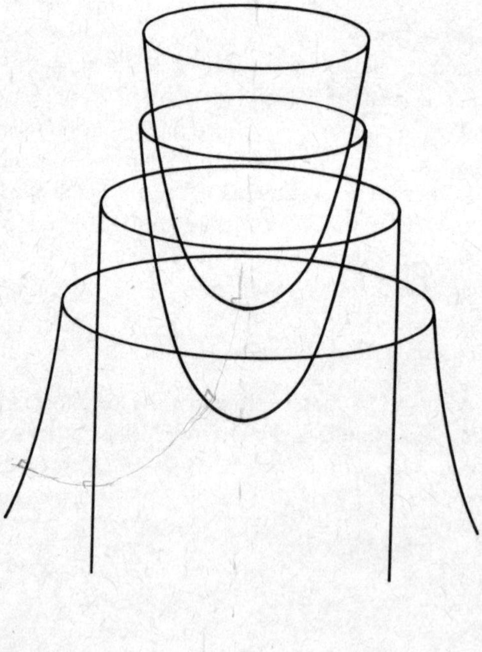

Figure 4

Therefore it is clear that the local charts (U,φ) define a C^r foliated manifold structure where the leaves are the connected components of the level sets $f^{-1}(c)$, $c \in N$. Let us look at a specific example. Let $f : \mathbb{R}^3 \longrightarrow \mathbb{R}$ be the submersion defined by $f(x_1,x_2,x_3) = \alpha(r^2)e^{x_3}$ where $r = \sqrt{x_1^2 + x_2^2}$ and $\alpha : \mathbb{R} \longrightarrow \mathbb{R}$ is a C^∞ function such that $\alpha(1) = 0$, $\alpha(0) = 1$ and if $t > 0$ then $\alpha'(t) < 0$. Let $\mathfrak{F}$ be the foliation of $\mathbb{R}^3$ whose leaves are the connected components of the submanifolds $f^{-1}(c)$, $c \in \mathbb{R}$.

The leaves of $\mathfrak{F}$ in the interior of the solid cylinder $C = \{(x_1,x_2,x_3) \mid x_1^2 + x_2^2 \leq 1\}$ are all homeomorphic to $\mathbb{R}^2$ and can be parametrized by $(x_1,x_2) \in D^2 \mapsto (x_1,x_2,\log(c/\alpha(r^2)))$, where $c > 0$. The boundary of C, $\partial C = \{(x_1,x_2,x_3) \mid x_1^2 + x_2^2 = 1\}$ is also a leaf. Outside C the leaves are all homeomorphic to cylinders (see fig. 4).

Example 2. Fibrations.

The fibers of a fiber space (E,π,B,F) define a foliation on E whose leaves are diffeomorphic to the connected components of F. A fibered space (E,π,B,F) consists of differentiable manifolds E,B,F and a submersion $\pi : E \longrightarrow B$ such that for every $b \in B$ there is an open neighborhood U_b of b and a diffeomorphism $\varphi_b : \pi^{-1}(U_b) \longrightarrow U_b \times F$ which makes the following diagram commute:

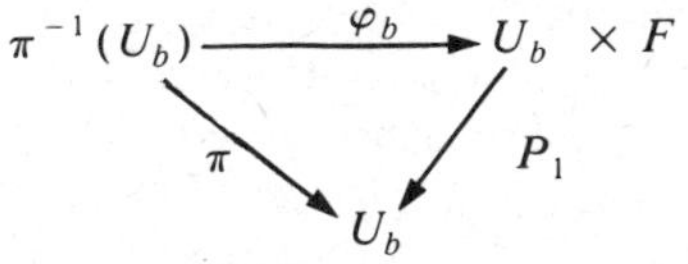

In the above diagram P_1 is the projection onto the first factor. The fibers of the fibration are the submanifolds $\pi^{-1}(b)$, $b \in B$.

An important example of this situation is given by the following theorem, whose proof the reader can find in [27].

Theorem (On Tubular Neighborhoods). *Let $N \subset M$ be a C^r submanifold with $r \geq 1$. There is an open neighborhood $T(N) \supset N$* and a *C^r submersion, $\pi : T(N) \longrightarrow N$, such that $\pi(q) = q$ for all $q \in N$. If the codimension of N is k, then $T(N)$ can be obtained in such a way that $(T(N),\pi,N,\mathbb{R}^k)$ is a fibered space.*

Example 3. The Reeb foliation of S^3, [46].

The following example plays an important role in the development of foliation theory.

Consider the submersion of Example 1, $f{:}D^2 \times R \longrightarrow \mathbb{R}$ given by $f(x_1,x_2,x_3) = \alpha(r) \cdot e^{x_3}$, where now $\alpha(r)$ is the function $\alpha(r) = \exp(-\exp(1/1 - r^2))$. The foliation defined by f has for leaves the graphs of the functions $x_3 =$

$= \exp(1/1 - r^2) + b$, $b \in \mathbb{R}$ and it extends to a C^∞ foliation of $\mathbb{R}^3$ whose leaves in the exterior of $D^2 \times \mathbb{R}$ are the cylinders $x_1^2 + x_2^2 = r^2, r > 1$.

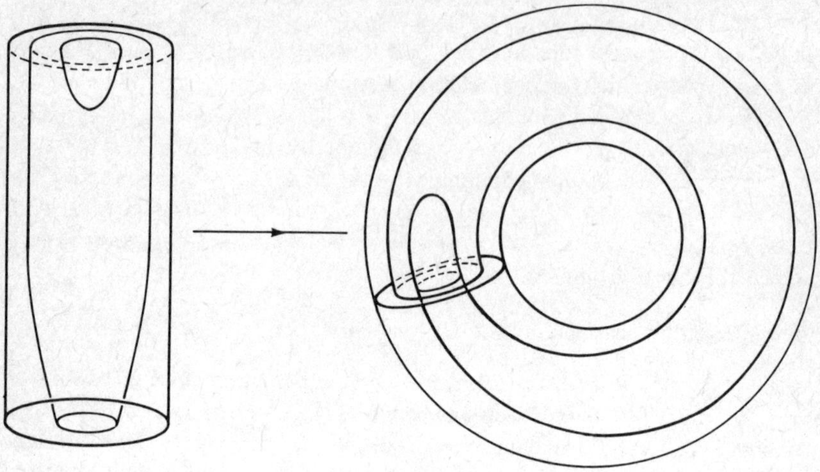

Figure 5

On $D^2 \times [0,1]$ we identify the boundary points in the following manner: $(x_1,x_2,0) \equiv (y_1,y_2,1)$ if and only if $(x_1,x_2) = (y_1,y_2)$. The quotient manifold $D^2 \times [0,1]/\equiv$ is diffeomorphic to $D^2 \times S^1$ and, since the foliation defined on $D^2 \times \mathbb{R}$ is invariant by translations along the x_3-axis (that is, these translations take leaves to leaves), it induces a C^∞ foliation, $\mathcal{R}$ on $D^2 \times S^1$. It is called the (*orientable*) *Reeb foliation* of $D^2 \times S^1$. In this foliation the boundary $\partial(D^2 \times S^1) = S^1 \times S^1$ is a leaf. Moreover, the other leaves are homeomorphic to $\mathbb{R}^2$ and they accumulate only on the boundary (see Fig. 5). It is easy to see that this foliation is not defined by a submersion. If, instead of the above identification, we consider on $D^2 \times [0,1]$ the relation $(x_1,x_2,0) \sim (y_1,y_2,1)$ if and only if $(x_1,x_2) = (y_1,-y_2)$ then the quotient $D^2 \times [0,1]/\sim$ will be a non-orientable three-dimensional manifold, K^3, whose boundary is diffeomorphic to the Klein bottle. Since this identification preserves the foliation of $D^2 \times \mathbb{R}$, this induces a foliation $\mathcal{R}$ of K^3 called the *non-orientable Reeb foliation* of K^3. The leaves of $\mathcal{R}$ in the interior of K^3 are all homeomorphic to $\mathbb{R}^2$ and the boundary of K^3 is a leaf.

From two Reeb foliations of $D^2 \times S^1$ we can get a C^∞ foliation of S^3 in the following way. The sphere $S^3 = \{(x_1,x_2,x_3,x_4) \in \mathbb{R}^4 \mid \sum_{i=1}^4 x_i^2 = 1\}$ can be considered as a union of two solid tori $T_i \simeq D^2 \times S^1$, $i = 1,2$ identified along the boundary by a diffeomorphism which takes meridians of ∂T_1 to parallels of ∂T_2 and vice versa. The solid torus T_1 can be defined by the equations $\sum_{i=1}^4 x_i^2 = 1$ and $x_1^2 + x_2^2 \leq 1/2$ and the solid torus T_2 by the equations

$\sum_{i=1}^{4} x_i^2 = 1$ and $x_1^2 + x_2^2 \geq 1/2$.

Another way of seeing this decomposition of S^3 into two solid tori is the following. Consider the stereographic projection $\pi : S^3 - P \longrightarrow \mathbb{R}^3$ where $P = (0,0,0,1)$ and $\pi(x)$ is the point of intersection with the plane $x_4 = 0$ of a ray containing P and x. Since π is a diffeomorphism, S^3 can be thought of as the union of $\mathbb{R}^3$ and the point P at infinity.

Consider now in the x_1x_3-plane the region S_1 bounded by the circles of radius 1 centered at $p = (2,0)$, $q = (-2,0)$ and $S_2 = \mathbb{R}^2 - S_1$.

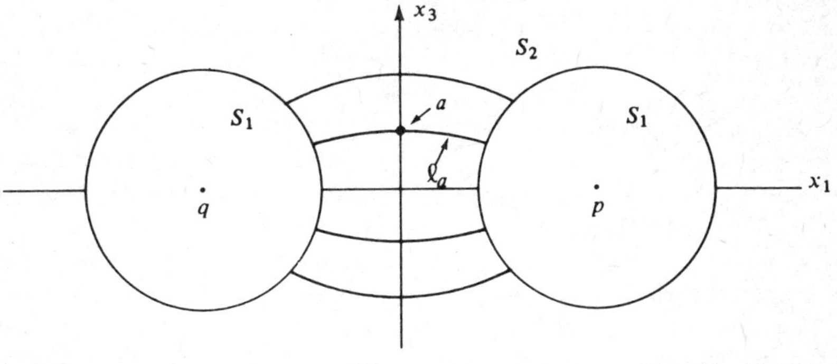

Figure 6

Taking this figure into $\mathbb{R}^3$ by rotating about the x_3-axis, we generate by the region S_1 a solid torus S_1'. Let $S_2' = \mathbb{R}^3 - S_1'$ and note that $\pi^{-1}(S_2') \cup P$ is also a solid torus. Indeed consider the circles in the x_1x_3-plane with center on the x_3-axis and which pass through the points p and q. Let ℓ_a be the connected component of one of these circles in S_2 which intersects the x_3-axis in the point a. Letting $\ell_\infty = \{(x_1,0) \mid x_1^2 \geq 9\}$, we get a foliation of S_2 whose leaves are the curves $\{\ell_a\}$, $a \in \mathbb{R} \cup \{\infty\}$. By rotating the x_1x_3-plane about the x_3-axis, we see that the curves $\ell_a (a \in \mathbb{R})$ generate disks D_a and ℓ_∞ generates a cylinder D_∞. Each D_a, $a \in \mathbb{R} \cup \{\infty\}$, meets the torus $\partial S_1'$ along a parallel.

It is now clear that by adding to $\mathbb{R}^3$ the point P of the stereographic projection, the x_3-axis together with P defines the axis of a solid torus T_2 which is foliated by disks $\pi^{-1}(D_a)$, $a \in \mathbb{R}$ plus the disk $D_P = \pi^{-1}(D_\infty) \cup P$.

If we denote by T_1 the solid torus $\pi^{-1}(S_1')$ we get that $S^3 = T_1 \cup T_2$ and the parallels of ∂T_1 coincide with the meridians of ∂T_2 and vice-versa.

Returning to the construction of the foliation of S^3, consider the foliation that results from glueing two Reeb foliations of T_1 and T_2 where $\partial T_1 = \partial T_2$ is a leaf. We obtain in this way a foliation of S^3 of codimension one called the *Reeb foliation of* S^3. This foliation is C^∞ and has one leaf homeomorphic to T^2. All the other leaves are homeomorphic to $\mathbb{R}^2$ and accumulate on the compact leaf.

Example 4. Vector fields without singularities.

A vector field on a manifold M is a map that associates to each point $p \in M$ a vector of the tangent space to M at p. An *integral curve of* X *through* $p \in M$ is a curve $\gamma : (a,b) \longrightarrow M$, with $\gamma(0) = p$, such that $X(\gamma(t)) = \gamma'(t)$ for any $t \in (a,b)$. Thus integral curves are, locally, solutions of the differential equation $dx/dt = X(x)$. The existence and uniqueness theorem for ordinary differential equations guarantees that under certain differentiability conditions on X, for instance if X is C^r, $r \geq 1$, there passes through any $p \in M$ an integral curve of X (of class C^r) which is unique in the sense that any two integral curves through the same point necessarily agree in the intersection of their domains of definition. Moreover, these integrals define a local flow at any point $p \in M$, i.e. there is a neighborhood U_p of p and an interval (α,β) such that for any $q \in U_p$ the integral curve through $q, \gamma_q(t)$ is defined for all $t \in (\alpha,\beta)$ and the map (local flow) $\varphi : (\alpha,\beta) \times U_p \longrightarrow M$ given by $\varphi(t,q) = \gamma_q(t)$ is of class C^r. Clearly, $\varphi(0,q) = q$ for any $q \in M$ and $\varphi(s,\varphi(t,q)) = \varphi(s+t,q)$ provided that $\varphi(t,q) \in U_p$ and $s,t,s+t \in (\alpha,\beta)$. Let X be a vector field on M without singular points. Let $i : B^{m-1} \longrightarrow M$, $i(0) = p$, be an imbedding of a small $m-1$ disc centered at $0 \in \mathbb{R}^m$, transverse to X everywhere. Since $X(p) \neq 0$, the map $\Phi : B^{m-1} \times (\alpha,\beta) \longrightarrow M$ given by $\Phi(x,t) = \varphi(t,i(x))$ has maximal rank at $(0,0) \in B^{m-1} \times (\alpha,\beta)$. Thus, by the inverse mapping theorem there is a neighborhood $V \subset M$ of p such that Φ^{-1}/V is a diffeomorphism onto a product neighborhood $B'^{m-1} \times (\alpha',\beta') \subset B^m \times (\alpha,\beta)$ of $(0,0)$. This map will be a local chart for the one dimensional foliation on M defined by the integral curves of X.

Example 5. Actions of Lie groups.

A Lie group is a group G which has a C^∞ differentiable manifold structure such that the map $G \times G \longrightarrow G$, $(x,y) \longmapsto xy^{-1}$ is C^∞. This last condition is equivalent to saying that the maps $(x,y) \longmapsto xy$ and $x \longmapsto x^{-1}$ are C^∞. An immersed C^∞ submanifold $H \subset G$ that is also a subgroup of G is called a *Lie subgroup of* G.

Examples of Lie groups

(1) The additive group $\mathbb{R}^n$.
(2) The group $\mathbb{C}^* = \mathbb{C} - \{0\}$, with the multiplication of complex numbers. The circle $S^1 = \{z \in \mathbb{C} \mid |z| = 1\}$ is a Lie subgroup of $\mathbb{C}^*$.
(3) The torus $T^n = S^1 \times \ldots \times S^1$ (n times) with the multiplication

$$(z_1, \ldots, z_n) \cdot (w_1, \ldots, w_n) = (z_1 \cdot w_1, \ldots, z_n \cdot w_n) .$$

(4) The group $GL(n,\mathbb{R})$ of all real $n \times n$ nonsingular matrices is a Lie group of dimension n^2. This can be seen by considering each element $A = (a_{ij}) \in GL(n,\mathbb{R})$ as a point of $\mathbb{R}^{n^2}$. So $GL(n,\mathbb{R})$ can be taken to be an open submanifold of $\mathbb{R}^{n^2}$.

(5) The orthogonal group $\mathcal{O}(n,\mathbb{R})$ which consists of $n \times n$ real matrices A such that $AA' = I$ is a compact Lie subgroup of $GL(n,\mathbb{R})$. In fact, if $\mathcal{S}(n,\mathbb{R})$ denotes the space of symmetric matrices, the map $f: GL(n,\mathbb{R}) \longrightarrow \mathcal{S}(n,\mathbb{R})$, $f(X) = XX'$ is well-defined and is C^∞. Moreover $\mathcal{O}(n,\mathbb{R}) = f^{-1}(I)$. One can verify that I is a regular value of f. Consequently, $\mathcal{O}(n,\mathbb{R})$ is a submanifold of $GL(n,\mathbb{R})$. If $\alpha_1,\ldots,\alpha_n$ are the rows of $A \in \mathcal{O}(n,\mathbb{R})$, the condition $AA' = I$ implies $|\alpha_1|^2 + \ldots + |\alpha_n|^2 = n$. So $\mathcal{O}(n,\mathbb{R})$ is compact.

A C^r action of a Lie group G on a manifold M is a map $\varphi : G \times M \longrightarrow M$ such that $\varphi(e,x) = x$ and $\varphi(g_1 \cdot g_2,x) = \varphi(g_1,\varphi(g_2,x))$ for any $g_1,g_2 \in G$ and $x \in M$. The *orbit of a point* $x \in M$ for the action φ is the subset $\mathcal{O}_x(\varphi) = \{\varphi(g,x) \in M \mid g \in G\}$. The *isotropy group of* $x \in M$ is the subgroup $G_x(\varphi) = \{g \in G \mid \varphi(g,x) = x\}$. It is clear that $G_x(\varphi)$ is closed in G. The map $\psi_x : G \longrightarrow M$ given by $\psi_x(g) = \varphi(g,x)$ induces the map $\overline{\psi}_x : G/G_x(\varphi) \longrightarrow M$, $\overline{\psi}_x(\overline{g}) = \psi_x(g)$ where $\overline{g} = g \cdot G_x(\varphi)$. Since $g_1^{-1}g_2 \in G_x(\varphi)$ if and only if $\varphi(g_1,x) = \psi_x(g_1) = \psi_x(g_2) = \varphi(g_2,x)$, we conclude that $\overline{\psi}_x$ is well-defined and injective. Further it can be shown that $G/G_x(\varphi)$ has a differentiable structure and that $\overline{\psi}$ is an injective immersion whose image is $\mathcal{O}_x(\varphi)$. (See Chapter VIII).

We say that $\varphi : G \times M \longrightarrow M$ is a *foliated action* if for every $x \in M$ the tangent space to the orbit of φ passing through x has fixed dimension k. When k is the dimension of G we say φ is *locally free*.

Proposition 1. *The orbits of a foliated action define the leaves of a foliation.*

Proof. Let $\varphi : G \times M \longrightarrow M$ be a foliated action whose orbits have dimension $k \geq 1$. Fixing $x_0 \in M$, let $E \subset T_eG$ be a subspace complementary to the tangent space of $G_{x_0}(\varphi)$ and $i_1 : B^k \longrightarrow G$, $i_1(0) = e$, an imbedding of a k-disk tangent to E at e. If $i_2 : B^{m-k} \longrightarrow M$, $i_2(0) = x_0$, denotes the imbedding of a small transverse section to the orbit $\mathcal{O}_{x_0}(\varphi)$ of x_0, we can define the map $\Phi : B^k \times B^{m-k} \longrightarrow M^m$ given by $\Phi = \varphi(i_1,i_2)$. Since $D\Phi(0,0) : \mathbb{R}^k \times \mathbb{R}^{m-k} \longrightarrow T_{x_0}M$ is an isomorphism, there is a neighborhood U of x_0 such that $\Phi^{-1} : U \longrightarrow B^k \times B^{m-k}$ is a diffeomorphism taking the orbit of $i_2(x) \in U \cap i_2(B^{m-k})$ to an open subset of the surface $B^k \times \{x\}$. This defines a local chart of the foliation by orbits of φ.

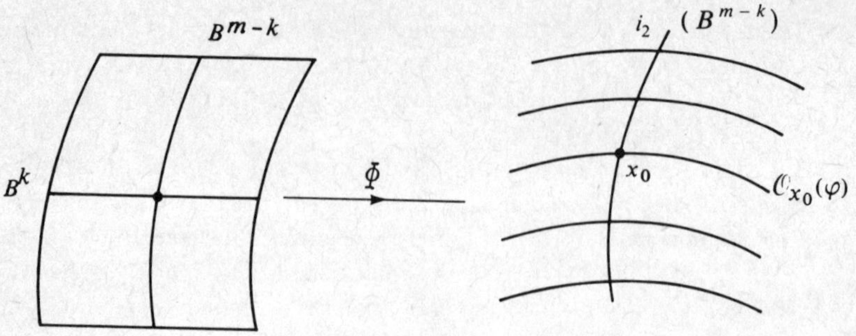

Figure 7

Remark 2. Given a Lie group G and a Lie subgroup $H \subset G$, the map $H \times G \longrightarrow G$, $(h,g) \longmapsto g \cdot h$ defines an action of H on G. The isotropy group of each element $g \in G$ is the identity. So this action is locally free and the orbits define a foliation of G whose leaves are all homeomorphic to H. The leaves of this foliation are imbedded if and only if H is closed in G. In fact, if H is not imbedded, there exists an $h \in H$ and V a neighborhood of h such that $V \cap H$ contains a countably infinite number of path-connected components. Let Σ be a transverse section to the orbits of the action passing through h. Then $\Sigma \cap H$ contains a countably infinite number of points. On the other hand, as is easy to see, the set of accumulation points of $\Sigma \cap H$ contains $\Sigma \cap H$, so $\overline{\Sigma \cap H} \subset \Sigma$ is perfect. From general topology we know that a perfect set is not countable, thus $\overline{\Sigma \cap H} - \Sigma \cap H$ is not empty. Thus H is not closed.

Conversely, if H is not closed there is a sequence h_n in H such that $h_n \longrightarrow g \notin H$. Hence there exists an open set V containing g such that $V \cap H$ has an infinite number of path-connected components. Let $U = g^{-1} \cdot V$. We have that $1 \in U \cap H$, the sequence $g_n = h_n^{-1} \cdot h_{n+1}$ is in H, converges to 1 and if $m \neq n$ then g_m and g_n are in distinct path-connected components of $H \cap U$. So H is not imbedded.

A specific example of this can be obtained by considering $G = \mathcal{O}(4, \mathbb{R})$ and H the subgroup generated by matrices A of the form

$$A = \begin{bmatrix} A_1 & 0 \\ 0 & A_2 \end{bmatrix} \text{ where } A_1 = \begin{bmatrix} \cos s & \sin s \\ -\sin s & \cos s \end{bmatrix} \text{ and } A_2 = \begin{bmatrix} \cos t & \sin t \\ -\sin t & \cos t \end{bmatrix}.$$

It is easy to see that H is isomorphic to T^2.

Remark 3. An action $\varphi : \mathbb{R} \times M \longrightarrow M$ ($\mathbb{R}$ the additive group of real numbers) is also called a flow on M. A flow φ on M satisfies the following properties:

(a) $\varphi(0,x) = x$ for all $x \in M$

(b) $\varphi(s + t,x) = \varphi(s,\varphi(t,x)), \quad s,t \in \mathbb{R},\ x \in M.$

To a C^r flow φ on $M(r \geq 1)$ it is possible to associate a C^{r-1} vector field by the formula: $X(x) = (d\varphi(t,x)/dt)\big|_{t=0}$. The field X so defined is such that $t \mapsto \varphi(t,x)$ is the integral of X which passes through x. In fact, from this definition we have $d\varphi(t,x)/dt = (d\varphi(s + t,x)/ds)\big|_{s=0} = (d(\varphi(s,\varphi(t,x))/ds)\big|_{s=0} = X(\varphi(t,x))$. Conversely, if X is a C^r vector field on M whose integrals are defined on $\mathbb{R}$, there exists a unique C^r flow, $\varphi : \mathbb{R} \times M \longrightarrow M$, such that $t \mapsto \varphi(t,x)$ is the integral of X with initial condition $\varphi(0,x) = x$. ([20]).

Proposition 1 in the case of flows is called the *flow-box theorem*.

§2. The leaves

Each leaf F of a C^r foliation $\mathfrak{F}$ has a C^r differentiable manifold structure induced by the charts of $\mathfrak{F}$. This structure, called the *intrinsic structure* of F, is constructed in the following manner. Given $p \in F$, let (U,φ) be a chart of $\mathfrak{F}$ such that $p \in U$ and $\varphi(U) = U_1 \times U_2 \subset \mathbb{R}^{n+s}$ where U_1 and U_2 are open balls in $\mathbb{R}^n$ and $\mathbb{R}^s$ respectively. Let α be the plaque of U which contains p. Setting $\varphi = (\varphi_1,\varphi_2)$ where $\varphi_1 : U \longrightarrow \mathbb{R}^n$, $\varphi_2 : U \longrightarrow \mathbb{R}^s$, we define $\overline{\varphi}$: $\alpha \longrightarrow \mathbb{R}^n$ by $\overline{\varphi} = \varphi_1|_\alpha$. It is clear that $\overline{\varphi} : \alpha \longrightarrow U_1 \subset \mathbb{R}^n$ is a homeomorphism since $\varphi(\alpha) = U_1 \times \{a\}$ for some $a \in U_2$. Now we prove that

$$\mathcal{B} = \{(\alpha,\overline{\varphi}) \mid \alpha \subset F \text{ is plaque of } U \text{ with } (U,\varphi) \in \mathfrak{F}\}$$

is a C^r atlas of dimension n for F.

It is enough to show that if $(\alpha,\overline{\varphi})$, $(\beta,\overline{\psi})$ are in $\mathcal{B}$ and $\alpha \cap \beta \neq \emptyset$ then $\overline{\varphi}(\alpha \cap \beta)$ and $\overline{\psi}(\alpha \cap \beta)$ are open in $\mathbb{R}^n$ and $\overline{\varphi} \circ \overline{\psi}^{-1} : \overline{\psi}(\alpha \cap \beta) \longrightarrow \overline{\varphi}(\alpha \cap \beta)$ is a C^r diffeomorphism. First we show that $\alpha \cap \beta$ is open in α and in β. Let (U,φ), (V,ψ) be in $\mathfrak{F}$ such that $\overline{\varphi} = \varphi_1|_\alpha$ and $\overline{\psi} = \psi_1|_\beta$. From condition $(*)$ in §1, $\varphi \circ \psi^{-1} : \psi(U \cap V) \longrightarrow \varphi(U \cap V)$ is given by $\varphi \circ \psi^{-1}(x,y) = (h_1(x,y),h_2(y)) \in \mathbb{R}^n \times \mathbb{R}^s$ with $(x,y) \in \mathbb{R}^n \times \mathbb{R}^s$.

In particular since $\alpha \cap \beta \neq \emptyset$ we have

$$(**) \qquad \varphi \circ \psi^{-1}(x,b) = (h_1(x,b),h_2(b)) = (h_1(x,b),a) .$$

Since $\psi(\beta \cap U) = \psi(U \cap V \cap \beta) = \psi(U \cap V) \cap (\mathbb{R}^n \times \{b\})$ and $\varphi(\alpha \cap V) = \varphi(U \cap V) \cap (\mathbb{R}^n \times \{a\})$, from $(**)$ we have

$$\varphi(\beta \cap U) = \varphi \circ \psi^{-1}(\psi(\beta \cap U)) = \varphi \circ \psi^{-1}(\psi(U \cap V) \cap (\mathbb{R}^n \times \{b\}))$$

$$\subset \varphi(U \cap V) \cap (\mathbb{R}^n \times \{a\}) = \varphi(\alpha \cap V)$$

i.e., $\beta \cap U \subset \alpha \cap V$. Analogously, $\alpha \cap V \subset \beta \cap U$, so $\alpha \cap \beta = \alpha \cap V = \beta \cap U$. This proves the claim.

Since $\overline{\varphi}$ and $\overline{\psi}$ are homeomorphisms we get that $\overline{\varphi}(\alpha \cap \beta)$ and $\overline{\psi}(\alpha \cap \beta)$ are open in $\mathbb{R}^n$. The map $\overline{\varphi} \circ \overline{\psi}^{-1} : \overline{\psi}(\alpha \cap \beta) \longrightarrow \overline{\varphi}(\alpha \cap \beta)$ is C^r since $\overline{\varphi} \circ \overline{\psi}^{-1}(x) = h_1(x,b)$ if $x \in \overline{\psi}(\alpha \cap \beta)$. Similarly $\overline{\psi} \circ \overline{\varphi}^{-1}$ is C^r, hence $\overline{\varphi} \circ \overline{\psi}^{-1}$ is a C^r diffeomorphism. This defines the intrinsic structure of F. It can be shown that the leaves have a countable basis. (See exercise 6).

We note that the topology of F associated to the atlas $\mathcal{B}$ defined above is such that the set of all the plaques α of $\mathfrak{F}$ with $\alpha \subset F$ constitute a basis of open sets of F. This topology, in general, does not coincide with the one induced naturally by the topology of M. The reason is that the leaf F can eventually meet the domain U of a chart $(U,\varphi) \in \mathfrak{F}$ in a sequence of plaques $(\alpha_n)_{n \in \mathbb{N}}$ which accumulate on a plaque $\alpha \subset F$, i.e., any neighborhood of α contains an infinite number of plaques of F and, hence F is not locally connected in the topology induced by M, while in the intrinsic topology, F is a manifold and hence locally connected. In Chapter III we will see some examples where this occurs.

We consider now the canonical inclusion $i : F \longrightarrow M$, $i(p) = p$, and F with its intrinsic manifold structure. It is easy to verify that i is a one-to-one C^r immersion. When i is an imbedding we say that F is an imbedded leaf. This occurs if and only if the intrinsic topology of F coincides with the topology induced by that of M. We summarize the above in the following:

Theorem 1. *Let M be a manifold foliated by a C^r n-dimensional foliation $\mathfrak{F}$. Each leaf F of $\mathfrak{F}$ has the structure of a C^r manifold of dimension n such that the domains of the local charts are plaques of $\mathfrak{F}$. The map $i : F \longrightarrow M$ defined by $i(p) = p$ is a C^r one-to-one immersion, where on F we take the intrinsic manifold structure. Further F is a C^r submanifold of M if and only if i is an imbedding.*

§3. Distinguished maps

As one would expect, the transverse structure plays an important role in the study of foliations. This structure is emphasized in the following alternate definition of foliation.

Definition. A $C^r (r \geq 1)$ codimension s foliation $\mathfrak{F}$ of M is defined by a maximal collection of pairs (U_i, f_i), $i \in I$, where the U_i's are open subsets of M and the f_i's, $f_i : U_i \longrightarrow \mathbb{R}^s$, are submersions satisfying

(1) $\cup_{i \in I} U_i = M$

(2) if $U_i \cap U_j \neq \varnothing$, there exists a local C^r diffeomorphism g_{ij} of $\mathbb{R}^s$ such that $f_i = g_{ij} \circ f_j$ on $U_i \cap U_j$.

The f_i's are called *distinguished maps of* $\mathfrak{F}$.

In this definition the plaques of $\mathfrak{F}$ in U_i are the connected components of the sets $f_i^{-1}(c)$, $c \in \mathbb{R}^s$.

We verify that this definition is equivalent to the definition of §1. To do this we need the following lemma.

Lemma 1. *Let $\mathfrak{F}$ be a foliation of a manifold M. There is a cover $\mathcal{C} = \{U_i \mid i \in I\}$ of M by domains of local charts of $\mathfrak{F}$ such that if $U_i \cap U_j \neq \emptyset$ then $U_i \cup U_j$ is contained in the domain of a local chart of $\mathfrak{F}$.*

Proof. We consider a cover of M by compact sets K_n where $K_n \subset \operatorname{int}(K_{n+1})$. For each $n \in N$ we fix a cover of K_n by domains of foliation charts of $\mathfrak{F}$, $\{V_i^n \mid i = 1, \ldots, k_n\}$. Let $\delta_n > 0$ be the Lebesgue number of this cover with respect to some fixed metric on M. We can assume that the sequence (δ_n) is decreasing.

It is now sufficient to take a cover of K_n by domains of foliation charts, $\{U_j^n \mid j = 1, \ldots, \ell_n\}$, such that the diameters of the U_j^n's are less than $\delta_n/2$ for all $j = 1, \ldots, \ell_n$. It is clear that if $U_i^n \cap U_j^n \neq \emptyset$ then $U_i^n \cup U_j^n \subset V_\mu^n$ for some $\mu \in \{1, \ldots, k_n\}$. Then $\mathcal{C} = \{U_j^n \mid j = 1, \ldots, \ell_n, n \in \mathbb{N}\}$. ■

Suppose now that M has an atlas $\mathfrak{F}$ which defines a codimension s foliation, according to the definition in §1. Consider a cover $\mathcal{C} = \{U_i \mid i \in I\}$ of M by domains of local charts of $\mathfrak{F}$, as in Lemma 1.

Given an open set $U_i \in \mathcal{C}$, $i \in I$, we have defined $\varphi_i : U_i \longrightarrow \mathbb{R}^n \times \mathbb{R}^s$ on $\mathfrak{F}$ such that $\varphi_i(U_i) = U_1^i \times U_2^i$ where U_1^i and U_2^i are open balls in $\mathbb{R}^n$ and $\mathbb{R}^s$ respectively. Let $p_2 : \mathbb{R}^n \times \mathbb{R}^s \longrightarrow \mathbb{R}^s$ be the projection on the second factor. Then $f_i = p_2 \circ \varphi_i : U_i \longrightarrow \mathbb{R}^s$ is a submersion and for every $c \in U_2^i$, $f_i^{-1}(c)$ is a plaque of U_i. If $U_i \cap U_j \neq \emptyset$, $U_i \cup U_j$ is contained in the domain of a chart $(V, \varphi) \in \mathfrak{F}$ with $\varphi(V) = V_1 \times V_2$. Hence, letting α and β be plaques of U_i and U_j with $\alpha \cap \beta \neq \emptyset$, we have that $\alpha \cup \beta$ is contained in a plaque γ of V. Then $\beta \cap U_i \subset \alpha$, which implies that $f_i(\beta \cap U_i)$ contains a unique point. This implies that given $y \in f_j(U_i \cap U_j)$, $f_i \circ f_j^{-1}(y)$ contains a unique point $g_{ij}(y)$. We leave it to the reader to verify that $g_{ij} : f_j(U_i \cap U_j) \longrightarrow f_i(U_i \cap U_j)$ is a C^r diffeomorphism. Suppose now that there exists a collection of pairs (U_i, f_i), $i \in I$, satisfying the definition of this section. Since for every $i \in I$, $f_i : U_i \longrightarrow \mathbb{R}^s$ is a C^r submersion, from the theorem of the local form of submersions, given $p \in U_i$, there exist V_1 and V_2 open balls of $\mathbb{R}^n$ and $\mathbb{R}^s$ respectively, with $V_2 \subset f_i(U_i)$, and a C^r local chart $\varphi : V \longrightarrow V_1 \times V_2$ with $V \subset U_i$, such that $f_i \circ \varphi^{-1} : V_1 \times V_2 \longrightarrow V_2$ is the projection on the second factor. The set of all the charts (V, φ) constructed in this manner is a C^r atlas of a foliation of M.

In fact, if (V, φ) and (W, ψ) are two charts as above such that $V \cap W \neq \emptyset$, then $V \subset U_i$ and $W \subset U_j$ with $i, j \in I$. On the other hand, $\varphi \circ \psi^{-1} : \psi(V \cap W) \longrightarrow \varphi(V \cap W)$ can be written as $\varphi \circ \psi^{-1}(x,y) = (h_1(x,y), h_2(x,y))$, $(x,y) \in \mathbb{R}^n \times \mathbb{R}^s$. So $h_2(x,y) = p_2 \circ \varphi \circ \psi^{-1}(x,y) = f_i \circ \varphi^{-1} \circ \varphi \circ \psi^{-1}(x,y) = f_i \circ \psi^{-1}(x,y) = g_{ij} \circ f_j \circ \psi^{-1}(x,y) = g_{ij}(y)$ only depends on y, as we wished.

Definition. Let N be a manifold. We say that $g : N \longrightarrow M$ is *transverse to* $\mathfrak{F}$ when g is transverse to all the leaves of $\mathfrak{F}$, i.e., if for every $p \in N$ we have

$$Dg(p) \cdot T_p(N) + T_q(\mathfrak{F}) = T_qM \qquad q = g(p)$$

where by $T_q(\mathfrak{F})$ we mean the tangent space to the leaf of $\mathfrak{F}$ which passes through q.

Theorem 2. *Let $\mathfrak{F}$ be a C^r foliation $(r \geq 1)$ of M and $g : N \longrightarrow M$ a C^r map. Then g is transverse to $\mathfrak{F}$ if and only if for each distinguished map (U,f) of $\mathfrak{F}$ the composition $f \circ g : g^{-1}(U) \longrightarrow \mathbb{R}^s$ is a submersion.*

Proof. Suppose g is transverse to $\mathfrak{F}$ and let (U,f) be a distinguished map of $\mathfrak{F}$. If $p \in U$ and $q = g(p)$, we have $T_qM = T_q(\mathfrak{F}) + Dg(p) \cdot T_p(N)$. Applying $Df(q) : T_qM \longrightarrow \mathbb{R}^s$ to both sides and using the chain rule, we get

$$\mathbb{R}^s = Df(q) \cdot T_qM = Df(q) \cdot T_q(\mathfrak{F}) + D(f \circ g)_p \cdot T_pN \ .$$

Since f is constant on the plaque of $\mathfrak{F}$ in U which passes through q, we have $Df(q) \cdot T_q(\mathfrak{F}) = \{0\}$, so $D(f \circ g)_p \cdot T_pN = \mathbb{R}^s$, i.e., $f \circ g$ is a submersion.

Conversely, consider $p \in N$ and a distinguished map (U,f) of $\mathfrak{F}$ such that $q = g(p) \in U$. From the hypothesis, we have that $Df(q) \circ Dg(p) \cdot T_pN = \mathbb{R}^s$, i.e., the restriction $Df(q) \mid Dg(p) \cdot T_pN$ is onto $\mathbb{R}^s$ and so $Dg(p) \cdot T_pN$ contains a subspace E of dimension s such that $Df(q) \mid E : E \longrightarrow \mathbb{R}^s$ is an isomorphism. Since $T_q(\mathfrak{F}) = (Df(q))^{-1}(0)$ it follows that $E \cap T_q(\mathfrak{F}) = \{0\}$, so $T_qM = E \oplus T_q(\mathfrak{F})$, since $\dim(E) = \operatorname{cod}(T_q(\mathfrak{F}))$. Hence $T_qM = Dg(p) \cdot T_pN + T_q(\mathfrak{F})$, as desired. ■

As an immediate consequence of this theorem we have the following

Theorem 3. *Let $\mathfrak{F}$ be a C^r foliation on M and $g : N \longrightarrow M$ a C^r map transverse to $\mathfrak{F}$. Then there exists a unique C^r foliation $g^*(\mathfrak{F})$ on N of codimension* cod $(\mathfrak{F})$ *whose leaves are the connected components of the sets $g^{-1}(F)$, F a leaf of $\mathfrak{F}$.*

Proof. Let $\{(U_i,f_i,g_{ij}) \mid f_i = g_{ij} \circ f_j,\ i,j \in I\}$ be a system of distinguished maps of $\mathfrak{F}$. Since $f_i \circ g = g_{ij} \circ f_j \circ g$ on $g^{-1}(U_i) \cap g^{-1}(U_j)$, it is evident that the collection $\{(g^{-1}(U_i), f_i \circ g, g_{ij}) \mid i,j \in I\}$ is a system of distinguished C^r maps of a foliation $g^*(\mathfrak{F})$ on N of codimension cod $(\mathfrak{F})$. On the other hand, if F is a leaf of $\mathfrak{F}$ and α is a plaque of $F \cap U_i$ for some $i \in I$, then the connected components of $g^{-1}(\alpha)$ are plaques of $g^*(\mathfrak{F})$ in $g^{-1}(U_i)$ and this implies that $g^{-1}(F)$ is a union of leaves of $g^*(\mathfrak{F})$. Since $\{g^{-1}(F) \mid F$ is a leaf of $\mathfrak{F}\}$ is a cover of N, it follows that the leaves of $g^*(\mathfrak{F})$ are the connected components of the sets $g^{-1}(F)$, F a leaf of $\mathfrak{F}$. ■

In particular when g is a submersion $g^*\mathfrak{F}$ is well-defined for any foliation $\mathfrak{F}$ of M.

§4. Plane fields and foliations

A field of k-planes on a manifold M is a map P which associates to each point $q \in M$ a vector subspace of dimension k of $T_q M$.

A field of 1-planes is also called a *line field*. For example, if X is a vector field without singularities on M, we can define a line field P on M by letting $P(q) = \mathbb{R} \cdot X(q)$, the subspace of dimension 1 of $T_q M$ generated by $X(q)$.

Conversely, if P is a line field on M, we can define a vector field without singularities on M by choosing at each point $q \in M$ a nonzero vector in $P(q)$. One says that such a vector field is tangent to P. We say that a line field on M is of class C^r when for each $q \in M$ there is a C^r vector field X defined in a neighborhood V of q such that $P(x) = \mathbb{R} \cdot X(x)$ for every $x \in V$. In general, a continuous line field on M does not have a continuous tangent vector field. The figure below illustrates some examples in $\mathbb{R}^2 - \{0\}$.

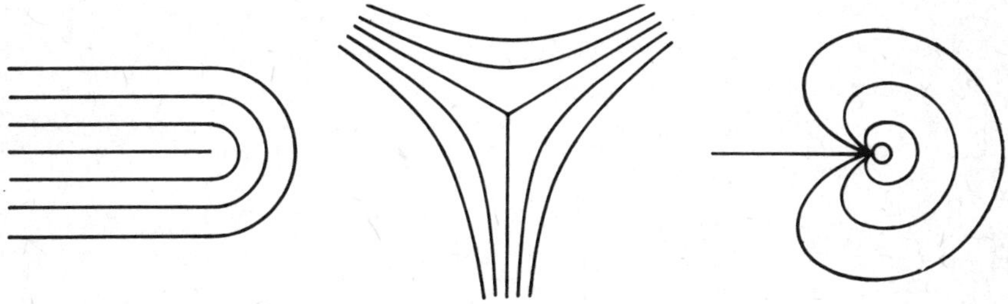

Figure 8

Analogous to the case of line fields, we say that a k-plane field P on M is of class C^r if for every $q \in M$ there exist k C^r vector fields $X^1, X^2, \ldots, X^k$, defined in a neighborhood V of q such that for every $x \in V$, $\{X^1(x), \ldots, X^k(x)\}$ is a basis for $P(x)$. A relevant fact is the following.

Proposition 2. *Every C^r $(r \geq 1)$ k-dimensional foliation $\mathfrak{F}$ on M defines a C^{r-1} k-plane field on M which will be denoted $T\mathfrak{F}$.*

Proof. Let $P(x) = T_x\mathfrak{F}$ be the subspace of $T_x M$ tangent at x to the leaf of $\mathfrak{F}$ which passes through x. Given $x_0 \in M$, let (U, φ), $x_0 \in U$, be a local chart of $\mathfrak{F}$. It is easy to see that for each $x \in U$, $P(x)$ is the subspace generated by vectors $(\partial/\partial x_i)(x) = (D\varphi(x))^{-1}(e_i)$, $i = 1, \ldots, k$ where e_i is the ith vector of the canonical basis of $\mathbb{R}^m$. Since φ is C^r, the fields $\partial/\partial x_i$, $i = 1, \ldots, k$, are C^{r-1}, and hence P is C^{r-1}. ■

In particular, if M does not admit a continuous k-plane field, then M does not possess a k-dimensional foliation. For example, the sphere S^5 does not have a continuous 2-plane field (see [53] page 142); so, there do not exist foliations of dimension 2 on S^5. A natural question then is the following. Given a k-plane field P on M, under what conditions does there exist a k-dimensional foliation $\mathfrak{F}$ such that, for each $q \in M$, $T_q\mathfrak{F} = P(q)$?

This question is answered by the theorem of Frobenius whose statement follows. Its proof is in the appendix.

Definition. One says that a plane field P is *involutive* if, given two vector fields X and Y such that, for each $q \in M$, $X(q)$ and $Y(q) \in P(q)$, then $[X, Y](q) \in P(q)$. (For the definition of the Lie bracket [,], see the appendix).

Theorem *(of Frobenius). Let P be a C^r k-plane field $(r \geq 1)$ on M. If P is involutive then there exists a C^r foliation $\mathfrak{F}$ of dimension k on M such that $T_q(\mathfrak{F}) = P(q)$ for every $q \in M$. Conversely, if $\mathfrak{F}$ is a C^r $(r \geq 2)$ foliation and P is the tangent plane field to $\mathfrak{F}$ then P is involutive.*

We say also that an involutive plane field is *completely integrable.*

In particular, if $k = 1$, P is always completely integrable. In this case the theorem reduces to the existence and uniqueness theorem for ordinary differential equations.

§5. Orientation

Given a vector space E of dimension $n \geq 1$, we say that two ordered bases of E, $\mathcal{B} = \{u_1, \ldots, u_n\}$ and $\mathcal{B}' = \{v_1, \ldots, v_n\}$, define the same orientation on E if the matrix for the change of basis, $A = (a_{ij})_{1 \leq i,j \leq n}$, defined by $v_i = \sum_{j=1}^{n} a_{ij} u_j$, has positive determinant. If B is the set of ordered bases of E, the relation "$\mathcal{B}$ and $\mathcal{B}'$ define the same orientation on E" is an equivalence relation on B, which has two equivalence classes, called *the orientations of E.* Let P be a continuous k-plane field on M. We will say that P is orientable if for each $x \in M$ it is possible to choose an orientation $\mathcal{O}(x)$ on $P(x)$ such that the map $x \mapsto \mathcal{O}(x)$ is continuous in the following sense. We consider a cover of M by open sets $(U_i)_{i \in I}$ such that for each $i \in I$, the restriction $P \mid U_i$ is defined by k continuous vector fields $X^1, \ldots, X^k$. For each $x \in U_i$, the bases $\mathcal{B}(x) = \{X^1(x), \ldots, X^k(x)\}$ and $\mathcal{B}'(x) = \{-X^1(x), X^2(x), \ldots, X^k(x)\}$ define two distinct orientations of $P(x)$, say $\mathcal{O}_i^+(x)$ and $\mathcal{O}_i^-(x)$. We say that the choice of $\mathcal{O}$ is continuous if $\mathcal{O} \mid U_i = \mathcal{O}_i^+$ for every i and if $U_i \cap U_j \neq \emptyset$ then $\mathcal{O}_i^+ = \mathcal{O}_j^+$ in the intersection.

If $k = \dim(M)$ and $P(x) = T_xM$ we say that M is orientable.

For example, a line field P on M is orientable if and only if there exists a continuous vector field X on M such that, for all $x \in M$, $P(x)$ is the subspace generated by $X(x)$. In figure 8 we see some examples of line fields that are not orientable.

§6. Orientable double coverings

The orientable double covering of a k-plane field P is defined in the following way: Let $\tilde{M} = \{(x, \mathcal{O}) \mid x \in M$ and $\mathcal{O}$ is one of the orientations of $P(x)\}$ and $\pi : \tilde{M} \longrightarrow M$ be the projection $\pi(x, \mathcal{O}) = x$. For each $x \in \tilde{M}$, $\pi^{-1}(x) = \{(x, \mathcal{O}), (x, -\mathcal{O})\}$, where $\mathcal{O}$ is one of the orientations of $P(x)$ and $-\mathcal{O}$ the other. On $\tilde{M}$ consider the topology whose basis of open sets is constructed as follows. Given $(x_0, \mathcal{O}_0) \in \tilde{M}$, let U be a neighborhood of x_0 on M where there is defined a continuous orientation $\mathcal{O}$ of $P \mid U$. We can assume $\mathcal{O}(x_0) = \mathcal{O}_0$. We define then a neighborhood V of $(x_0, \mathcal{O}_0)$ as $V = \{(x, \mathcal{O}(x)) \mid x \in U\}$. With this topology, $\pi : \tilde{M} \longrightarrow M$ is a two-sheeted covering (that is, $\pi^{-1}(x)$ is a set of two elements for every $x \in M$). Since π is a local homeomorphism we can define a C^∞ differentiable structure on $\tilde{M}$, compatible with the topology co-induced by π and such that π is a C^∞ local diffeomorphism (see example 6 of Chapter I).

The orientable double covering of P is by definition the k-plane field $\pi^*(P)$ given by $\pi^*(P)_x = D\pi(x)^{-1} P(\pi(x))$.

Example. Consider on the torus T^2 the non-orientable field P tangent to the foliation $\mathfrak{F}$ sketched in Figure 9a.

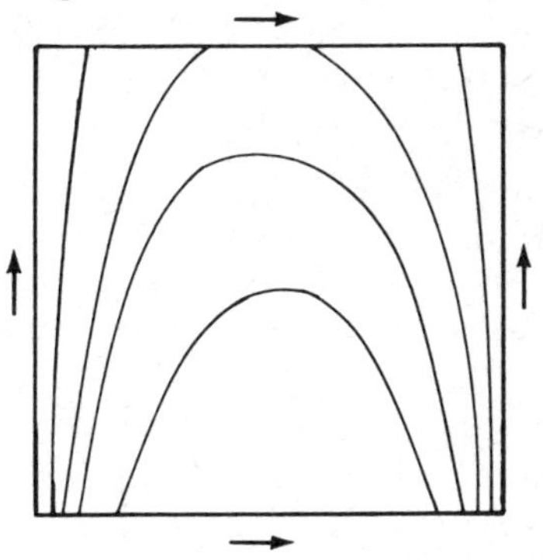

Figure 9a

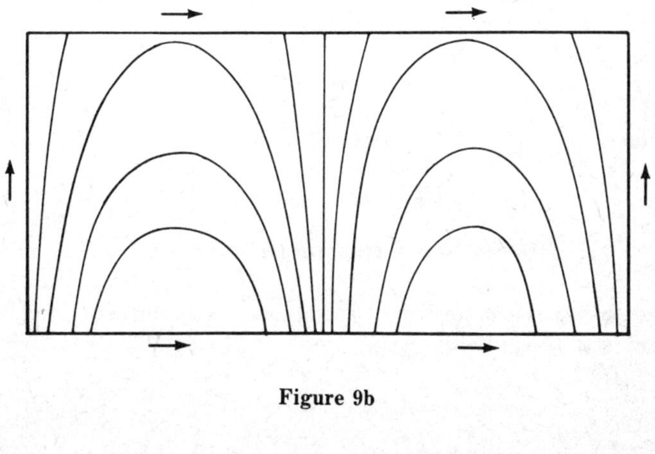

Figure 9b

In this figure, T^2 is the square $[0,1] \times [0,1]$ with sides identified according to the arrows. The vertical sides constitute a unique leaf of $\mathfrak{F}$. The double covering of P is still defined on T^2 and looks like figure 9b. In this figure, T^2 is the rectangle $[0,2] \times [0,1]$ with the sides identified.

Theorem 4. *Suppose M is connected and let P be a continuous k-plane field on M. Let $(\tilde{M}, \pi, \pi^*(P))$ be the double covering of P. Then*

(a) $\pi^*(P)$ is orientable.
(b) $\tilde{M}$ is connected if and only if P is not orientable.

Proof. (a) By definition $\pi^*(P)(x,\mathcal{O}) = (D\pi(x,\mathcal{O}))^{-1} \cdot (P(x))$, if $(x,\mathcal{O}) \in \tilde{M}$. We define an orientation $\mathcal{O}^*$ for $\pi^*(P)$ by saying that the ordered basis $\{X^1(x,\mathcal{O}), \ldots, X^k(x,\mathcal{O})\}$ of $\pi^*(P)(x,\mathcal{O})$ is in $\mathcal{O}^*(x)$ if the basis $\{D\pi(x,\mathcal{O}) \cdot X^1(x,\mathcal{O}), \ldots, D\pi(x,\mathcal{O}) \cdot X^k(x,\mathcal{O})\}$ is in $\mathcal{O}$. One easily verifies that $\mathcal{O}^*$ is a continuous orientation of $\pi^*(P)$.

(b) Suppose $\tilde{M}$ is connected. Given $x_0 \in M$, consider a continuous path $\gamma^* : [0,1] \longrightarrow \tilde{M}$ such that $\gamma^*(0) = (x_0,\mathcal{O}_0)$ and $\gamma^*(1) = (x_0, -\mathcal{O}_0)$, where $\mathcal{O}_0$ is a fixed orientation on $P(x_0)$ and $-\mathcal{O}_0$ is the opposite orientation. Let $\gamma = \pi \circ \gamma^*$. Then γ is a closed path in M with $\gamma(0) = \gamma(1) = x_0$. Suppose P were orientable with continuous orientation $\mathcal{O}$. We can suppose that $\mathcal{O}(x_0) = \mathcal{O}_0$. However, in this case the path $\tilde{\gamma}(t) = (\gamma(t), \mathcal{O}(\gamma(t)))$ satisfies $\pi \circ \tilde{\gamma}(t) = \gamma(t)$ for every $t \in [0,1]$ and $\tilde{\gamma}(0) = (x_0,\mathcal{O}_0) = \gamma^*(0)$. Therefore it is easy to see that we must have $\tilde{\gamma}(t) = \gamma^*(t)$ for every $t \in [0,1]$ and hence $\gamma^*(1) = (x_0,\mathcal{O}_0)$, a contradiction.

Conversely, suppose $\tilde{M}$ is disconnected. If $\tilde{M}_1$ is a connected component of $\tilde{M}$, we must have $\pi(\tilde{M}_1) = M$, since π is a local diffeomorphism and M is connected. This implies that $\tilde{M}$ has exactly two connected components $\tilde{M}_1$ and $\tilde{M}_2$.

In fact, if there were another connected component $\tilde{M}_3$, for every $x \in M$, $\pi^{-1}(x)$ would contain at least three points, which does not happen. It is also true that for every $x \in M$, if $(x,\mathcal{O}) \in \tilde{M}_1$ then $(x, -\mathcal{O}) \in \tilde{M}_2$. Then, $\pi \mid \tilde{M}_1$ is bijective and so $h = \pi \mid \tilde{M}_1 : \tilde{M}_1 \longrightarrow M$ is a diffeomorphism. Since $\pi^*(P)$ is orientable, it follows that the restriction $\pi^*(P) \mid \tilde{M}_1$ is orientable and we can define a continuous orientation on P by $h^{-1} : M \longrightarrow \tilde{M}_1$, as was done in part (a) of the proof. ■

Corollary. *If M is simply connected then every k-plane field is orientable* $(1 \leq k \leq \dim(M))$. *In particular, M is orientable.*

§7. Orientable and transversely orientable foliations

In the theory of foliations, more useful than the notion of orientability is that of *transverse orientability*.

Let P be a k-plane field on M. We say that $\tilde{P}$ is a field complementary to P or transverse to P if for every $x \in M$ we have $P(x) + \tilde{P}(x) = T_x M$ and $P(x) \cap \tilde{P}(x) = \{0\}$. It is a plane field of codimension k. If P is C^r it is possible to define a complementary C^r field as follows. Fix on M a Riemannian metric $\langle\ ,\ \rangle$. Let $P^\perp(x) = \{v \in T_x M \mid \langle u, v\rangle_x = 0$ for all $u \in P(x)\}$. It is clear that $P^\perp$ is a plane field complementary to P. We leave it to the reader to verify that $P^\perp$ is C^r.

Definition. Let P be a continuous k-plane field. We say that P is tranversely orientable if there exists a field complementary to P which is continuous and orientable.

Proposition 3. *If P is tranversely orientable, any continuous plane field complementary to P is orientable.*

Proof. It is enough to show the following: let $P^\perp$ be a plane field orthogonal to P and let $\tilde{P}$ be a plane field complementary to P. Then $P^\perp$ is orientable if and only if $\tilde{P}$ is orientable. For each $q \in M$ let $A_q : T_q M \longrightarrow P^\perp(q)$ be the orthogonal projection. Given $u \in T_q M$, we can write $u = u^1 + u^2$ where $u^1 \in P(q)$ and $u^2 = A_q(u) \in P^\perp(q)$ are unique. If X is a continuous vector field on M, it is easy to see that $Y(q) = A_q(X(q))$ is also a continuous field. Moreover if $F \subset T_q M$ is a plane complementary to $P(q)$ then the restric tion $A_q \mid F : F \longrightarrow P^\perp(q)$ is an isomorphism. It follows that $A_q \mid \tilde{P}(q) : \tilde{P}(q) \longrightarrow P^\perp(q)$ induces a bijection A_q^* between the orientations of $P^\perp(q)$ and the orientations of $\tilde{P}(q)$: if $\mathcal{O}$ is an orientation of $P^\perp(q)$ and $\{u_1, \ldots, u_{n-k}\} \in \mathcal{O}$ then $A_q^*(\mathcal{O})$ is the orientation determined by the ordered basis $\{A_q^{-1}(u_1), \ldots, A_q^{-1}(u_{n-k})\}$. Therefore $\mathcal{O}$ is a continuous orientation of $\mathcal{P}^\perp$ if and only if $A^*(\mathcal{O})$, defined by $A^*(\mathcal{O})(q) = A_q^*(\mathcal{O})$ is a continuous orientation of $\tilde{P}$. ■

Theorem 5. *Let P be a C^r k-plane field on M. The following properties are true:*

(a) *if P is orientable and transversely orientable, then M is orientable,*
(b) *if M is orientable then P is orientable if and only if it is transversely orientable.*

Proof. (a) Let $\mathcal{O}$ be a continuous orientation of P and $\mathcal{O}^\perp$ a continuous orientation of $P^\perp$. We define a continuous orientation $\overline{\mathcal{O}}$ of M in the following way: given $q \in M$ consider $\{u_1, \ldots, u_k\} \in \mathcal{O}(q)$ and $\{u_{k+1}, \ldots, u_n\} \in \mathcal{O}^\perp(q)$. Define $\overline{\mathcal{O}}(q)$ to be the equivalence class of the ordered basis $\{u_1, \ldots, u_n\}$. It is easily verified that $\overline{\mathcal{O}}$ is a continuous orientation of M.

(b) Suppose M and P are orientable. Let $\overline{\mathcal{O}}$ and $\mathcal{O}$ be continuous orientations of M and P respectively. We define a continuous orientation $\mathcal{O}^\perp$ of $P^\perp$ in the following manner: given $q \in M$, let $\{u_1, \ldots, u_k\} \in \mathcal{O}(q)$. We say $\{u_{k+1}, \ldots, u_n\} \in \mathcal{O}^\perp(q)$ if $\{u_1, \ldots, u_n\} \in \overline{\mathcal{O}}(q)$. It is easily shown that $\mathcal{O}^\perp$ is a continuous orientation of $P^\perp$. If M and $P^\perp$ are orientable, one proves similarly that P is orientable. ■

Definition. A C^r $(r \geq 1)$ foliation $\mathfrak{F}$ is orientable if the plane field tangent to $\mathfrak{F}$ is orientable. Similarly, $\mathfrak{F}$ is transversely orientable if the plane field tangent to $\mathfrak{F}$ is transversely orientable. We denote the plane field tangent to $\mathfrak{F}$ by $T\mathfrak{F}$.

If $\mathfrak{F}$ is not orientable, we can consider the orientable double cover of $T\mathfrak{F}$, $\pi^*(T\mathfrak{F})$, where $\pi : \tilde{M} \longrightarrow M$ is as before. Since π is a local diffeomorphism, we can define the foliation $\pi^*(\mathfrak{F})$ on $\tilde{M}$. Evidently $T(\pi^*(\mathfrak{F})) = \pi^*(T\mathfrak{F})$. So, $\pi^*(\mathfrak{F})$ is orientable.

Analogously, if $\mathfrak{F}$ is not transversely orientable, we can consider the orientable double cover $(\pi^\perp)^*(T\mathfrak{F}^\perp)$ of $T\mathfrak{F}^\perp$ where $T\mathfrak{F}^\perp$ is a plane field complementary to $T\mathfrak{F}$ and $\pi^\perp : \tilde{M} \longrightarrow M$ is the covering projection. The foliation $(\pi^\perp)^*(\mathfrak{F})$ is transversely orientable. This construction is used in the proofs of many theorems in the theory of foliations in order to reduce the arguments to the case that $\mathfrak{F}$ is orientable and transversely orientable.

The transverse orientability of a foliation $\mathfrak{F}$ of co-dimension k and class C^r, $r \geq 1$, can be expressed in terms of a collection of distinguished maps, $\mathfrak{D} = \{f_i : U_i \longrightarrow \mathbb{R}^k\}$ with $\cup_{i \in I} U_i = M$. Recall that if $U_i \cap U_j \neq \varnothing$ then there exists a diffeomorphism $g_{ij} : f_j(U_i \cap U_j) \longrightarrow f_i(U_i \cap U_j)$ such that $f_i = g_{ij} \circ f_j$. We say that $\mathfrak{D}$ is a *coherent system of distinguished maps* when, for every $i, j \in I$, $g_{ij} : f_j(U_i \cap U_j) \longrightarrow f_i(U_i \cap U_j)$ preserves the canonical orientation of $\mathbb{R}^k$. In other words, if $\{e_1, \ldots, e_k\}$ is the canonical basis of $\mathbb{R}^k$ then for every $i, j \in I$ and every $q \in f_j(U_i \cap U_j)$ the basis $\{Dg_{ij}(q) \cdot e_1, \ldots, Dg_{ij}(q) \cdot e_k\}$ has the same orientation as $\{e_1, \ldots, e_k\}$. This is equivalent to saying that $\det(Dg_{ij}(q)) > 0$ for every $q \in f_j(U_i \cap U_j)$.

Theorem 6. *Let $\mathfrak{F}$ be a C^r $(r \geq 1)$ foliation. Then $\mathfrak{F}$ is transversely orientable if and only if $\mathfrak{F}$ possesses a coherent system of distinguished maps.*

Proof. Suppose $\mathfrak{F}$ is transversely orientable. Let $\mathcal{O}^\perp$ be a continuous orientation of $P^\perp$, the plane field orthogonal to $\mathfrak{F}$. Let $\mathfrak{D} = \{(U_i, f_i, g_{ij}) \mid i, j \in I\}$ be a system of distinguished maps of $\mathfrak{F}$. From $\mathfrak{D}$ we are going to construct a coherent system $\overline{\mathfrak{D}} = \{(\overline{U}_i, \overline{f}_i, \overline{g}_{ij} \mid i, j \in J\}$ for $\mathfrak{F}$. For each $i \in I$ and each $q \in U_i$, let $A_q^i = Df_i(q) \mid P^\perp(q) : P^\perp(q) \longrightarrow \mathbb{R}^k$. Since $P^\perp(q)$ is orthogonal to $\mathfrak{F}$ at q, A_q^i is an isomorphism for every $i \in I$ and every $q \in U$. Let $\{u_1, \ldots, u_k\} \in \mathcal{O}^\perp(q)$. In the case that the basis $\{A_q^i(u_1), \ldots, A_q^i(u_k)\}$ has the same orientation as the canonical basis of $\mathbb{R}^k$, we define $\overline{f}_i(q) = f_i(q)$. In the other case we define $\overline{f}_i(q) = R(f_i(q))$ where $R : \mathbb{R}^k \longrightarrow \mathbb{R}^k$ is the reflection $R(x_1, \ldots, x_k) = (-x_1, x_2, \ldots, x_k)$. Since $\mathcal{O}^\perp$ is a continuous orientation, we have that if $\overline{f}_i(q) = f_i(q)$ for some $q \in U_i$ then $\overline{f}_i = f_i$ on U_i (since

U_i is connected). Analogously if $\overline{f}_i(q) = R(f_i(q))$ for some $q \in U_i$, then $\overline{f}_i = R \circ f_i$. The transition diffeomorphisms are defined by

(i) $\overline{g}_{ij} = g_{ij}$ in the case that $\overline{f}_i = f_i$ and $\overline{f}_j = f_j$
(ii) $\overline{g}_{ij} = g_{ij} \circ R$ in the case that $\overline{f}_i = f_i$ and $\overline{f}_j = R \circ f_j$
(iii) $\overline{g}_{ij} = R \circ g_{ij}$ in the case that $\overline{f}_i = R \circ f_i$ and $\overline{f}_j = f_j$
(iv) $\overline{g}_{ij} = R \circ g_{ij} \circ R$ in the case that $\overline{f}_i = R \circ f_i$ and $\overline{f}_j = R \circ f_j$.

One easily verifies that $\overline{f}_i = \overline{g}_{ij} \circ \overline{f}_j$ for any $i \neq j$ such that $U_i \cap U_j \neq \varnothing$. Moreover if $\overline{A}^i_q = D\overline{f}_i(q) \mid P^{\perp}(q)$, $i \in I$, $\overline{A}^i_q$ takes a basis of $P^{\perp}(q)$ in $\mathcal{O}^{\perp}(q)$ to a basis of $\mathbb{R}^k$ with the same orientation as the canonical basis. Then, if $x = \overline{f}_j(q)$, $D\overline{g}_{ij}(x) = \overline{A}^i_q \circ (\overline{A}^j_q)^{-1}$ takes the canonical basis of $\mathbb{R}^k$ to a basis with the same orientation, as we wished.

Conversely, if $\mathcal{F}$ has a coherent system of distinguished maps $\mathcal{D} = \{f_i : U_i \longrightarrow \mathbb{R}^k \mid i \in I\}$, consider on $P^{\perp}$ the orientation $\mathcal{O}^{\perp}$ induced by $A^i_q = Df_i \mid P^{\perp}$ and by the canonical orientation of $\mathbb{R}^k$: $\{u_1 ,\ldots, u_k\} \in \mathcal{O}^{\perp}(q)$ if and only if $\{A^i_q(u_1) ,\ldots, A^i_q(u_k)\}$ has the orientation of the canonical basis of $\mathbb{R}^k$. One easily verifies that $\mathcal{O}^{\perp}$ is a continuous orientation of $P^{\perp}$. ■

Notes to Chapter II

(1) Suspension of diffeomorphisms

A specific example of a foliation of dimension one, induced by a vector field without singularities, is obtained by the *suspension of a* C^r diffeomorphism $(r \geq 1)$ $f : N \longrightarrow N$ defined as follows.

On the product $N \times \mathbb{R}$ consider the equivalence relation $\sim$ defined by $(x,t) \sim (x',t')$ if and only if $t - t' = n \in \mathbb{Z}$ and $x' = f^n(x)$ (f^n is defined by $f^0 = id$, $f^n = f \circ \ldots \circ f$ (n times) if $n \geq 1$ and $f^n = f^{-1} \circ \ldots \circ f^{-1}$ ($|n|$ times) if $n \leq -1$). So, if $g(x,t) = (f^{-1}(x), t+1)$, then g is a C^r diffeomorphism of $N \times \mathbb{R}$ and $(x,t) \sim (x',t')$ if and only if $(x',t') = g^n(x,t)$ for some $n \in \mathbb{Z}$. Let $M = N \times \mathbb{R}/\sim$ be the quotient space, with the quotient topology, and $\pi : N \times \mathbb{R} \longrightarrow M$ the projection of the equivalence relation.

Given $y = \pi(x,t) \in M$, let $V = N \times (t - 1/2, t + 1/2)$ and $U = \pi(V)$. It is easily verified that $\pi^{-1}(U) = \cup_{n \in \mathbb{Z}} V_n$, where $V_n = g^n(V)$ and $\pi \mid V_n : V_n \longrightarrow U$ is a homeomorphism for every $n \in \mathbb{Z}$. One concludes then that $\pi : N \times \mathbb{R} \longrightarrow M$ is a covering map. On the other hand, for every $n \in \mathbb{Z}$, one has that $g^n = (\pi \mid V_n)^{-1} \circ \pi \mid V : V \longrightarrow V_n$ is a C^r diffeomorphism. Hence it is possible to induce on M a C^r manifold structure such that π is a local C^r diffeomorphism and $\dim(M) = \dim(N) + 1$ (see example 6 in Chapter I).

On $N \times R$, consider the foliation $\mathcal{F}_0$ whose leaves are the lines $\{x\} \times \mathbb{R}$, $x \in N$, which is tangent to the vector field $X^0(x,t) = (0,1)$ on $N \times \mathbb{R}$. This foliation and the field X^0, are invariant under the diffeomorphism g, that is, $g^*(\mathcal{F}_0) = \mathcal{F}_0$ and $g^*(X^0) = (dg)^{-1} \cdot (X^0 \circ g) = X^0$). Under these conditions, it is easy to see that there exists a foliation $\mathcal{F}$ and a field X on M such that $\mathcal{F}_0 = \pi^*(\mathcal{F})$ and $X^0 = \pi^*(X)$. The integral curves of X are the leaves of $\mathfrak{F}$. The foliation $\mathfrak{F}$ is called the *suspension of the diffeomorphism* f and the field X, the *suspension field of* f.

Let us see the relation that exists between f and $\mathfrak{F}$. Let $N_0 = \pi(N \times \{0\})$. Then N_0 is an imbedded subvariety of M, diffeomorphic to N by $h = \pi \mid N \times \{0\} \longrightarrow N_0$. Given $y = \pi(x,0) \in N_0$, let F_y be the leaf of $\mathfrak{F}$ which passes through y. Then F_y can be parametrized by $\gamma(t) = \pi(x,t)$, which is the orbit of X with initial conditions $\gamma(0) = y$. This orbit returns to N_0 at $t = 1$ in the point $\pi(x,1) = \pi(f(x),0) = h \circ f \circ h^{-1}(y)$, so we can define a map of "first return," $\tilde{g} : N_0 \longrightarrow N_0$, by $\tilde{g} = h \circ f \circ h^{-1}$. The map $\tilde{g}$ is a C^r diffeomorphism of N_0 conjugate to f ($\tilde{g} = h \circ f \circ h^{-1}$). This map is also known as the *Poincaré map* of the section N_0 and of the field X. The topological and algebraic properties of f translate to analogous properties of $\mathfrak{F}$. For example, if $x \in N$ is a periodic point of f of period k (that is, $f^{\ell}(x) = x$ if and only if $\ell = j \cdot k$ for some $j \in \mathbf{Z}$) then the leaf of $\mathfrak{F}$ containing $y = h(x)$ is a closed curve that cuts N_0 k times. If x is not a periodic point of f, then the leaf of $\mathfrak{F}$ through $h(x)$ is not compact and cuts N_0 in the set $\{h(f^n(x)) \mid n \in \mathbf{Z}\}$. If the orbit of x for f, $\{f^n(x) \mid n \in \mathbf{Z}\}$, is dense in N, the leaf of $\mathfrak{F}$ through $h(x)$ is dense in M.

As a specific example of this situation, consider the diffeomorphism $f_\alpha : S^1 \longrightarrow S^1$ defined by $f_\alpha(z) = e^{2\pi i\alpha} \cdot z$. This diffeomorphism is a rotation of angle $2\pi\alpha$. The suspension of f_α is a foliation defined on $M = T^2$, which we denote by $\mathfrak{F}_\alpha$. If $\alpha = p/q \in Q$, where p and q are relatively prime, then all the points of S^1 are periodic for f_α, which means that the leaves of $\mathfrak{F}_\alpha$ are all homeomorphic to S^1 and they cut $N_0 = \pi(S^1 \times \{0\})$ q times. If $\alpha \notin \mathbf{Q}$ all the leaves of $\mathfrak{F}_\alpha$ are dense in T^2 (see exercise 13).

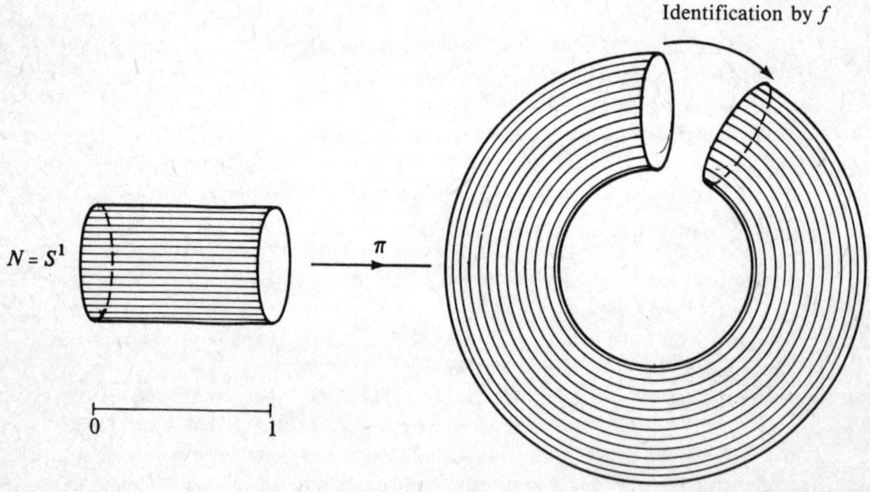

Figure 10

The method of suspension will be generalized in Chapter V. It is used, as we will see, to construct examples of foliations exhibiting specific recurrence phenomena.

(2) The Hopf fibration

Let $S^{2n-1} = \{(z_1, \ldots, z_n) \in \mathbb{C}^n \mid \sum_{i=1}^n |z_i|^2 = 1\}$, $n \geq 2$. Given an n-tuple of integers $k = (k_1, \ldots, k_n)$, where $k_i \neq 0$ for all $i = 1, \ldots, n$, we can associate an action $\varphi_k : S^1 \times S^{2n-1} \longrightarrow S^{2n-1}$ letting

$$\varphi_k(u, z_1, \ldots, z_n) = (u^{k_1} z_1, \ldots, u^{k_n} z_n) .$$

It is easy to verify that the vector field X_k defined on S^{2n-1} by

$$X_k(z_1, \ldots, z_n) = (ik_1 z_1, \ldots, ik_n z_n)$$

is tangent to the orbits of φ_k and hence φ_k is locally free, since $X_k(z) \neq 0$ for every $z \in S^{2n-1}$. The leaves of the foliation $\mathfrak{F}_k$, induced by φ_k, are all homeomorphic to S^1. In the case that $k_1 = k_2 = \cdots = k_n = 1$, the foliation $\mathfrak{F}_k$ is known as the *Hopf fibration.* In this case, for every $z \in S^{2n-1}$ the isotropy group of z is $\{1\}$ and, for this reason, it is possible to prove that the orbit space of φ_k has a differentiable manifold structure. This manifold is diffeomorphic to the *complex projective space* $\mathbb{C}P^{n-1}$, which is defined as being the quotient space of $\mathbb{C}^n - \{0\}$ by the equivalence relation which identifies two points $z, w \in \mathbb{C}^n - \{0\}$ if there is a $\lambda \in \mathbb{C}^*$ such that $z = \lambda w$. The projective space $\mathbb{C}P^{n-1}$ has real dimension $2n - 2$, and since the projection for this equivalence $\pi : S^{2n-1} \longrightarrow \mathbb{C}P^{n-1}$ is a submersion, $\mathfrak{F}_k = \pi^*(\mathfrak{F}_0)$, where $\mathfrak{F}_0$ is the foliation of $\mathbb{C}P^{n-1}$ by points.

In the case $n = 2$, $\mathbb{C}P^{n-1}$ is diffeomorphic to S^2 and the projection π can be defined, for example, by the expression:

$$\pi(z_1, z_2) = (|z_1|^2 - |z_2|^2, 2\mathrm{Re}(z_1 \bar{z}_2), 2\mathrm{Im}(z_1 \bar{z}_2)) \in S^2 .$$

(3) Holomorphic vector fields

A holomorphic flow is a holomorphic action $\Phi : \mathbb{C} \times M \longrightarrow M$ where M is a complex manifold (locally diffeomorphic to $\mathbb{C}^m$ by a holomorphic diffeomorphism). The flow Φ induces a holomorphic field Z on M defined by $Z(p) = (d\Phi(T,p)/dT)\big|_{T=0}$. If $Z(p) = 0$ then $\Phi(T,p) = p$ for every $T \in \mathbb{C}$, or, the orbit of p is the point p. In the case $Z(p) \neq 0$, the orbit of Φ through p has complex dimension 1 (2 real).

Consider for example the flow on $\mathbb{C}^n$ defined by

$$\Phi(T, z) = \exp(T \cdot A) \cdot z, \qquad T \in \mathbb{C}, \qquad z \in \mathbb{C}^n ,$$

where $A \in \mathfrak{D}(n, \mathbb{C})$, the set of complex $n \times n$ matrices.

The field induced by Φ is the linear field $z \longmapsto A \cdot z$. If A is nonsingular, Φ defines a foliation of two real dimensions in $\mathbb{C}^n - \{0\}$.

Let us look at the case where $A = \mathrm{diag}(\lambda_1, \ldots, \lambda_n)$, where $\lambda_i \notin \mathbb{R}\lambda_j$ if $i \neq j$. In this case, the eigenspaces E_i, relative to the eigenvalues λ_i, are invariant under the field A, $(A(E_i) = E_i)$. So, for $i = 1, \ldots, n$, $E_i - \{0\}$ is an orbit of Φ homeomorphic to $\mathbb{R} \times S^1$. The orbits of Φ through points outside $E_1 \cup \ldots \cup E_n$ are all homeomorphic to $\mathbb{C}$. Let $\mathfrak{K}(A) \subset \mathbb{C}$ be the closed convex hull of the set $\{\lambda_1, \ldots, \lambda_n\}$. $\mathfrak{K}(A)$ is the smallest polygonal convex set which contains $\{\lambda_1, \ldots, \lambda_n\}$. If $0 \notin \mathfrak{K}(A)$ it is possible to prove that there exists an $r_0 \in \mathbb{C} - \{0\}$ such that $\lim_{t \to +\infty} \Phi(t \cdot r_0, z) = 0$ for all $z \in \mathbb{C}^n$. This means that there exists a sphere S of real dimension $2n - 1$, transverse to the trajectories of the real "subflow" $X_t(z) = \Phi(t \cdot r_0, z)$. So the intersection of orbits of Φ with S define a foliation $\mathfrak{F}$ of (real) dimension 1 on S. The intersection of each eigenspace E_i with S is a closed curve. Any other leaf of $\mathfrak{F}$ is homeomorphic to $\mathbb{R}$. For example, if $n = 2$, $\mathfrak{F}$ has two leaves homeomorphic to S^1 and any other has as its limit set the union of these two.

In the case that $A = \operatorname{diag}(1, \ldots, 1)$, the identity on $\mathbb{C}^n$, one sees that the foliation $\mathfrak{F}$ is the Hopf fibration of S^{2n-1}.

Another example of a foliation generated by a holomorphic differential equation is the following.

Consider the holomorphic differential equation

(1) $$y'' = f(x, y, y') \qquad x \in \mathbb{C}$$

where the derivatives of y are taken with respect to the parameter x. Setting $z = y'$, equation (1) is transformed to the system

(2) $$y' = z$$

$$z' = f(x, y, z)$$

on $\mathbb{C}^3$. This equation induces a holomorphic vector field Z on $\mathbb{C}^3$ given by

(3) $$Z(x, y, z) = (1, z, f(x, y, z))$$

whose integral surfaces contain the solutions of (2). The field Z can be considered as a plane field of real dimension 2 on $\mathbb{C}^3$, completely integrable and transverse to the fibers $x =$ constant.

(4) Turbulization of a foliation

Let $\mathfrak{F}$ be a codimension one foliation of a three manifold M. Suppose there exists an imbedding

$$\varphi : S^1 \times D^2 \longrightarrow M$$

such that:

(i) $\varphi \mid S^1 \times \{x\}$ is transverse to $\mathfrak{F}$ for each $x \in D^2$
(ii) $\varphi(\theta \times D^2)$ is contained in a leaf for all $\theta \in S^1$.

In this case it is possible to modify $\mathfrak{F}$ in the solid torus $\varphi(S^1 \times D^2)$ without changing the foliation on the outside, as follows:

In the solid cylinder $\mathbb{R} \times D^2$ consider the C^∞ foliation defined by the cylinder $x_1^2 + x_2^2 = 1/2$ plus the graphs of the functions $x_3 = \alpha(x_1^2 + x_2^2) + b$, $b \in \mathbb{R}$, on the points (x_1, x_2) with $x_1^2 + x_2^2 \neq 1/2$.

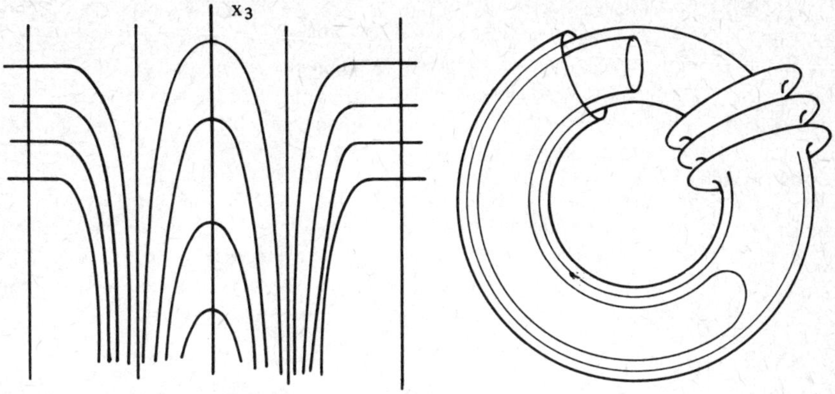

Figure 11

Here $\alpha(r) = 0$ in a neighborhood of $r = 1$ and $(d^n\alpha(r)/dr^n) \longrightarrow -\infty$ when $r \longrightarrow 1/2$ for all $n \geq 0$. Since this foliation is invariant under translation along the x_3-axis, it defines a folaition of $S^1 \times D^2$ which in turn induces via φ a foliation $\mathfrak{F}'$ of M with one more compact leaf. In the region $\varphi(S^1 \times D_{1/2})$, the foliation $\mathfrak{F}'$ has a Reeb component, where $D_{1/2}$ is the disk of radius $1/2$.

We say $\mathfrak{F}'$ is obtained from $\mathfrak{F}$ by turbulization.

(5) Foliations of three-manifolds

We will sketch here the proof of the theorem of Lickorish [25], Novikov-Zieschang [40]: every compact, orientable manifold of dimension three admits a foliation of codimension one. The proof of [40] is based on the following result on the topology of three-manifolds.

Theorem (H. Wallace [58]). *Let M be a compact, connected, orientable manifold of dimension three. There exist imbedded, disjoint solid tori T_i in M and T_i' in S^3, $i = 1, \ldots, n$, such that $M - \cup_{i=1}^{n} T_i$ and $S^3 - \cup_{i=1}^{n} T_i'$ are diffeomorphic.*

One shows then that the tori T_i' can be taken transverse to a Reeb foliation of S^3. Using the turbulization process, we obtain a C^∞ foliation of $S^3 - \cup_{i=1}^{n} T_i'$ where each $\partial T_i'$ in the boundary is a leaf diffeomorphic to T^2. By the theorem of Wallace this induces a foliation of $M - \cup_{i=1}^{n} T_i$ where the ∂T_i's are leaves. This foliation is completed to a foliation of M by putting a Reeb foliation in each solid torus T_i.

Remark. Using Alexander decomposition, B. Lawson ([24]) proved in 1970 the existence of codimension-one foliations on the spheres S^{2^k+3}, $k = 1,2,3$. This result was generalized to any odd-dimensional sphere by Durfee [9] and Tamura [54].

Later, and with different methods, W. Thurston ([55],[56]) proved that any plane field of codimension one on a compact orientable manifold of dimension n is homotopic to a completely integrable plane field, that is, tangent to a foliation.

In the case $n = 3$ this result was already known [60]. In particular, one has that any compact manifold with zero Euler characteristic admits a foliation of codimension one.

(6) Geodesic flows

Let M be a compact Riemannian manifold. A unit vector $v \in T_pM$ determines a unique geodesic γ passing through p in the direction of v. We denote by $\varphi_t(p,v)$ the point of M at distance t from p measured along γ in the direction of v. Since the vector $(d\varphi_s(p,v)/ds)\big|_{s=t}$ has length 1, the map $\Phi_t : T^1M \longrightarrow T^1M$ given by $\phi_t(p,v) = (\varphi_t(p,v),\ (d\varphi_s(p,v)/ds)\big|_{s=t})$ is well-defined on the fiber space T^1M of unit tangent vectors to M. (See example 3, chapter V). Setting $\Phi_t(p,v) = \Phi_{-t}(p,-v)$ for $t < 0$, we obtain a flow $\Phi : \mathbb{R} \times T^1M \longrightarrow T^1M$, given by $\Phi(t,(p,v)) = \Phi_t(p,v)$. This is called the geodesic flow of M.

On manifolds of negative sectional curvature, the geodesic flows has the following property. There is a decomposition of the tangent fiber space $T(T^1M)$ invariant under $D\Phi_t (t \in \mathbb{R})$:

$$T(T^1M) = E^s \oplus E^u \oplus X$$

where $\oplus$ denotes the Whitney sum and X the space generated by the velocity vector of the geodesic flow, and there are positive constants $a,b,\lambda > 0$ such that

$$\|D\Phi_t(p,v)\cdot\xi\| \le ae^{-\lambda t}\|\xi\| \quad \text{if} \quad t > 0 \quad \text{and} \quad \xi \in E^s$$

$$\|D\Phi_t(p,v)\cdot\xi\| \le be^{\lambda t}\|\xi\| \quad \text{if} \quad t < 0 \quad \text{and} \quad \xi \in E^u\ .$$

The plane fields generated by $E^u \oplus X$ and $E^s \oplus X$ are completely integrable, i.e., they define two foliations which intersect transversely along orbits of Φ. The study of these foliations, taking into account the dynamical properties of the system, proves that the set of periodic orbits of Φ is dense in T^1M, that is, given $(p,v) \in T^1M$ and $\epsilon > 0$ there exists $(p',v') \in T^1M$, ϵ-close to (p,v) such that the geodesic which passes through p' in the direction of v' is closed. This result is due to Anosov [2]. We also recommend [3] and [4].

III. THE TOPOLOGY OF THE LEAVES

We saw in the previous chapter that the leaves of a C^r foliation inherit a C^r differentiable manifold structure immersed in the ambient manifold. In this chapter we will study the topological properties of these immersions, giving special emphasis to the asymptotic properties of the leaves.

§1. The space of leaves

Let M^m be a foliated manifold with a foliation $\mathfrak{F}$ of dimension $n < m$. The *space of leaves of* $\mathfrak{F}$, $M/\mathfrak{F}$, is the quotient space of M under the equivalence relation R which identifies two points of M if they are on the same leaf of $\mathfrak{F}$. From the definition of leaf it is clear that this relation coincides with the relation R defined in §1 of Chapter II. On $M/\mathfrak{F}$ take the quotient topology. The topology of $M/\mathfrak{F}$ is in general very complicated, possibly being non-Hausdorff as in the case of the Reeb foliation of S^3, or of the foliation of $\mathbb{R}^2$ as in figure 6 of Chapter I.

Let $A \subset M$. The *saturation of* A *in* $\mathfrak{F}$ is by definition the set $\mathfrak{F}(A) = \{x \in M \mid xRy \text{ for some } y \in A\}$. If $\pi : M \longrightarrow M/\mathfrak{F}$ is the projection for the quotient, we have $\mathfrak{F}(A) = \pi^{-1}(\pi(A)) = \cup_{x \in A} F_x$, where by F_x we denote the leaf of $\mathfrak{F}$ containing x.

Theorem 1. *The projection* π *is an open map, or, the saturation* $\mathfrak{F}(A)$ *of an open subset* A *of* M *is open.*

Proof. Let $p \in \mathfrak{F}(A)$ and F the leaf of $\mathfrak{F}$ through p. Then $F \cap A \neq \emptyset$, and if $q \in A \cap F$ there is a path of plaques $\alpha_1, \ldots, \alpha_k$ such that $q \in \alpha_1$ and $p \in \alpha_k$.

Suppose that each α_j is a plaque of U_j with $(U_j,\varphi_j) \in \mathfrak{F}$ and let $\varphi_j(U_j) = U_j' \times U_j''$ where U_j' and U_j'' are open disks in $\mathbb{R}^n$ and $\mathbb{R}^{m-n}$ respectively, $j = 1,\ldots,k$. Suppose that for some $j \in \{1,\ldots,k\}$ there is an $x \in \alpha_j$ which has an open neighborhood $V \subset \mathfrak{F}(A) \cap U_j$. Since $\varphi_j : U_j \longrightarrow U_j' \times U_j''$ is a homeomorphism, $\varphi_j(V)$ is open in $U_j' \times U_j''$, so if $\pi_2 : U_j' \times U_j'' \longrightarrow U_j''$ is the projection onto the second factor, $\pi_2^{-1}(\varphi_j^2(V))$ is open in $U_j' \times U_j''$. On the other hand $W = \varphi_j^{-1}(\pi_2^{-1}(\varphi_j^2(V)))$ is open and $\alpha_j \subset W \subset \mathfrak{F}(A)$ and so α_j is in the interior of $\mathfrak{F}(A)$. It suffices now to note that since A is open, α_1 is in the interior of $\mathfrak{F}(A)$, so $\alpha_1 \cap \alpha_2$ is also in the interior of $\mathfrak{F}(A)$ and from the previous argument, there is a neighborhood W of α_2 such that $\alpha_2 \subset W \subset \mathfrak{F}(A)$. Repeating this process $k-1$ times, one proves inductively that α_k is in the interior of $\mathfrak{F}(A)$ and hence that $\mathfrak{F}(A)$ is open. ■

In general the projection π is not closed. When this occurs, each leaf F of $\mathfrak{F}$ is a closed subset of M, since $F = \pi^{-1}(\pi(x))$ where $x \in F$. The foliation in figure 6 of Chapter I is an example where π is not closed, although each leaf of $\mathfrak{F}$ is a closed subset of $\mathbb{R}^2$.

Definition. We say that a set $A \subset M$ is *invariant* or *saturated in* $\mathfrak{F}$ when the saturation of A in $\mathfrak{F}$ is A, that is, $\pi^{-1}(\pi(A)) = A$.

Theorem 2. *Let $A \subset M$ be an invariant subset in $\mathfrak{F}$. Then its interior $\mathring{A}$, its closure $\overline{A}$ and its boundary ∂A are also invariant in $\mathfrak{F}$.*

Proof. The set $\mathring{A}$ is characterized as being the largest open set contained in A, that is, if B is an open set such that $\mathring{A} \subset B \subset A$ then $B = \mathring{A}$. Since π is open we have that $\pi^{-1}(\pi(\mathring{A})) = B$ is open. Further, $\mathring{A} \subset B \subset A$, since A is invariant, hence $\mathring{A} = B = \pi^{-1}(\pi(\mathring{A}))$. Observe now that if A is invariant then $M - A$ is also, so $\operatorname{int}(M - A)$ is. On the other hand $\operatorname{int}(M - A) = M - \overline{A}$, or, $M - \overline{A}$ is invariant and hence $\overline{A}$ is invariant. Since $\partial A = \overline{A} - \mathring{A}$, it follows that ∂A is also invariant.

§2. Transverse uniformity

Let Σ be a submanifold of M. We say that Σ is transverse to $\mathfrak{F}$ when Σ is transverse to every leaf of $\mathfrak{F}$ that it meets. When $\dim(\Sigma) + \dim(\mathfrak{F}) = \dim(M)$ we say Σ is a *transverse section of* $\mathfrak{F}$.

Given $p \in M$ there is always a transverse section of $\mathfrak{F}$ passing through p. In fact it suffices to consider a local chart $(U,\varphi) \in \mathfrak{F}$ with $p \in U$, $\varphi(U) = U_1 \times U_2 \subset \mathbb{R}^n \times \mathbb{R}^s$, $\varphi(p) = (c_1,c_2)$ and take $\Sigma = \varphi^{-1}(\{c_1\} \times U_2)$. Since $\{c_1\} \times U_2$ is transverse to the plaques $U_1 \times \{c\}$, $c \in U_2$ it is clear that Σ is a transverse section of $\mathfrak{F}$.

Theorem 3 (Transverse uniformity of $\mathfrak{F}$). *Let F be a leaf of $\mathfrak{F}$. Given $q_1,q_2 \in F$, there exist transverse sections of $\mathfrak{F}$, Σ_1 and Σ_2, with $q_i \in \Sigma_i$ $(i = 1,2)$ and a C^r diffeomorphism, $f : \Sigma_1 \longrightarrow \Sigma_2$ such that for any leaf F' of $\mathfrak{F}$, one has $f(F' \cap \Sigma_1) = F' \cap \Sigma_2$.*

Proof. Let $q_1,q_2 \in F$. Consider a path of plaques α_1 ,..., α_k joining q_1 and q_2, that is $q_1 \in \alpha_1$ and $q_2 \in \alpha_k$. Suppose that, for each j, α_j is a plaque in $(U_j,\varphi_j) \in \mathfrak{F}$. By Lemma 1 in Chapter II we can suppose that if $U_i \cap U_j \neq \varnothing$ then $U_i \cup U_j$ is contained in some open set, the domain of a local chart of $\mathfrak{F}$. Let $\varphi_j(U_j) = U_1^j \times U_2^j \subset \mathbb{R}^n \times \mathbb{R}^s$. Fix points $p_j \in \alpha_j \cap \alpha_{j+1}$, $1 \leq j \leq k-1$, $p_0 = q_1$ and $p_k = q_2$. For each $j = 1$,..., k, set $\varphi_j(p_j) = (x_j,y_j)$ and consider the transverse disk $D_j = \varphi_j^{-1}(\{x_j\} \times U_2^j)$. For $j = 0$, let $\varphi_1(p_0) = (x_0,y_0)$ and $D_0 = \varphi_1^{-1}(\{x_0\} \times U_2^1)$.

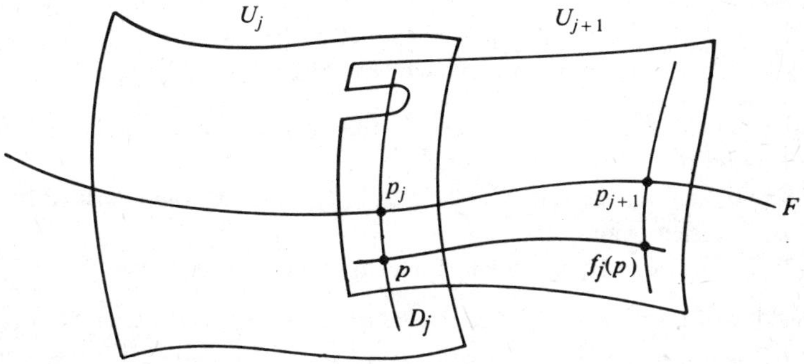

Figure 1

Since for each $j = 0$,..., $k-1$, $U_j \cup U_{j+1}$ is contained in a chart of $\mathfrak{F}$, there is a disk B_j of dimension s with $p_j \in B_j \subset D_j \cap U_{j+1}$ such that each plaque of U_{j+1} cuts B_j at most one time. We can then define an injective map $f_j : B_j \longrightarrow D_{j+1}$ calling $f_j(p)$, the point of intersection of the plaque of U_{j+1} which passes through p with D_{j+1}. So we have that $f_j(p_j) = p_{j+1}$ and that $f_j : B_j \longrightarrow f_j(B_j) \subset D_{j+1}$ is a C^r diffeomorphism. From the construction it is immediate that, for every leaf F' of $\mathfrak{F}$, $f_j(F' \cap B_j) = F' \cap f_j(B_j)$ if $j = 0,1$,..., $k-1$. Finally take a disk Σ_1 such that $p_0 \in \Sigma_1 \subset B_0 \cap f_0^{-1}(B_1) \cap \ldots \cap f_0^{-1}(f_1^{-1}(\ldots f_{k-1}^{-1}(D_k) \ldots))$ and define $f : \Sigma_1 \longrightarrow D_k$ by $f(p) = f_{k-1}(f_{k-2}(\ldots f_1(f_0(p)) \ldots))$. It is clear that f is a diffeomorphism on $\Sigma_2 = f(\Sigma_1) \subset D_k$ and that for every leaf F' of $\mathfrak{F}$ we have $f(F' \cap \Sigma_1) = F' \cap \Sigma_2$, as we wanted.

Let us see some consequences of this theorem.

Theorem 4. *Let $\mathfrak{F}$ be a foliation of M, F a leaf of $\mathfrak{F}$ and Σ a transverse section of $\mathfrak{F}$ such that $\Sigma \cap F \neq \emptyset$. We have three possibilities:*

(1) $\Sigma \cap F$ is discrete and in this case F is an imbedded leaf.
(2) The closure of $\Sigma \cap F$ in Σ contains an open set.
This occurs if and only if the closure $\overline{F}$ has non-empty interior and $\text{int}(\overline{F}) = \overline{F} - \partial \overline{F}$ is an open set which contains F. In this case we say F is locally dense.
(3) $\overline{\Sigma \cap F}$ is a perfect set (i.e., without isolated points) with empty interior. In this case we say F is an exceptional leaf.

Proof. Suppose $p \in \Sigma \cap F$ and $q \in F$, $q \neq p$. By Theorem 3 there exist transverse disks Σ_1 and Σ_2 with $p \in \Sigma_1 \subset \Sigma$, $q \in \Sigma_2$ and a diffeomorphism $f : \Sigma_1 \longrightarrow \Sigma_2$ such that, for every leaf F' of $\mathfrak{F}$, $f(F' \cap \Sigma_1) = F' \cap \Sigma_2$. In particular $f(F \cap \Sigma_1) = F \cap \Sigma_2$, therefore $F \cap \Sigma_1$ is homeomorphic to $F \cap \Sigma_2$ and $\overline{F \cap \Sigma_1}$ is homeomorphic to $\overline{F \cap \Sigma_2}$ (by f). In the case $q \in \Sigma$ we can take $\Sigma_2 \subset \Sigma$. From Theorem 3 we have the following possibilities:

(1) p is an isolated point of $\Sigma \cap F$. In this case $\Sigma \cap F$ consists entirely of isolated points, i.e., is discrete,
(2) p is an interior point of $\overline{\Sigma \cap F}$ in Σ. In this case all the points of $\Sigma \cap F$ are interior points of $\overline{\Sigma \cap F}$,
(3) $\overline{\Sigma \cap F}$ has empty interior in Σ and $\Sigma \cap F$ is not discrete.

Consider alternative (1). In this case, if $q \in F$ and (U, φ) is a foliation chart of $\mathfrak{F}$ with $q \in U$ and $\varphi(U) = U_1 \times U_2$ then $D^s = \varphi^{-1}(\{x\} \times U_2)$ is a transverse disk to $\mathfrak{F}$, so $F \cap D^s$ is a discrete set which contains q. Therefore, there is a smaller disk $\tilde{U}_2 \subset U_2$ such that $D = \varphi^{-1}(\{x\} \times \tilde{U}_2)$ cuts F only in q, or $\varphi^{-1} \mid U_1 \times \tilde{U}_2 : U_1 \times \tilde{U}_2 \longrightarrow M$ is a C^r imbedding such that $F \cap \varphi^{-1}(U_1 \times \tilde{U}_2)$ contains only one plaque of U; hence F is a C^r submanifold of M. Conversely, if F is a C^r submanifold of M one proves easily that $F \cap \Sigma$ is discrete.

Let us look at alternative (2). Let $q \in F$ and $(U, \varphi) \in \mathfrak{F}$, $q \in U$, with $\varphi(U) = U_1 \times U_2$, as above. Let $D^s = \varphi^{-1}(\{x\} \times U_2)$. Then q is an interior point of $\overline{F \cap D^s}$, so there is a disk $\tilde{U}_2 \subset U_2$ such that $D = \varphi^{-1}(\{x\} \times \tilde{U}_2) \subset \overline{F \cap D^s}$. Therefore $\overline{F}$ contains the open set $A = \varphi^{-1}(U_1 \times \tilde{U}_2)$. On the other hand, $\overline{F}$ is invariant under $\mathfrak{F}$ so $\overline{F}$ contains the saturation of A, $\mathfrak{F}(A)$, which is open, by Theorem 2. Hence, $F \subset \mathfrak{F}(A) \subset \overline{F}$ and $\mathfrak{F}(A) = \overline{F} - \partial \overline{F}$ as we wanted.

Alternative (3) is the complement of the other two and this concludes the proof. ■

As a consequence of this theorem, the immersed submanifold F, sketched in the figure below, is not a leaf of any foliation defined on an open set of $\mathbb{R}^2$.

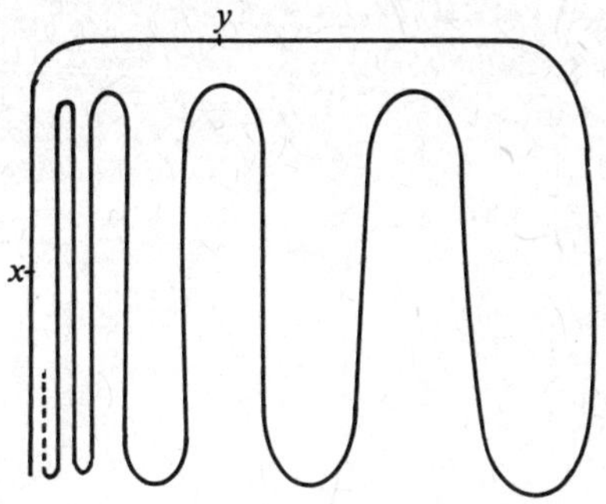

Figure 2

The reason is that a point x, as in the figure, is an accumulation point of points of $\Sigma \cap F$, where Σ is a transverse segment passing through x, while the same does not happen with the point y.

§3. Closed leaves

Theorem 5. *Let F be a leaf of a foliation $\mathfrak{F}$ on M. The following statements are equivalent:*

(a) *F is a closed leaf,*
(b) *if (U,φ) is a foliation chart of $\mathfrak{F}$ such that $\overline{U}$ is compact then $U \cap F$ contains a finite number of plaques of U,*
(c) *the immersion $i : F \longrightarrow M$, $i(x) = x$, is proper* (*). *In particular, if F is closed then F is imbedded.*

Proof. Let F be a closed leaf of $\mathfrak{F}$ and (U,φ) a foliation chart of $\mathfrak{F}$ with $\varphi(U) = U_1 \times U_2 \subset \mathbb{R}^n \times \mathbb{R}^s$ such that $U \cap F \neq \emptyset$. Let $D \subset \overline{D} \subset U_2$ be a disk such that $\Sigma = \varphi^{-1}(\{x\} \times D)$ is a transverse section with $\overline{\Sigma} \subset U$ and $\Sigma \cap F \neq \emptyset$. Since the plaques of $U \cap F$ are open and disjoint in the intrinsic topology of F, $U \cap F$ contains at most a countable number of plaques and therefore $\overline{\Sigma} \cap F$ is countable. Since F is closed, $\overline{F \cap \Sigma} = F \cap \overline{\Sigma}$ is countable. However, we know from general topology that a perfect set is uncountable, hence F necessarily satisfies alternative (1) of Theorem 4, so, $\overline{\Sigma} \cap F$ is discrete and hence finite since $\overline{\Sigma}$ is compact. We conclude therefore that $F \cap \varphi^{-1}(U_1 \times D)$

(*) A continuous map $f : X \longrightarrow Y$ is proper if for every compact $K \subset Y$, $f^{-1}(K)$ is compact.

contains a finite number of plaques of U. Taking $\overline{U}$ compact, this implies that $F \cap U$ contains a finite number of plaques. Conversely, suppose that for each foliation chart (U, φ) of $\mathfrak{F}$ with $\overline{U}$ compact, $U \cap F$ contains a finite number of plaques of U. In this case, if p_n, $n \in \mathbb{N}$, is a sequence in F with $\lim_n p_n = = p \in M$, we take a foliation chart (U, φ) of $\mathfrak{F}$ such that $p \in U$ and $U \cap F$ contains a finite number of plaques. Then there exists an n_0 and a plaque α of U such that $p_n \in \alpha$ for all $n \geq n_0$. So $p \in \alpha$ and hence $p \in F$, i.e., F is a closed leaf of $\mathfrak{F}$.

We will show that (a) implies (c). If F is a closed leaf of $\mathfrak{F}$, then $i : F \longrightarrow M$ is an imbedding and the intrinsic topology of F coincide with the topology induced by the topology of M. Thus if $K \subset M$ is compact, $K \cap F = i^{-1}(K)$ is compact in F, so the immersion $i : F \longrightarrow M$ is proper.

We will now prove that (c) implies (b). If $i : F \longrightarrow M$ is proper, given a foliation chart (U, φ) of $\mathfrak{F}$ such that U is compact, $i^{-1}(U)$ is compact in F; so, $U \cap F$ contains a finite number of plaques, which shows that F is a closed leaf.

§4. Minimal sets of foliations

Let M be a manifold with a foliation $\mathfrak{F}$. A subset $\mu \subset M$ is *minimal* if it satisfies the following properties:

(a) μ is closed, nonempty and invariant,
(b) if $\mu' \subset \mu$ is a closed, invariant subset then either $\mu' = \emptyset$ or $\mu' = \mu$.

Examples.

(1) Every closed leaf of $\mathfrak{F}$ is a minimal subset.
(2) The only minimal set of the Reeb foliation of S^3 is the compact leaf homeomorphic to T^2.
(3) Let $\mathfrak{F}_\alpha$ be the foliation of T^2 obtained by a suspension of the rotation $f_\alpha : S^1 \longrightarrow S^1$, $f_\alpha(z) = e^{2\pi i \alpha} \cdot z$. If α is irrational, all the leaves of $\mathfrak{F}_\alpha$ are dense in T^2 and, therefore T^2 is the only minimal set of $\mathfrak{F}_\alpha$.

Theorem 6. *Every foliation of a compact manifold has a minimal set.*

Proof. Let M be a compact manifold and $\mathfrak{F}$ a foliation of M. Denote by $\mathcal{C}$ the collection of nonempty compact subsets of M invariant in $\mathfrak{F}$. It is clear that $\mathcal{C}$ is nonempty. If F is a leaf of $\mathfrak{F}$ then $\overline{F} \in \mathcal{C}$. Consider on $\mathcal{C}$ the partial ordering induced by set inclusion. Given a sequence $\mu_1 \supset \mu_2 \supset \ldots$ of elements of $\mathcal{C}$, it is clear that $\mu = \bigcap_{i=1}^{\infty} \mu_i$ is nonempty, invariant, compact and $\mu \subset \mu_i$ for every $i \in \mathbb{N}$. By Zorn's lemma, $\mathcal{C}$ contains minimal elements. A minimal element of $\mathcal{C}$ is obviously a minimal subset of $\mathfrak{F}$. ■

Theorem 7. *Let μ be a minimal subset of $\mathfrak{F}$. The following properties hold:*

(a) *Every leaf of $\mathfrak{F}$ contained in μ is dense in μ.*
(b) *If M is connected and μ has nonempty interior then $\mu = M$.*
(c) *Let $p = \operatorname{cod}\mathfrak{F}$ and Σ a p-dimensional disk transverse to $\mathfrak{F}$ such that $\mu \cap \Sigma \neq \emptyset$. If μ is not a closed leaf then $\mu \cap \Sigma$ is a perfect set.*
(d) *Moreover, if $p = 1$, $\partial\Sigma \cap \mu \neq \emptyset$ and μ has empty interior but is not a closed leaf, then $\mu \cap \Sigma$ is homeomorphic to a Cantor set. In this case we say μ is an* exceptional minimal set.

Proof.

(a) Let F be a leaf of $\mathfrak{F}$ contained in μ. We have then that $\overline{F}$ is closed, $\overline{F} \neq \emptyset$ and $\overline{F} \subset \mu$, since μ is invariant. As μ is minimal, $\overline{F} = \mu$.

(b) Since μ has nonempty interior, there exists an open subset $A \subset \mu$. By (a), all leaves of μ meet A, so $\mathfrak{F}(A) = \mu$. Since $\mathfrak{F}(A)$ is open in M, μ is closed and open in M, which is connected, so $\mu = M$.

(c) Given $x \in \Sigma \cap \mu$, let F be the leaf of $\mathfrak{F}$ through x. Since μ is not a closed leaf, there exists a leaf $F' \neq F$, contained in μ. Taking a local chart of $\mathfrak{F}$ containing x and using (a), one concludes that x is an accumulation point of $F' \cap \Sigma$.

(d) By (c), $\mu \cap \Sigma$ is a compact, perfect subset with empty interior in Σ. The proof is reduced to the following lemma, which can be found in [12], p. 123. ■

Lemma. *Let $K \subset \mathbb{R}$ be a compact perfect subset with empty interior. Then there exists a homeomorphism $h : K \longrightarrow \mathcal{C}$ where $\mathcal{C}$ is the Cantor set in* $[0,1]$.

Notes to Chapter III

(1) Foliated structures of the plane.

In Example 5 of Chapter I we saw a foliation of $\mathbb{R}^2$ whose leaf space is a non-Hausdorff manifold of dimension one, equivalent to the manifold obtained by taking two copies of the real line and identifying points with the same negative coordinate.

One can say that this space is obtained by branching over $\mathbb{R}$, at the point $0 \in \mathbb{R}$, a ray in the positive direction. If we repeat this process at all the rationals, we get a manifold called a simple feather for which $\mathbb{R}$ represents the shaft and the branched rays the barbs.

A *double feather* is obtained by substituting each barb by a simple feather. If now this process is taken to the limit substituting all the barbs by simple feathers one gets a *complete feather*. In this space the nonseparable points form a dense set.

It is possible to show that there exist foliations of $\mathbb{R}^2$ whose leaf spaces are simple feathers. This process is sketched in figure 3.

Taking, in the figure, the curve L as axis, the process is repeated by substituting a simple feather for a barb. Substituting simple feathers for all the barbs by this process, one obtains a foliation of $\mathbb{R}^2$ whose leaf space is a complete feather. This is called Wazewski's example [59]. The following theorems are known about foliations of the plane.

Theorem (Kaplan [22]). *To each continuous foliation of the plane it is possible to associate a continuous real-valued function defined on $\mathbb{R}^2$ which does not have either a local maximum or a local minimum and is constant on each leaf.*

Theorem (Kamke [21]). *If $\mathfrak{F}$ is a C^∞ foliation and U is bounded then there exists a C^∞ map $f : U \longrightarrow \mathbb{R}$ which is constant on the leaves of $\mathfrak{F}/U$ and such that $df \neq 0$ at each point of U.*

Theorem (Wazewski [59]). *There is a C^∞ foliation of $\mathbb{R}^2$ such that each C^1 function on $\mathbb{R}^2$, constant on the leaves, is constant.*

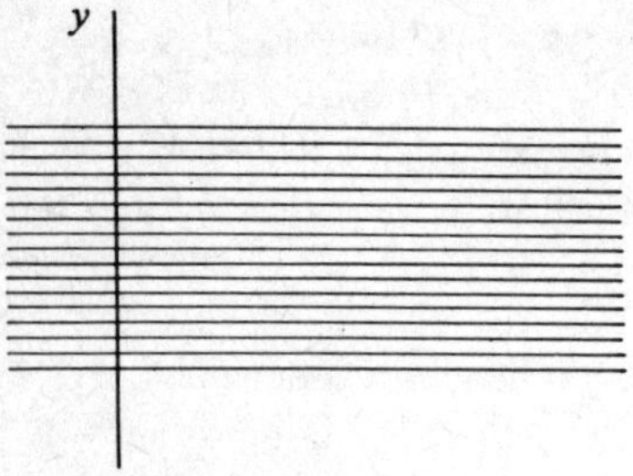

trivial foliation

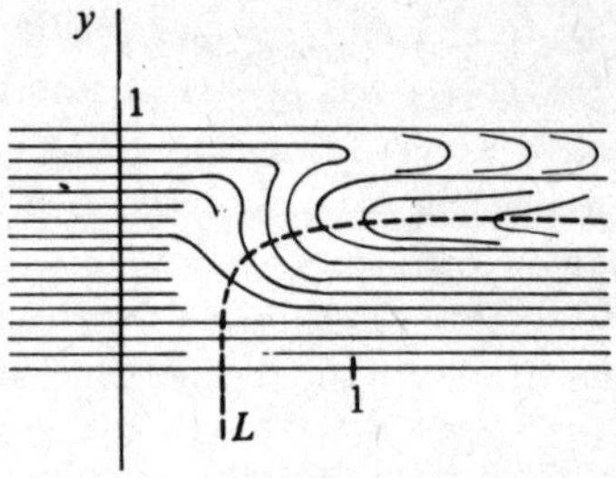

insertion of a barb at $y = 1$

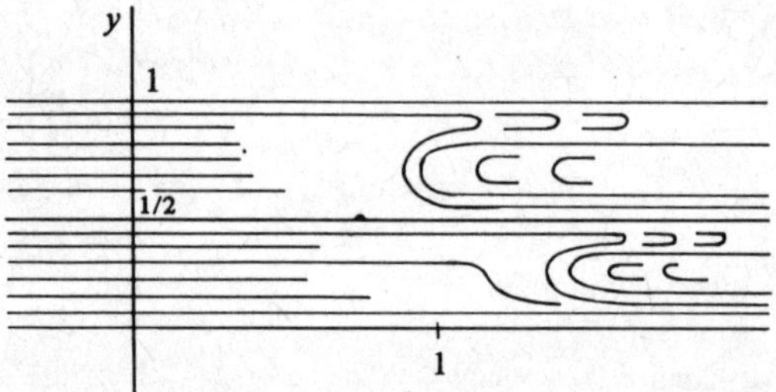

insertion at $y = \frac{1}{2}$

Figure 3

These theorems were proved again by Haefliger and Reeb in [16] and [47], by means of an analysis of leaf spaces of foliations of the plane. They observed that the leaf space of a C^r foliation of $\mathbb{R}^2$ is, in general, a non-Hausdorff manifold, of dimension one, simply connected and of class C^r. Kaplan's theorem follows from this and from the fact that on a simply connected, one-dimensional manifold (Hausdorff or not), there always exists a continuous strictly monotonic function. Kamke's theorem follows from the fact that the leaf space of $\mathfrak{F}/U$ has a C^∞ (Hausdorff) manifold structure. Finally, Wazewski's theorem follows from the fact that, on a complete feather, there exists a C^∞ differentiable structure such that any C^1 function defined on it is necessarily constant.

(2) Minimal sets of foliations of codimension 1.

The study of minimal sets began with H. Poincaré who studied the behavior of orbits of analytic diffeomorphisms of the circle. The concept of rotation number, which he introduced, can be defined by the limit $\rho(f) = (\lim_{n \to \infty} (\tilde{f}^n(x) - x)/n) \pmod 1$, where $\tilde{f}: \mathbb{R} \to \mathbb{R}$ is a diffeomorphism which covers f, that is, $\pi(\tilde{f}(x)) = f(\pi(x))$ where $\pi(x) = e^{2\pi i x}$. One shows that the limit always exists and is independent of x. Poincaré proved that $\rho(f)$ is rational if and only if f has a periodic point. In terms of foliations, this means that a foliation $\mathfrak{F}$ of T^2 has no compact leaves if and only if it is equivalent to the foliation defined by the suspension of a diffeomorphism of S^1 with irrational rotation number.

In 1916 Bohl ([4]) observed, without constructing them that there exist foliations of T^2 having exceptional minimal sets. In 1924 Kneser ([23]) constructed examples of such foliations of class C^0.

The theory of foliations on T^2 has as one of its highest points the theorems of A. Denjoy ([7] and [8]). In 1932, he constructed examples of C^1 foliations on T^2 with exceptional minimal sets. Further he proved that if $\mathfrak{F}$ is a C^2 foliation of T^2 and μ is a minimal set of $\mathfrak{F}$ then $\mu = T^2$ or μ is a compact leaf ($\mu \approx S^1$). In particular if $\mathfrak{F}$ does not have compact leaves, Denjoy's theorem implies that $\mathfrak{F}$ is topologically equivalent(*) to a foliation obtained by a suspension of an irrational rotation of the circle (see note 1 in Chapter II). In note 3 of this chapter we construct Denjoy's example. Later results on C^r diffeomorphisms of the circle can be found in: M. R. Herman, Sur la conjugaison différentiable des diffeomorphismes du cercle à des rotations. Publ. IHES no 49 (1979).

The results of Denjoy motivated the study of minimal sets of foliations of codimension one on compact manifolds of dimension ≥ 3. In Chapter V we will see an example, due to Saksteder ([49], 1964) of a C^∞ codimension one foliation of $M_2 \times S^1$ (M_2 = surface of genus two) which has an exceptional minimal set.

Later, Raymond showed, using Fuchsian groups, the existence of foliations of S^3 with exceptional minimal sets.

(3) Denjoy's example.

Denjoy's example is obtained by suspending a C^1 diffeomorphism $f: S^1 \to S^1$ which has an exceptional minimal set. A subset $\mu \subset S^1$ is called *minimal* for f if μ is closed, nonempty, $f(\mu) = \mu$ and for all $\mu' \subset \mu$ with the above three properties we have $\mu' = \mu$. A minimal set is called *exceptional* when it is perfect and has empty interior, i.e., homeomorphic to a subset of the Cantor set on S^1.

The construction of a diffeomorphism with an exceptional minimal set will be done in the following manner. Given an irrational rotation $R_\alpha(e^{2\pi i t}) = e^{2\pi i (t+\alpha)}$, cut the circle S^1 at all the points of an orbit $\{\theta_n \mid n \in \mathbb{Z}\}$ of R_α. In each cut insert a segment J_n of length ℓ_n where $\sum_{n=-\infty}^{\infty} \ell_n = L < \infty$. We obtain in this manner a new circle longer than the first. The set $S^1 - \bigcup_{n \in \mathbb{Z}} J_n = \mu$ will be homeomorphic to the Cantor set. The diffeomorphism f will be constructed in such a way that f/μ' will be conjugate to

(*) We say two foliations $\mathfrak{F}$ and $\mathfrak{F}'$ on M are topologically equivalent when there exists a homeomorphism $h: M \to M$ which takes leaves of $\mathfrak{F}$ to leaves of $\mathfrak{F}'$.

$R_\alpha/(S^1 - \{\theta_n \mid n \in \mathbf{Z}\})$ where $\mu' = \mu - \cup_{n\in\mathbf{Z}} J_n$. From this fact one sees that every orbit of f in μ will be dense in μ.

In order to construct f formally, we construct a C^1 increasing diffeomorphism, $\tilde{f}$: $\mathbb{R} \longrightarrow \mathbb{R}$ such that $\tilde{f}(t) = t + \varphi(t)$ where φ is periodic of period 1. As it is known, such a map $\tilde{f}$ induces a C^1 diffeomorphism $f : S^1 \longrightarrow S^1$ which can be defined by the formula

$$f(e^{2\pi it}) = e^{2\pi i\tilde{f}(t)} .$$

(A) Construction of the minimal set μ.

Fix $\alpha \in (0,1) - \mathbb{Q}$ and $y_0 \in (0,1)$. Given $n \in \mathbf{Z}$ the number $y_0 + n\alpha$ is not an integer so $y_0 + n\alpha \in (m, m+1)$ for some $m \in \mathbf{Z}$. Set $y_n = y_0 + n\alpha - m \in (0,1)$. Since α is irrational the sequence $\{y_n\}_{n\in\mathbf{Z}}$ is dense in $[0,1]$ and moreover $y_m \neq y_n$ if $n \neq m$. Let $\{\ell_n\}_{n\in\mathbf{Z}}$ be a sequence of positive numbers satisfying the following properties: (1) $\sum_{-\infty}^{\infty} \ell_n = L < \infty$, (2) $\sum_{|n|\to\infty} \ell_{n+1}/\ell_n = 1$ and (3) $1/3 < \ell_{n+1}/\ell_n < 5/3$ for all $n \in \mathbf{Z}$. For example the sequence $\ell_n = (n^2+1)^{-1}$ satisfies these properties.

From the sequences $\{y_n\}_{n\in\mathbf{Z}}$ and $\{\ell_n\}_{n\in\mathbf{Z}}$ we will construct a set $K \subset (0,1+L)$, homeomorphic to the Cantor set and which will essentially be the minimal set of Denjoy's example. The idea for the construction of K is, basically, that for each $n \in \mathbf{Z}$ one cuts the interval $[0,1]$ at the point y_n and inserts an interval of length ℓ_n. Formally we proceed in the following manner.

Consider the two sequences $\alpha_n = y_n + \sum_{0<y_i<y_n} \ell_i$ and $\beta_n = \alpha_n + \ell_n$. Let $I_n = (\alpha_n, \beta_n)$. It is clear that for every $n \in \mathbf{Z}$ one has $0 < \alpha_n < \beta_n < 1 + L$, so $I_n \subset (0, 1+L)$. Moreover one can verify that if $y_n < y_m$ then $\beta_n < \alpha_m$ and hence $I_n \cap I_m = \emptyset$. Set $K = [0,1+L] - \cup_{n\in\mathbf{Z}} I_n$. It is clear that K is compact.

Lemma 1. *K is perfect and has empty interior.*

Proof. With this goal we define two auxiliary functions. Define h^- and $h^+ : [0,1] \longrightarrow [0,1+L]$ by $h^-(y) = y + \sum_{0<y_i<y} \ell_i$ and $h^+(y) = y + \sum_{0<y_i\leq y} \ell_i$. Using that the series $\sum_{n\in\mathbf{Z}} \ell_n$ converges, one easily proves that h^- and h^+ are monotonically increasing that h^- is lower semi-continuous (i.e. ${}_{s<y}\lim_{s\to y} h^-(s) = h^-(y)$) and that h^+ is upper semi-continuous (i.e., ${}_{s>y}\lim_{s\to y} h^+(s) = h^+(y)$). Moreover $h^-(y_n) = \alpha_n$ and $h^+(y_n) = \beta_n$, $n \in \mathbf{Z}$, and if $y \notin \{y_n \mid n \in \mathbf{Z}\}$ then $h^-(y) = h^+(y)$.

K has empty interior. Suppose $[a,b] \subset K$ is such that $a, b \in \partial K$. Since $a \in \partial K$, there is a sequence $\{\beta_{n_k}\}_{k\in\mathbf{Z}}$ such that $\beta_{n_k} < \beta_{n_{k+1}} < a$ for all k and $\lim_{k\to\infty} \beta_{n_k} = a$. Since $I_n \cap I_m = \emptyset$ if $n \neq m$ we have that $\beta_{n_k} < \alpha_{n_{k+1}} < \beta_{n_{k+1}} < a$ so $\lim_{k\to\infty} \alpha_{n_k} = a$. Since the sequence $\{\alpha_{n_k}\}_{k\in\mathbb{N}}$ is increasing, the corresponding sequence $\{y_{n_k}\}_{k\in\mathbb{N}}$ is also increasing, so $\lim_{k\to\infty} y_{n_k} = y$ for some $y \in [0,1]$ and hence $a = \lim_{k\to\infty} \alpha_{n_k} = \lim_{k\to\infty} h^-(y_{n_k}) = h^-(y)$. Analogously one concludes that $b = h^+(y')$ for some $y' \in [0,1]$. In the case $y = y_n$ for some n we have $a = \alpha_n$ so $a = b$ since $K \cap I_n = \emptyset$. In the other case $y \notin \{y_n \mid n \in \mathbf{Z}\}$ and so $h^-(y) = h^+(y) = a$. Hence, $h^+(y) = a \leq b = h^+(y')$ and thus we conclude that $y \leq y'$ because h^+ is increasing. On the other hand if $y < y'$ there exists $y_n \in (y, y')$ since $\{y_n\}_{n\in\mathbb{N}}$ is dense in $[0,1]$ so $a = h^+(y) < h^+(y_n) < h^+(y') = b$ or else $[a,b] \cap I_n \neq \emptyset$, a contradiction. We conclude that $y = y'$ so $a = b$ and hence K has empty interior.

K is perfect. Let $x \in K$. In the case $x = \alpha_n$ for some $n \in \mathbf{Z}$, let $\{y_{n_k}\}_{k \in \mathbf{N}}$ be a sequence such that $y_{n_k} < y_n$ for all $k \in \mathbf{N}$ and $\lim_{k \to \infty} y_{n_k} = y_n$. We have then that $\alpha_n = \lim_{k \to \infty} h^-(y_{n_k}) = \lim_{k \to \infty} \alpha_{n_k}$ and therefore α_n is an accumulation point of K. In the case $x = \beta_n$ for some $n \in \mathbf{Z}$, with an analogous argument, one proves that $\beta_n = \lim_{k \to \infty} \beta_{n_k}$ where $\beta_{n_k} > \beta_n$ for all $k \in \mathbf{N}$ and therefore β_n is an accumulation point of K. On the other hand if $x \notin \{\alpha_n, \beta_n \mid n \in \mathbf{Z}\}$, it is easy to see that $x = \lim_{k \to \infty} \alpha_{n_k}$ for some sequence $\{\alpha_{n_k}\}_{k \in \mathbf{N}}$, since K has empty interior. This proves that K is perfect. ■

Now set $\tilde{X} = K - \{\alpha_n, \beta_n \mid n \in \mathbf{Z}\}$ and $\tilde{Y} = [0,1] - \{y_n \mid n \in \mathbf{Z}\}$. Define $\tilde{h} : \tilde{Y} \to \tilde{X}$ by $\tilde{h}(y) = h^+(y) = h^-(y)$, $y \in \tilde{Y}$.

Lemma 2. *$\tilde{h}$ is a homeomorphism onto $\tilde{X}$.*

Proof. $\tilde{h}$ is continuous since h^- is lower semicontinuous and h^+ is upper semicontinuous. On the other hand $\tilde{h}$ is increasing since h^- is. The inverse of $\tilde{h}$ is given by

$$\tilde{h}^{-1}(x) = x - \sum_{0<\alpha_i<x} \ell_i = x - \sum_{0<\beta_i<x} \ell_i, \qquad x \in \tilde{X},$$

which, as one can verify directly, is continuous on $\tilde{X}$. ■

We consider now the map $\psi(x) = (1 + L)^{-1}x$. The homeomorphism ψ takes $K \subset [0, 1 + L]$ to the set $\psi(K) \subset [0,1]$, which is homeomorphic to the Cantor set. Set $\tilde{\mu} = \psi(K) + \mathbf{Z}$. By an abuse of notation we designate $\psi(I_n)$ by I_n, $\psi(\alpha_n)$ by α_n and $\psi(\beta_n)$ by β_n. With this notation we have $[0,1] - \tilde{\mu} = \cup_{n \in \mathbf{N}} I_n$.

(B) Construction of $\tilde{f}$:

Lemma 3. *There exists a continuous function $\varphi : \mathbf{R} \to \mathbf{R}$ satisfying the following properties:*

(a) *φ is positive and periodic of period* 1,

(b) $\varphi \mid \tilde{\mu} \equiv 1$,

(c) $\int_{\alpha_n}^{\beta_n} \varphi(\tau)\, d\tau = \beta_{n+1} - \alpha_{n+1} = (1 + L)^{-1} \ell_{n+1}$.

Proof. It is enough to define φ on $\cup_{n \in \mathbf{Z}} I_n$. Using this we extend φ to $\mathbf{R}$, letting $\varphi(t + m) = \varphi(t)$ if $m \in \mathbf{Z}$ and $\varphi \mid \tilde{\mu} \equiv 1$. We then define $\varphi(t) = 1 + \lambda_n(\beta_n - t) \cdot (t - \alpha_n)$ for $t \in I_n$ where

$$\lambda_n = \frac{6}{(\beta_n - \alpha_n)^2} \left[\frac{\beta_{n+1} - \alpha_{n+1}}{\beta_n - \alpha_n} - 1 \right] = \frac{6(1 + L)^2}{\ell_n^2} \left[\frac{\ell_{n+1}}{\ell_n} - 1 \right].$$

Condition (c) is immediately verified. We verify that φ is positive and continuous. For $t \in I_n$, we see that

$$|\varphi(t) - 1| = |\lambda_n(\beta_n - t)(t - \alpha_n)| \le \frac{|\lambda_n|}{4} (\beta_n - \alpha_n)^2 = \frac{3}{2} \left| \frac{\ell_{n+1}}{\ell_n} - 1 \right| < 1$$

since $1/3 < \ell_{n+1}/\ell_n < 5/3$. So φ is positive.

On the other hand

$$\sup_{t\in I_n} |\varphi(t) - 1| = \frac{3}{2}\left|\frac{\ell_{n+1}}{\ell_n} - 1\right| ,$$

so $\lim_{|n|\to\infty} (\sup_{t\in I_n} |\varphi(t) - 1|) = 0$, since $\lim_{|n|\to\infty} \ell_{n+1}/\ell_n = 1$. This implies that φ is continuous on $\tilde{\mu} \cap [0,1]$ and thus continuous on $\mathbb{R}$. ■

Now let $c = (1 + L)^{-1}(\alpha + \sum_{0<y_i<\alpha} \ell_i)$ where α is the irrational number used in definition of the sequence $\{y_n\}_{n\in\mathbb{Z}}$. Define $\tilde{f}: \mathbb{R} \to \mathbb{R}$ by $\tilde{f}(x) = c + \int_0^x \varphi(\tau)d\tau$. It is clear that $\tilde{f}$ is a C^1 diffeomorphism.

Lemma 4. *The following properties are true:*

(a) $\tilde{f}(t) = t + \varnothing(t)$ *where* $\varnothing$ *is periodic of period* 1,

(b) *for each* $n \in \mathbb{Z}$ *such that* $y_n + \alpha < 1$, *one has that* $\tilde{f}(\alpha_n) = \alpha_{n+1}$ *and* $\tilde{f}(\beta_n) = \beta_{n+1}$,

(c) *for each* $n \in \mathbb{Z}$ *such that* $y_n + \alpha > 1$, *one has that* $\tilde{f}(\alpha_n) = 1 + \alpha_{n+1}$ *and* $\tilde{f}(\beta_n) = 1 + \beta_{n+1}$.

Proof. Let m be the Lebesgue measure on $\mathbb{R}$. We have then that

$$\tilde{f}(\alpha_n) - c = \int_0^{\alpha_n} \varphi(\tau)d\tau = \int_{K_n} \varphi(\tau)d\tau + \int_{L_n} \varphi(\tau)d\tau$$

where $K_n = \tilde{\mu} \cap [0,\alpha_n]$ and $L_n = \cup_{0<\alpha_j<\alpha_n} I_j$. Since $\int_{I_j} \varphi(\tau)d\tau = (1 + L)^{-1}\ell_{j+1}$, one sees that $\int_{L_n} \varphi(\tau)d\tau = (1 + L)^{-1}\sum_{0<\alpha_j<\alpha_n} \ell_{j+1}$. On the other hand

$$\int_{K_n} \varphi(\tau)d\tau = \int_{K_n} d\tau = m(\tilde{\mu} \cap [0,\alpha_n]) = \alpha_n - (1 + L)^{-1} \sum_{0<\alpha_j<\alpha_n} \ell_j$$

since $\tilde{\mu} \cap [0,\alpha_n] = [0,\alpha_n] - \cup_{0<\alpha_j<\alpha_n} I_j$ and $m(I_j) = (1 + L)^{-1}\ell_j$. Recalling that $\sum_{0<y_i<y_n} \ell_j = \sum_{0<\alpha_j<\alpha_n} \ell_j$ and that after the change of scale ψ, $\alpha_n = (1 + L)^{-1}\cdot(y_n + \sum_{0<y_j<y_n} \ell_j)$ we have finally

$$\tilde{f}(\alpha_n) = (1 + L)^{-1}(y_n + \alpha + \sum_{0<y_i<\alpha} \ell_i + \sum_{0<y_j<y_n} \ell_{j+1}) .$$

Suppose $y_n + \alpha < 1$. In this case, from the definition of the sequence $\{y_j\}_{j\in\mathbb{Z}}$, one sees that $y_n + \alpha = y_{n+1}$ and $y_j + \alpha = y_{j+1}$ for every j such that $y_j \leq y_n$.

We have then that $\sum_{0<y_j<y_n} \ell_{j+1} = \sum_{\alpha<y_j<y_{n+1}} \ell_j$ and therefore $\tilde{f}(\alpha_n) = (1 + L)^{-1}(y_{n+1} + \sum_{0<y_j<y_{n+1}} \ell_j) = \alpha_{n+1}$. On the other hand $\tilde{f}(\beta_n) = \tilde{f}(\alpha_n) + \int_{I_n} \varphi(\tau)d\tau = \alpha_{n+1} + \beta_{n+1} - \alpha_{n+1} = \beta_{n+1}$. In an analogous manner one proves that $\tilde{f}(\alpha_n) = 1 + \alpha_{n+1}$ and $\tilde{f}(\beta_n) = 1 + \beta_{n+1}$, in the case $y_n + \alpha > 1$. We leave the details to the reader. In order to prove (a) it is enough to observe that φ is periodic of period 1 and that

$$\int_t^{t+1} \varphi(\tau)d\tau = \int_0^1 \varphi(\tau)d\tau = \int_{\mu\cap[0,1]} \varphi(\tau)d\tau + \int_{\bigcup_{n\in\mathbb{Z}} I_n} \varphi(\tau)d\tau =$$

$$= (1 + L)^{-1} + (1 + L)^{-1} \sum_{j=-\infty}^{\infty} \ell_j = 1 .$$

Therefore $\tilde{f}(t + 1) - \tilde{f}(t) = 1$ for all $t \in \mathbf{R}$, so $\tilde{f}(t) = t + \varnothing(t)$ where $\varnothing$ is periodic of period 1.

Now let $\mu = \{e^{2\pi it} \mid t \in \tilde{\mu}\}$, $a_n = e^{2\pi i\alpha_n}$, $b_n = e^{2\pi i\beta_n}$, $J_n = \{e^{2\pi it} \mid t \in I_n\}$ and $f(e^{2\pi it}) = e^{2\pi i\tilde{f}(t)}$. It is clear that $f : S^1 \longrightarrow S^1$ is well-defined and is a C^1 diffeomorphism of S^1. Further $\mu \subset S^1$ is homeomorphic to the Cantor set and $\mu = S^1 - \cup_{n\in\mathbf{Z}} J_n$.

Lemma 5. *μ is a minimal set of f.*

Proof. Conditions (b) and (c) of the preceding lemma imply that $f(J_n) = J_{n+1}$ for all $n \in \mathbf{Z}$ and hence $f(\mu) = S^1 - \cup_{n\in\mathbf{Z}} f(J_n) = \mu$, μ is invariant under f. In order to prove that μ is a minimal set for f it is enough to prove that the orbit of θ under f, $\mathcal{O}(\theta) = \{f^n(\theta) \mid n \in \mathbf{Z}\}$, is dense in μ for all $\theta \in \mu$.

1st *case*. $\theta = a_k$ (or b_k) for some $k \in \mathbf{Z}$. In this case $\mathcal{O}(a_k) = \{a_n \mid n \in \mathbf{Z}\}$ which is dense in μ since $S^1 - \mu = \cup_{n\in\mathbf{Z}} J_n$, μ is homeomorphic to the Cantor set and for all $n \in \mathbf{Z}$, a_n is an endpoint of J_n.

2nd *case* - general case. Consider the rotation by angle $2\pi\alpha$, $R_\alpha(e^{2\pi it}) = e^{2\pi i(t+\alpha)}$. For each $n \in \mathbf{Z}$ set $\theta_n = e^{2\pi iy_n}$. It is easy to see that $\theta_n = R_\alpha^n(\theta_0)$ for all $n \in \mathbf{Z}$. Consider also the homeomorphism $\tilde{h} : \tilde{Y} \longrightarrow \tilde{X} \subset [0, 1 + L]$. This map induces a homeomorphism

$h : Y = (S^1 - \{\theta_n \mid n \in \mathbf{Z}\}) \longrightarrow S^1 - \{a_n, b_n \mid n \in \mathbf{Z}\}) = X$ setting $h(e^{2\pi it}) = e^{2\pi i(1+L)^{-1}\tilde{h}(t)}$, $t \in \tilde{Y}$. Observe that Y is invariant under R_α and X is invariant under f. The proof resumes now to show that the following diagram commutes

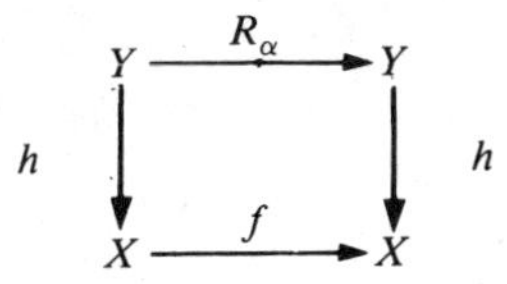

Indeed, the fact that all the orbits of R_α are dense implies that all the orbits of f in X are dense in X, so in μ. The commutativity of the diagram is equivalent to showing that $\tilde{f}(\tilde{h}(y)) - \tilde{h}(y + \alpha) \in \mathbf{Z}$ for all $y \in Y$. This can be verified directly using the formulas given above. We leave the details to the reader.

IV. HOLONOMY AND THE STABILITY THEOREMS

In this chapter $\mathfrak{F}$ denotes a foliation of codimension n and class C^r, $r \geq 1$, of a manifold M^m. Our objective is to study the behavior of the leaves near a fixed compact leaf F. By the transverse uniformity of $\mathfrak{F}$ it is sufficient to study the first returns of leaves to a small transverse section Σ of dimension n passing through a point $p \in F$. For each closed path γ in F passing through p, these returns can be expressed by a local C^r diffeomorphism of Σ, f_γ, with $f_\gamma(p) = p$ and where for $x \in \Sigma$ sufficiently near p, $f_\gamma(x)$ is the first return "over γ" of the leaf of $\mathfrak{F}$ which passes through x.

If γ' is another closed path in F through p, homotopic to γ (with fixed endpoints) then it happens that f_γ and $f_{\gamma'}$ coincide on a common neighborhood of $p \in \Sigma$, that is, $f_{\gamma'}$ and f_γ have the same germ at $p \in \Sigma$. So it follows that the map $\gamma \mapsto f_\gamma$ induces a homomorphism

$$\pi_1(F,p) \longrightarrow G(\Sigma,p)$$

from the fundamental group of F at p (the group of homotopy classes of closed paths) and the group of germs of C^r diffeomorphisms of Σ at p. This representation is called the holonomy of F at p. This notion was introduced by Ehresmann in [10].

It is clear that if $\pi_1(F,p)$ is small the recurrence of leaves near F will be small. The *Theorem of Local Stability* expresses this in the following manner: if F is a compact leaf of $\mathfrak{F}$ with finite fundamental group, then there exists a neighborhood V of F, saturated by leaves of $\mathfrak{F}$, such that every leaf in V is compact with finite fundamental group.

The *Global Stability Theorem* offers the following global version: if $\operatorname{cod} \mathfrak{F} = 1$, M is compact and connected and there is a compact leaf with finite fundamental group then all the leaves are compact and have finite fundamental group. These theorems are due to G. Reeb [46].

§1. Holonomy of a leaf

Let $\gamma : I = [0,1] \longrightarrow F$ be a continuous path and Σ_0, Σ_1 small transverse sections to $\mathfrak{F}$ of dimension n passing through $p_0 = \gamma(0)$ and $p_1 = \gamma(1)$ respectively. We will define a local map between Σ_0 and Σ_1 "along" the leaves of $\mathfrak{F}$, "over" the path γ taking p_0 to p_1.

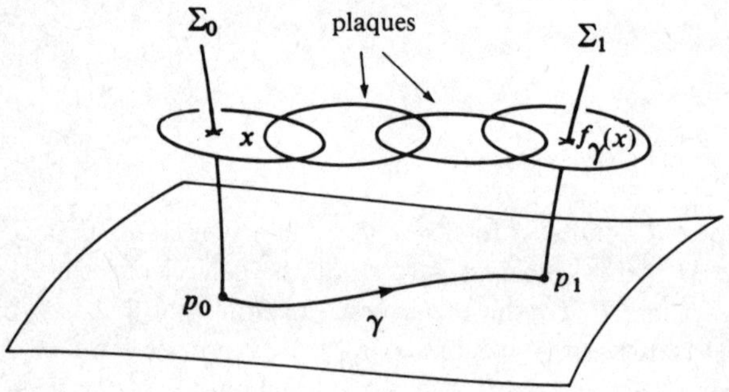

Figure 1

According to Lemma 1 of Chapter II, there exists a sequence of local charts $(U_i)_{i=0}^{k}$ and a partition of $[0,1]$, $0 = t_0 < \dots < t_{k+1} = 1$ such that (a) if $U_i \cap U_j \neq \emptyset$ then $U_i \cup U_j$ is contained in a local chart of $\mathfrak{F}$ and (b) $\gamma([t_i, t_{i+1}]) \subset U_i$ for all $0 \leq i \leq k$. We will say then that $(U_i)_{i=0}^{k}$ is a *chain subordinated* to γ.

For each $0 < i < k+1$ fix a transverse section to $\mathfrak{F}$, $D(t_i) \subset U_{i-1} \cap U_i$, homeomorphic to an n-dimensional disk, passing through $\gamma(t_i)$. Also set $D(0) = \Sigma_0$ and $D(1) = \Sigma_1$. Then for each $x \in D(t_i)$ sufficiently near $\gamma(t_i)$ the plaque of U_i passing through x meets $D(t_{i+1})$ in a unique point $f_i(x)$.

The domain of the map f_i contains a disk $D_i \subset D(t_i)$ containing $\gamma(t_i)$.

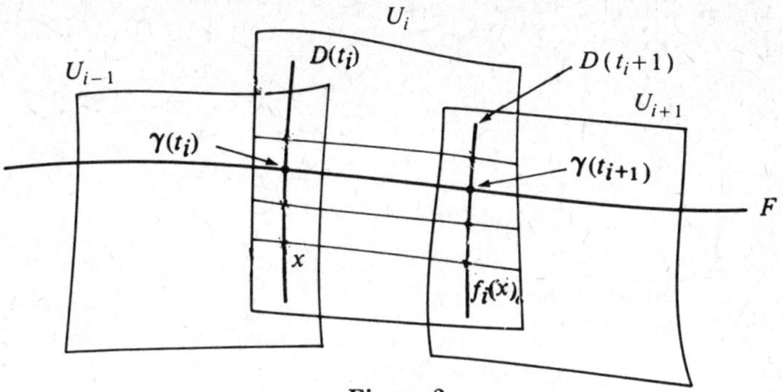

Figure 2

So it is clear that the composition

$$f_\gamma = f_k \circ f_{k-1} \circ \ldots \circ f_0$$

is well-defined in a neighborhood of $p_0 \in \Sigma_0$. We will call f_γ the holonomy map associated to γ.

Lemma 1. *The map f_γ is independent of the disks $(D(t_i))_{i=1}^k$ chosen and of the chain subordinated to γ, that is, if $\tilde{f}_\gamma$ and f_γ are two holonomy maps associated to the same path then $\tilde{f}_\gamma$ and f_γ coincide on the intersection of the two domains.*

Proof. First fix the chain subordinated to γ, and suppose that $(D(t_i))_{i=1}^k$ and $(B(t_i))_{i=1}^k$ are two collections of disks as above with $\gamma(t_i) \in D(t_i) \cap B(t_i)$ for every i. Given $x \in D(t_i)$ near $\gamma(t_i)$ we can define $\beta_i(x) \in B(t_i)$ as the intersection of the plaque of U_i which passes through x and $B(t_i)$. We have then defined a homeomorphism $\beta_i : \hat{D}_i \longrightarrow \hat{B}_i$ between open sets $\hat{D}_i \subset D(t_i)$ and $\hat{B}_i \subset B(t_i)$ containing $\gamma(t_i)$. Let $f_i : D_i' \longrightarrow D(t_{i+1})$ and $g_i : B_i' \longrightarrow B(t_{i+1})$ be as before. Since $U_i \cup U_{i+1} \subset V$, a local chart of $\mathfrak{F}$, each plaque of V meets each transverse disk in at most one point; then, the following diagram commutes

$$\begin{array}{ccc} \hat{D}_i \cap D_i' & \xrightarrow{\ f_i\ } & \hat{D}_{i+1} \\ {\scriptstyle \beta_i}\downarrow & & \downarrow{\scriptstyle \beta_{i+1}} \\ \hat{B}_i \cap B_i' & \xrightarrow[\ g_i\ ]{} & \hat{B}_{i+1} \end{array}$$

Restricting the maps f_i, g_i and β_i, $0 \leq i \leq k$, to smaller domains, we have that

$$f_\gamma = f_k \circ f_{k-1} \circ \ldots \circ f_0$$
$$= \beta_{k+1}^{-1} \circ g_k \circ \beta_k \circ \beta_k^{-1} \circ g_{k-1} \circ \ldots \circ \beta_1 \circ \beta_1^{-1} \circ g_0 \circ \beta_0 = g_\gamma$$

since β_0 and β_{k+1} are identity maps. With an analogous argument one proves that f_γ is independent of the partition $0 = t_0 < t_1 < \ldots < t_{k+1} = 1$. Now we take $(U_i)_{i=0}^{k}$ and $(U_i')_{i=0}^{k'}$, two chains subordinated to γ. If for every $i \in \{0, \ldots, k'\}$, $U_i' \subset U_j$ for some j, it is immediate that f_γ and f_γ' coincide on the intersection of their domains. The argument is similar to the one above. In case this does not occur, by Lemma 1 of Chapter II, we can construct a chain subordinated to γ, $(U_i'')_{i=0}^{k''}$ which refines $\{U_i \cap U_j'\}_{1 \leq i \leq k}^{1 \leq j \leq k'}$, that is, such that for every $i \in \{0, \ldots, k''\}$, $U_i'' \subset U_j \cap U_\ell'$ for some $j \in \{0, \ldots, k\}$ and $\ell \in \{0, \ldots, k'\}$. This implies that $f_\gamma = f_\gamma''$ and $f_\gamma' = f_\gamma''$ on the intersection of their domains, so $f_\gamma = f_\gamma'$.

Remark 1. The following facts follow from the definition of f_γ :

(a) $f_\gamma(p_0) = p_1$,
(b) if $\gamma^{-1}(t) = \gamma(1-t)$ then $f_{\gamma^{-1}} = (f_\gamma)^{-1}$
(c) if $\mathfrak{F}$ is of class $C^r (r \geq 1)$. The intermediate maps $f_i (0 \leq i \leq k)$ are C^r so f_γ is C^r. Since f_γ has an inverse, also C^r, f_γ is a C^r diffeomorphism,
(d) if $\tilde{\gamma}$ is obtained from γ by a small deformation in the leaf, with endpoints fixed, then $f_{\tilde{\gamma}} = f_\gamma$ in a neighborhood of $p_0 \in \Sigma_0$. One proves this observing that, for $\tilde{\gamma}$ near γ, there is a chain that is simultaneously subordinated to γ and $\tilde{\gamma}$,
(e) if $\Delta_0 \subset U_0$ and $\Delta_1 \subset U_k$ are other transverse disks to $\mathfrak{F}$, centered at p_0 and p_1, the projections along the plaques of U_0 and U_k define C^r diffeomorphisms, $\varphi_0 : \Delta_0 \longrightarrow \Sigma_0$ and $\varphi_1 : \Delta_1 \longrightarrow \Sigma_1$. The new holonomy transformation, $\Delta_0 \supset \Delta_0' \xrightarrow{g_\gamma} \Delta_1$ satisfies

$$g_\gamma(x) = \varphi_1^{-1} \circ f_\gamma \circ \varphi_0(x), \qquad x \in \Delta_0 .$$

In particular, if γ is a closed path, $\Sigma_0 = \Sigma_1$ and $\Delta_0 = \Delta_1$, then g_γ and f_γ are conjugate, that is, $g_\gamma = \varphi^{-1} \circ f_\gamma \circ \varphi$ where $\varphi = \varphi_0 = \varphi_1$.

Definition 1. Let X, Y be topological spaces and $x \in X$. On the set of maps $X \supset V \xrightarrow{f} Y$, where V is a neighborhood of x, we introduce the equivalence relation R : fRg if there is a neighborhood W of x such that $f \mid W = g \mid W$. The equivalence class of f is called the *germ of f at x*.

When $X = Y$ the set $G(X,x)$ of germs of local homeomorphisms which leave x fixed is a group with the multiplication germ $(f) \circ$ germ $(g) =$ germ $(f \circ g)$

where the domain of $f \circ g$ is the intersection of the domain of g with g^{-1} (domain of f).

Theorem 1. *Let $\gamma_i : I \longrightarrow M$, $i = 0,1$ be paths contained in a leaf F of $\mathfrak{F}$ such that $\gamma_i(0) = p_0$, $\gamma_i(1) = p_1$, $i = 0,1$. Let Σ_0, Σ_1 be transverse sections to F at p_0, p_1; $\Sigma_0 \supset D_i \xrightarrow{f_{\gamma_i}} \Sigma_1$ holonomy maps associated to γ_i and φ_{γ_i}, the germ of f_{γ_i} at p_0.*

(1) *If $\gamma_0 \simeq \gamma_1$ rel$(0,1)$ then $\varphi_{\gamma_0} = \varphi_{\gamma_1}$.*

(2) *If $p_0 = p_1$ and $\Sigma_0 = \Sigma_1$ then the transformation $\gamma \mapsto \varphi_{\gamma^{-1}}$ induces a homomorphism*

$$\Phi : \pi_1(F, p_0) \longrightarrow G(\Sigma_0, p_0), \qquad \Phi([\gamma]) = \varphi_{\gamma^{-1}}$$

from the fundamental group of F at p_0 to the group of germs of C^r diffeomorphisms of Σ_0 which leave p_0 fixed.

Proof. Let $H : I \times I \longrightarrow F$ be a homotopy between γ_0 and γ_1, i.e., $H(t,0) = \gamma_0(t)$, $H(t,1) = \gamma_1(t)$ and $H(0,s) = p_0$, $H(1,s) = p_1$ for all $s,t \in I$.

For each path $\gamma_s : I \longrightarrow F$, $\gamma_s(t) = H(t,s)$, there is a chain subordinated to γ_s. By continuity, each chain is also subordinated to any $\gamma_{s'}$ with s' near s. Consequently, we can find a collection of chains $(C_i)_{i=1}^m$ and a partition of $I : 0 = s_0 < s_1 < \ldots < s_m = 1$ such that for any $i = 1, \ldots, m$, C_i is subordinate to all the paths γ_s, $s_{i-1} \le s \le s_i$. This implies by (d) of Remark 2 that germ$(f_{\gamma_{s_{i-1}}}) =$ germ$(f_{\gamma_{s_i}})$ for all $i = 1, \ldots, m$. So $\varphi_{\gamma_0} = \varphi_{\gamma_1}$. For this reason, when $p_0 = p_1$ the map $\Phi : [\gamma] \longrightarrow \varphi_{\gamma^{-1}}$ is well-defined. Finally, if C_λ, C_μ are chains subordinated to λ^{-1} and μ^{-1} then $C_\lambda \cup C_\mu$ is a chain subordinated to the product $\lambda^{-1} * \mu^{-1}$ of the paths λ and μ. This says that

$$\Phi([\mu] \cdot [\lambda]) = \varphi_{\lambda^{-1} * \mu^{-1}} = \varphi_{\mu^{-1}} \circ \varphi_{\lambda^{-1}} = \Phi([\mu]) \cdot \Phi([\lambda])$$

and so Φ is a homomorphism between the groups $\pi_1(F, p_0)$ and $G(\Sigma_0, p_0)$. ■

Definition 2. The subgroup $\mathrm{Hol}(F, p_0) = \Phi(\pi_1(F, p_0))$ of $G(\Sigma_0, p_0)$ is called the *holonomy group of F at p_0*. Given $p_0, p_1 \in F$, any path $\alpha : I \longrightarrow F$ with $\alpha(0) = p_0$ and $\alpha(1) = p_1$ induces an isomorphism

$$\alpha^* : \mathrm{Hol}(F, p_0) \longrightarrow \mathrm{Hol}(F, p_1)$$

where $\alpha^*(\Phi[\mu]) = \varphi_\alpha \circ \Phi[\mu] \circ \varphi_{\alpha^{-1}}$. In this way we can speak of the *holonomy group of F* as any group isomorphic to $\mathrm{Hol}(F, p_0)$. To simplify notation, we will at times identify an element of the holonomy group with a particular representation.

Example 1. In the Reeb foliation of S^3 the fundamental group of the compact leaf is isomorphic to $\mathbb{Z} \oplus \mathbb{Z}$. Its holonomy group can be represented by two C^∞ diffeomorphisms $f, g : \mathbb{R} \longrightarrow \mathbb{R}$, $f(0) = g(0) = 0$ such that $f(x) = x$ for $x \geq 0$ and $f(x) < x$ if $x < 0$; $g(x) = x$ if $x \leq 0$ and $g(x) < x$ for $x > 0$ (see figure 3).

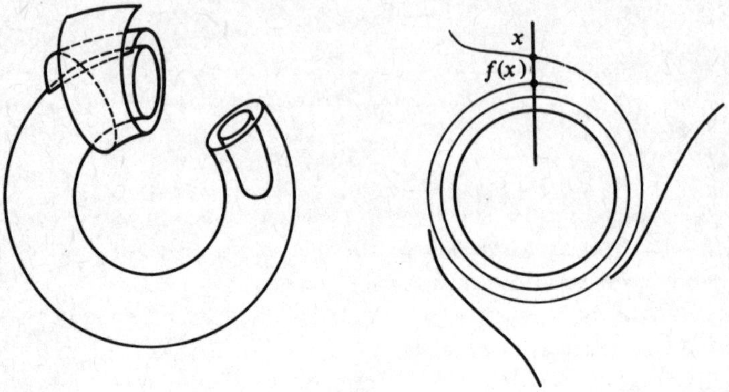

Figure 3

Remark 2. The holonomy of a leaf F can also be defined using a fibration transverse to F. With this goal in mind we first prove the following lemma.

Lemma 2. *Let F be a leaf of $\mathfrak{F}$ and $K \subset F$ a compact subset. There exist neighborhoods $U \supset W$ of K, U open in M and W open in F and a C^r retraction $\pi : U \longrightarrow W$ such that for any $x \in W$, $\pi^{-1}(x)$ is transverse to $\mathfrak{F}/U$.*

Proof. Since $K \subset F$ is compact there exists an imbedded surface W formed by a finite union of plaques of $\mathfrak{F}$, such that $K \subset W$. Let $\pi : \tilde{W} \longrightarrow W$ be a C^r tubular neighborhood of W. Since each fiber $\pi^{-1}(y)$, $y \in W$, meets W transversally and at y only, it is clear that if $x \in \pi^{-1}(y)$ is sufficiently near y, then $\pi^{-1}(y)$ meets the leaf of $\mathfrak{F}$ through x transversally at x. We can then obtain a neighborhood $U \subset \tilde{W}$ such that for all $y \in U \cap F$, $\pi^{-1}(y)$ meets $\mathfrak{F}/U$ transversally. ■

Now let $\gamma : I \longrightarrow F$ be a continuous curve with $\gamma(0) = p_0$ and $\gamma(1) = p_1$. Let U be a neighborhood of $\gamma(I)$ in M, W a neighborhood of $\gamma(I)$ in F and $\pi : U \longrightarrow W$ as in Lemma 2. Given $x \in \pi^{-1}(p_0)$ sufficiently near p_0, there is a unique path $\tilde{\gamma} : I \longrightarrow M$, contained in the leaf F_x of $\mathfrak{F}$ which passes through x such that $\tilde{\gamma}(0) = x$ and $\pi(\tilde{\gamma}(t)) = \gamma(t)$ for all $t \in I$. The path $\tilde{\gamma}$ is a lift of γ to the leaf F_x along the fibers of π. The map $f_\gamma(x) = \tilde{\gamma}(1)$ is well-defined for x near p_0 and coincides with the holonomy map associated to γ (see figure 4).

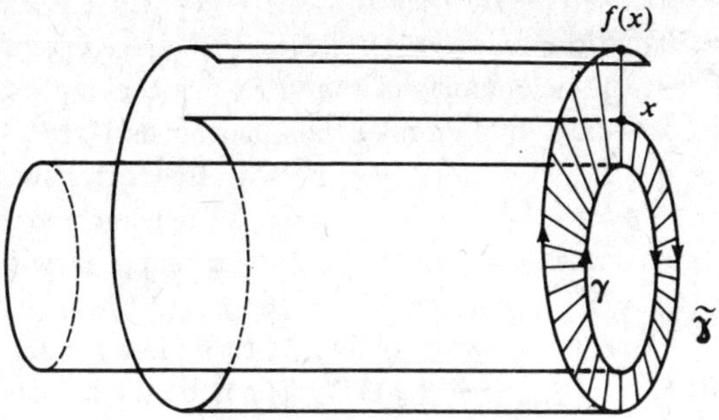

Figure 4

§2. Determination of the germ of a foliation in a neighborhood of a leaf by the holonomy of the leaf

Let $\mathfrak{F}$ be a foliation of a manifold M, F a leaf of $\mathfrak{F}$ and p_0 a point of F as above. The holonomy of F induces a local action of $\pi_1(F, p_0)$ on Σ_0 defined as follows: to each $[\gamma] \in \pi_1(F, p_0)$ and $x \in \Sigma_0$ sufficiently near p_0, we associate the point $f_\gamma(x) \in \Sigma_0$. When F is compact, this action characterizes the foliation in a neighborhood of F. More precisely, let $(M, \mathfrak{F})$, $(M', \mathfrak{F}')$ be foliations of codimension n and class C^r, $r \geq 1$, and $F \subset M$, $F' \subset M'$ be compact leaves. We say *the holonomies of F and F' are C^s conjugate* $(s \leq r)$ when there exist transverse sections to F and F', Σ_0 and Σ_0', $p_0 \in \Sigma_0 \cap F$, $p_0' \in \Sigma_0' \cap F'$ and a homeomorphism $h : F \cup \Sigma_0 \longrightarrow F' \cup \Sigma_0'$ such that $h(p_0) = p_0'$, $h \mid F$ and $h \mid \Sigma_0$ are C^s diffeomorphisms (a homeomorphism if $s = 0$) and for each $[\gamma] \in \pi_1(F, p_0)$, one has

$$h \circ f_\gamma \circ h^{-1}(x') = f_{h \circ \gamma}(x')$$

for every $x' \in \Sigma_0'$ sufficiently near p_0'.

Theorem 2. *Let F and F' be compact leaves of $\mathfrak{F}$ and $\mathfrak{F}'$ respectively. The holonomies of F and F' are C^s conjugate if and only if there exist neighborhoods $V \supset F$, $V' \supset F'$ and a C^s diffeomorphism $H : V \longrightarrow V'$, $H(F) = F'$, taking leaves of $\mathfrak{F}/V$ to leaves of $\mathfrak{F}'/V'$ (homeomorphism if $s = 0$). In this case we say $\mathfrak{F}$ and $\mathfrak{F}'$ are locally equivalent on F and F' and H is a local equivalence.*

Proof. Suppose first that there exists the local equivalence H. Fix $p_0 \in F$ and let Σ_0 and $\Sigma_0' = H(\Sigma_0)$ be transverse sections passing through p_0 and $p_0' =$ $= H(p_0)$. Let $[\gamma] \in \pi_1(F, p_0)$ and $\mathcal{C} = (U_i)_{i=0}^k$ be a chain subordinated to γ. The open sets $U_i' = H(U_i)$ define a chain $\mathcal{C}' = (U_i')_{i=0}^k$ subordinated to $H \circ \gamma$. Thus it follows that the holonomy transformations $\Sigma_0 \supset W_0 \xrightarrow{f_\gamma} \Sigma_0$

and $\Sigma_0' \supset W_0' \xrightarrow{f_{H\circ\gamma}} \Sigma_0'$ satisfy $H \circ f_\gamma \circ H^{-1}(x) = f_{H\circ\gamma}(x')$ if $x' \in \Sigma_0'$ is sufficiently near p_0'. Then the holonomies of F and F' are conjugate.

Suppose now that the holonomies of F and F' are conjugate. Since F and F' are compact, there exist, by Lemma 2, neighborhoods V of F, V' of F' and retractions $\pi : V \longrightarrow F$ and $\pi' : V' \longrightarrow F'$ with fibers transverse to $\mathfrak{F}$ and $\mathfrak{F}'$ respectively. Let $h : F \cup \Sigma_0 \longrightarrow F' \cup \Sigma_0'$ be a homeomorphism which conjugates the holonomies of F and F'. We can suppose without loss of generality that $\Sigma_0 = \pi^{-1}(p_0)$ and $\Sigma_0' = (\pi')^{-1}(h(p_0))$. Given $p \in F$, $p \neq p_0$, let $\gamma : I \longrightarrow F$ be a curve joining p_0 to p. Let $f_\gamma : D_{p_0} \subset \pi^{-1}(p_0) \longrightarrow \pi^{-1}(p)$ and $f'_{h(\gamma)} : D'_{h(p_0)} \longrightarrow (\pi')^{-1}(h(p))$ be the holonomy transformations of γ and $h \circ \gamma$. For $x \in f_\gamma(D_p)$ we define $H(x) = f'_{h(\gamma)}(h(f_{\gamma^{-1}}(x)))$ (see figure 5).

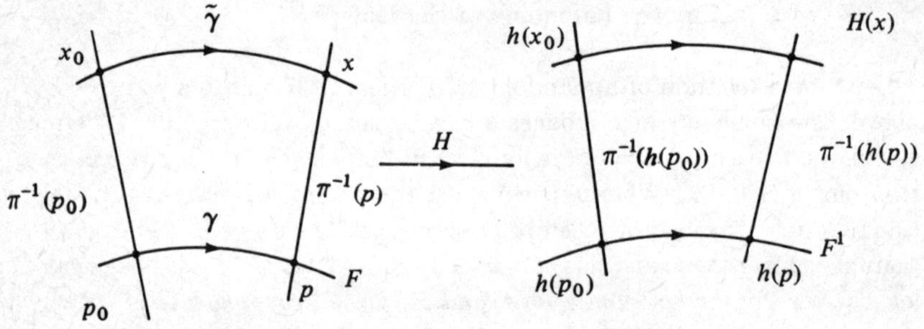

Figure 5

This definition is independent of the path γ chosen. In fact, if $\mu : I \longrightarrow F$ is another path with $\mu(0) = p_0$ and $\mu(1) = p$, we have by hypothesis that $h \circ f_{\mu\circ\gamma^{-1}} = f'_{h(\mu\circ\gamma^{-1})} \circ h$, consequently $f'_{h(\mu)} \circ h \circ f_{\mu^{-1}} = f'_{h(\gamma)} \circ h \circ f_{\gamma^{-1}} = H$.

Now we see the domain of H. Given $p \in F$ let U_p be a coordinate neighborhood of $\mathfrak{F}$ such that, for every $y \in U_p \cap F$, $\pi^{-1}(y)$ meets all the plaques of U_p and $U_p \cap F$ contains only one plaque. Fixing a curve γ which goes from p_0 to p, we can suppose that for every $z \in U_p$, the plaque of U_p through z cuts $\pi^{-1}(p)$ at a point z_1 in the domain of $f_{\gamma^{-1}}$. Let $U'_{h(p)}$, be a coordinate neighborhood of $\mathfrak{F}$ such that every plaque of $U'_{h(p)}$, cuts $(\pi')^{-1}(h(p))$ in a point of the domain of $f'_{h(\gamma^{-1})}$. Reducing, if necessary, U_p and $U'_{h(p)}$ we can also suppose that $h(f_{\gamma^{-1}}(\pi^{-1}(p) \cap U_p)) = f'_{h(\gamma^{-1})}((\pi')^{-1}(h(p)) \cap U'_{h(p)})$ and $h(U_p \cap F) = U'_{h(p)} \cap F'$. With these properties H is defined on U_p and $H(U_p) = U'_{h(p)}$. Since F is compact we can obtain a neighborhood of F, $U = U_{x_1} \cup \ldots \cup U_{x_k}$, where H is defined.

From the construction, it is immediate that $H : U \longrightarrow H(U) = U'$ is a C^s diffeomorphism (or homeomorphism if $s = 0$) and that H takes leaves of $\mathfrak{F}/U$ to leaves of $\mathfrak{F}'/U'$.

§3. Global trivialization lemma

Denote by $\mathfrak{F}$ a foliation of class C^r, $r \geq 1$, and codimension n on a manifold M^m.

Lemma 3. *Let $\gamma : I \longrightarrow M$ be a continuous, simple (injective) path whose image is contained in a leaf F of $\mathfrak{F}$. There is a neighborhood $V \supset \gamma(I)$ and a C^r diffeomorphism*

$$h : D^{m-n} \times D^n \longrightarrow V$$

such that $h^*\mathfrak{F}$ is a foliation whose leaves are the surfaces $P^{-1}(y)$ where $P : D^{m-n} \times D^n \longrightarrow D^n$ is the projection $P(x,y) = y$.

Proof. Let $(U_i)_{i=0}^k$ be a cover of $\gamma(I)$ by coordinate neighborhoods such that if $U_i \cap U_j \neq \varnothing$ then $U_i \cap U_j$ is a coordinate neighborhood of $\mathfrak{F}$. Let $\pi : U = \bigcup_{i=0}^k U_i \longrightarrow W \subset F$ be a retraction as in Lemma 2. Since γ has no self-intersections we can suppose that for each i, U_i meets at most U_{i-1} and U_{i+1} and $U_0 \cap U_k = \varnothing$. Let $p_0 = \gamma(0)$, $p_1 = \gamma(1)$ and $\Sigma_0 = \pi^{-1}(p_0)$, $\Sigma_1 = \pi^{-1}(p_1)$. It is clear that each leaf of $\mathfrak{F}/U$ cuts Σ_0 at most once. The holonomy map $\Sigma_0 \supset D_0 \xrightarrow{f_\gamma} \Sigma_1$ is well-defined on a disk D_0 with $p_0 \in D_0$. Let V be the saturation of D_0 by leaves of $\mathfrak{F}/\bigcup_{i=0}^k U_i$ and $f : V \longrightarrow D_0$ the map which to each $x \in V$ associates the intersection of the leaf $\mathfrak{F}/V$ passing through x with D_0. Since $\mathfrak{F}$ is C^r, f is also C^r. One easily verifies that f is a submersion.

Let $A \subset F$ be the leaf of $\mathfrak{F}/V$ containing γ. It is clear that by reducing A if necessary we can suppose that A is homeomorphic to the disk D^{m-n} by a diffeomorphism $k_1 : D^{m-n} \longrightarrow A$. Let $k_2 : D^n \longrightarrow D_0$ be a diffeomorphism such that $k_2(0) = p_0$. If $\mathfrak{F}$ is C^r we can take k_1 and k_2 as C^r diffeomorphisms. Define $h : D^{m-n} \times D^n \longrightarrow V$ letting $h(x,y) = \pi^{-1}(k_1(x)) \cap f^{-1}(k_2(y))$. Then h is a homeomorphism which takes the surface $P^{-1}(y) = D^{m-n} \times \{y\}$, $y \in D^n$, in the leaf $f^{-1}(k_2(y))$ of $\mathfrak{F}/V$. If $\mathfrak{F}$ is C^r, $r \geq 1$, it is easy to see that h is a C^r diffeomorphism, whose inverse is given by $h^{-1}(p) = (k_1^{-1}(\pi(p)), k_2^{-1}(f(p)))$.

When $\operatorname{cod}(\mathfrak{F}) = 1$, this result can be improved, as is shown in the following lemma which is used in later chapters.

Lemma 4. *Let F be a leaf of a transversally oriented foliation on M of codimension 1, and $f_0 : K \longrightarrow F$ a continuous map homotopic to a constant, K compact and path-connected. There is a continuous family of maps $f_t : K \longrightarrow M$, $t \in I = [0,1]$, such that $f_t(K)$ is contained in a leaf F_t. For $x \in K$ fixed, the curve $t \longrightarrow f_t(x)$ is normal to $\mathfrak{F}$.*

Proof. Let $\mathfrak{N}$ be a normal vector field to $\mathfrak{F}$ and denote by $\mathfrak{N}_t(x)$ the parametrized orbit of x. We can suppose that f_0 is homotopic to the constant map $C : K \longrightarrow F$, $C(x) = y_0$ where $y_0 = f_0(x_0)$ for $x_0 \in K$. Given a path $\gamma : I \longrightarrow K$, $\gamma(0) = x_0$, $\gamma(1) = x$, the path $\tilde{\gamma} : I \longrightarrow F$, $\tilde{\gamma} = f_0 \circ \gamma$ induces a holonomy map

$$J_{y_0} \supset V \xrightarrow{g_{\tilde{\gamma}}} J_y, \qquad y_0 \in V, \qquad y = f_0(x),$$

where J_y denotes an integral segment of $\mathfrak{N}$. For $t \geq 0$ sufficiently small, define $f_t(x) = g_{\tilde{\gamma}}(\mathfrak{N}_t(y_0))$. This definition is independent of γ. In fact, let $\mu : I \longrightarrow K$ be another path, $\mu(0) = x_0$, $\mu(1) = x$ and $H : K \times I \longrightarrow F$ the homotopy between f_0 and C, that is, $H(x,0) = f_0(x)$, $H(x,1) = y_0$ for all $x \in K$. Then the map $\tilde{H} : I \times I \longrightarrow F$, $\tilde{H}(s,u) = H(\gamma * \mu^{-1}(s),u)$ is a homotopy between the path $\tilde{\gamma} * \tilde{\mu}^{-1}(s)$ and the constant path $\tilde{C}(s) = y_0$. So $\tilde{\gamma} \simeq \tilde{\mu}$ and the germs of $g_{\tilde{\gamma}}$ and $g_{\tilde{\mu}}$ at y_0 coincide.

By the compactness of K, there is an interval $[0,\epsilon]$ such that $f_t(x)$, $t \in [0,\epsilon]$ is well-defined for all $x \in K$. Then it suffices to parametrize the integral curve J_{y_0} in order to obtain the same for all $t \in [0,1]$.

§4. The local stability theorem ([10],[46])

Theorem 3. *Let $\mathfrak{F}$ be a C^1 codimension n foliation of a manifold M and F a compact leaf with finite holonomy group. There exists a neighborhood U of F, saturated in $\mathfrak{F}$, in which all the leaves are compact with finite holonomy groups. Further, we can define a retraction $\pi : U \longrightarrow F$ such that, for every leaf $F' \subset U$, $\pi \mid F' : F' \longrightarrow F$ is a covering with a finite number of sheets and, for each $y \in F$, $\pi^{-1}(y)$ is homeomorphic to a disk of dimension n and is tranverse to $\mathfrak{F}$. The neighborhood U can be taken to be arbitrarily small.*

Proof. Since F is compact, there is a covering $(U_i)_{i=0}^k$ of F by coordinate neighborhoods, where $\bar{U}_i$ is compact, and diminishing the U_i's if necessary we can by Lemma 2 define a retraction $\pi : \cup_{i=0}^k U_i \longrightarrow F$ of class C^1 such that for any $x \in F$, $\pi^{-1}(x)$ is transverse to $\mathfrak{F} \mid \cup_{i=0}^k U_i$. In each U_i consider $D_i = \pi^{-1}(x_i)$ where $x_i \in U_i \cap F$. By hypothesis there exist closed paths $\gamma_j : I \longrightarrow F$, $\gamma_j(0) = \gamma_j(1) = x_0$, $j = 1 ,..., m$, and corresponding holonomy maps $f_j : \tilde{W}_j \subset D_0 \longrightarrow D_0$ such that $\mathrm{Hol}(F,x_0) = \{f_0 = 1, f_1 ,..., f_m\}$. Consequently, if $W = \cap_{j=1}^m \tilde{W}_j$, the domains of all the maps contain W. Set $V = \cup_{i=0}^k U_i$.

We will now show that there exists $W_0 \subset W$, a neighborhood of x_0, such that, if $y \in W_0$, then the leaf $\tilde{F}_y$ of the restriction $\mathfrak{F} \mid V$ which passes through y does not meet the boundary of V. This will imply in particular that if F_y is a leaf of $\mathfrak{F}$ through y then $F_y = \tilde{F}_y$. Moreover, if $\mathfrak{F}(W_0)$ is the saturation of W_0 in $\mathfrak{F}$ then $\mathfrak{F}(W_0) \subset V$. Observe first that if $y \in W$ then $\tilde{F}_y \cap D_0$

$\subset \{f_j(y) \mid j = 0, \ldots, m\}$. In fact, if there is a $z \in \tilde{F}_y \cap D_0$ such that $z \neq f_j(y)$, $j = 0, \ldots, m$, there exists a curve $\tilde{\gamma} : I \longrightarrow \tilde{F}_y$ such that $\tilde{\gamma}(0) = y$, $\tilde{\gamma}(1) = z$ and $\tilde{\gamma}(I) \subset V$. Since $\pi(y) = \pi(z) = x_0$, the curve $\gamma = \pi \circ \tilde{\gamma}$ induces an element $f_\gamma \in \mathrm{Hol}(F,x_0)$ such that $f_\gamma(y) = z \neq f_j(y)$, $j = 0, \ldots, m$, which is absurd.

The global trivialization lemma shows that for all $i = 0, \ldots, k$ there is a neighborhood W_i of x_0 in D_0 such that, for all $y \in W_i$, $\tilde{F}_y \cap D_i$ contains at most $m + 1$ points. Let $W_0' = W \cap W_1 \cap \ldots \cap W_k \subset D_0$ and for $y \in W_0'$, $\tilde{F}_y \cap D_i = \{q(i,j,y) \mid 1 \leq j \leq \ell_i\}$ where $\ell_i \leq m + 1$. Let $P(i,j,y)$ be the plaque of U_i through $q(i,j,y)$.

Let d be a metric on M. We have $d(\partial V, F) = \delta > 0$. Let $\delta_i(y) = \max_{1 \leq j \leq \ell_i} d(x_i, P(i,j,y))$, for $y \in W_0'$ and $1 \leq i \leq k$.

For $i = 0$ we have

$$\lim_{y \to x_0} \max_{1 \leq j \leq \ell_0} d(x_0, q(0,j,y)) \leq \lim_{y \to x_0} \max_{0 \leq j \leq m} d(x_0, f_j(y)) = 0.$$

Then $\lim_{y \to x_0} \delta_0(y) = 0$. Analogously $\lim_{y \to x_0} \delta_i(y) = 0$ for $1 \leq i \leq k$. We conclude therefore that there is a neighborhood $W_0 \subset W_0'$ of x_0, such that for all $y \in W_0$ and $i = 0, \ldots, k$, $\delta_i(y) \leq \delta/2$. We have then that for all $y \in W_0$ and $i = 0, \ldots, k$ the plaques of $\tilde{F}_y \cap U_i$ do not meet ∂V, so $F_y \subset V$. The above argument also shows that for all $y \in W_0$ the leaf $F_y = \tilde{F}_y$ is compact. In fact, for all $y \in W_0$, F_y is covered by a finite number of plaques $P(i,j,y)$, $0 \leq i \leq k$, $1 \leq j \leq \ell_i \leq m + 1$; so F_y is compact. In particular the saturation $U = \mathfrak{F}(W_0) \subset V$ is a neighborhood of F and $\pi = \pi \mid U : U \longrightarrow F$ is a retraction such that $\pi \mid F' : F' \longrightarrow F$ is a covering map with a finite number of sheets for each leaf $F' \subset U$.

We have yet to prove that the holonomy group of the leaves $F' \subset U$ is finite. It suffices to note that for all $j = 0, \ldots, m$, $f_j : W_0 \longrightarrow W_0$ is defined and therefore $\mathrm{Hol}(F') \subset \{f_0, \ldots, f_m\}$, since every closed curve in F' is the lift of a closed curve in F.

Corollary. *Let $\mathfrak{F}$ be a C^1, codimension n foliation of a manifold M and F a compact leaf with finite fundamental group. Then, there exists a neighborhood U of F, saturated by $\mathfrak{F}$, in which all the leaves are compact with finite fundamental group.*

Proof. First, F has finite holonomy. So it follows from Theorem 3 that there exists a neighborhood U of F saturated by $\mathfrak{F}$ such that if $F' \subset U$ is a leaf then $\pi \mid F' : F' \longrightarrow F$ is a covering. Since $(\pi \mid F')_* : \pi_1(F') \longrightarrow \pi_1(F)$ is injective, we have that $\pi_1(F')$ is finite.

§5. Global stability theorem. Transversely orientable case

In this section $\mathfrak{F}$ will denote a codimension one, transversely orientable, C^r $(r \geq 1)$ foliation on a compact connected manifold M.

Theorem 4 (global stability). *Suppose $\mathfrak{F}$ has a compact leaf F with finite fundamental group. Then all the leaves are diffeomorphic to F. Moreover there is a C^r submersion $f: M \longrightarrow S^1$ such that the leaves of $\mathfrak{F}$ are the sets $f^{-1}(\theta)$, $\theta \in S^1$.*

Since F has finite holonomy group, we begin by studying the properties of these groups.

Lemma 5. *Let* Hom $(\mathbb{R},0)$ *be the group of germs at* $0 \in \mathbb{R}$ *of homeomorphisms which leave* $0 \in \mathbb{R}$ *fixed. Let G be a finite subgroup of* Hom $(\mathbb{R},0)$. *Then G has at most two elements. If G has two elements, then one of them reverses orientation and the other is the identity.*

Proof. Let $f \in G$. We claim that $f^2 = 1$, the identity of Hom $(\mathbb{R},0)$. Consider a representation of f, which we will denote by f also. We have then that $f:(a,b) \longrightarrow (c,d)$ where $a < 0 < b$, $c < 0 < d$ and $f(0) = 0$.

Suppose initially that $f(0,b) \subset (0,+\infty)$. If there exists a $y \in (0,b)$ such that $f(y) < y$, we have that for all $k > 0$, $0 < f^k(y) < f^{k-1}(y) < \ldots < f(y) < y$ so the points of the sequence $\{f^k(y) \mid k \in \mathbb{N}\}$ are distinct points and therefore $f^k \neq f^j$ if $k \neq j$. This contradicts the hypothesis of G being finite. Analogously we cannot have $f(y) > y$. We conclude therefore that $f(y) = y$ for all $y \in (0,b)$. By an analogous argument $f(y) = y$ if $y \in (a,0)$ and therefore $f = 1$. In the case that $f(0,b) \subset (-\infty,0)$, that is, f reverses orientation, we have $f^2(0,b) \subset (0,+\infty)$, so $f^2 = 1$. We conclude therefore that every element of G has order ≤ 2. In particular G is abelian.

Suppose there exists $f,g \in G$ such that $f,g \neq 1$. Let (a,b) be the intersection of the domains of f and g with the image of g. Since $f,g \neq 1$, we have $g(0,b) \subset (a,0)$ and $f \circ g(0,b) \subset (0,+\infty)$; so $f \circ g = 1$ and hence $f = g^{-1}$. Since $g^2 = 1$, we have $f = g^{-1} = g$, as desired.

Corollary. *Let F be a compact leaf of a codimension 1 foliation $\mathfrak{G}$ such that $\pi_1(F)$ is finite. Then* Hol (F,x_0), *$x_0 \in F$, contains at most two elements. If $\mathfrak{G}$ is transversely orientable, then* Hol $(F,x_0) = \{1\}$.

Proof. By Lemma 5, it suffices to prove that Hol $(F,x_0) = \{1\}$, in the case that $\mathfrak{G}$ is transversely orientable. In this case, there is a vector field X on M that is nonzero and transverse to $\mathfrak{G}$. Given $x \in F$, take as a transverse section to $\mathfrak{G}$ through x a small segment Σ_x of the orbit of X through x. So, if

$f : \Sigma'_{x_0} \subset \Sigma_{x_0} \longrightarrow \Sigma_{x_0}$ is an element of $\text{Hol}(F, x_0)$, we have $f(x_0) = x_0$ and by the above lemma, $f = 1$ if and only if f preserves the orientation of Σ_{x_0}, that is, if and only if $Df(x_0) \cdot X(x_0) = \lambda_1 X(x_0)$ where $\lambda_1 > 0$ (in fact $\lambda_1 = 1$). Let $\gamma : [0,1] \longrightarrow F$ be a closed curve such that f is the element of $\text{Hol}(F, x_0)$ corresponding to $[\gamma]$. For each $t \in [0,1]$, there is a holonomy map $f_t : \Sigma'_t \subset \Sigma_0 \longrightarrow \Sigma_{\gamma(t)}$ and we have $Df_t(x_0) \cdot X(x_0) = \lambda(t) X(\gamma(t))$ where $\lambda : [0,1] \longrightarrow \mathbb{R}$ is a continuous nonvanishing function. Since $\lambda(0) = 1$, we have $\lambda_1 = \lambda(1) > 0$ as desired.

Lemma 6. *Let $\mathfrak{G}$ be a codimension one, C^r, $(r \geq 1)$ foliation. Suppose F is a compact leaf of $\mathfrak{G}$ such that $\#(\text{Hol}(F, x_0)) = 1$. Then there exists an open neighborhood $V(F)$ of F in M, saturated by $\mathfrak{G}$, and a C^r diffeomorphism $h : (-1,1) \times F \longrightarrow V(F)$ such that the leaves of $\mathfrak{G}$ in $V(F)$ are the sets of the type $h(\{t\} \times F)$, $t \in (-1,1)$, letting $F = h(\{0\} \times F)$. In particular $V(F) - F$ has two connected components.*

Proof. By the local stability theorem there is a neighborhood $V = V(F)$ of F, saturated by $\mathfrak{G}$, and a retraction $\pi : V \longrightarrow F$ such that, for every $x \in F$, $\pi^{-1}(x)$ is a segment transverse to $\mathfrak{G}$. Moreover if $F' \subset V$ is a leaf of $\mathfrak{G}$ then $1 \leq \#(F' \cap \pi^{-1}(x)) \leq \#(\text{Hol}(F, x_0)) = 1$, so $\#(F' \cap \pi^{-1}(x)) = 1$. Consider a C^r parametrization $\sigma : (-1,1) \longrightarrow \pi^{-1}(x_0)$ such that $\sigma(0) = x_0$ and $\sigma'(t) \neq 0$ for all $t \in (-1,1)$. Denote by F_t the leaf of $\mathfrak{G}$ in V which cuts $\pi^{-1}(x_0)$ at $\sigma(t)$. The map $t \mapsto F_t$ is evidently one-to-one. Consider now the map $g : V \longrightarrow (-1,1) \times F$ defined by $g(y) = (g_1(y), \pi(y))$ where $g_1(y)$ is the unique element of $(-1,1)$ such that $y \in F_{g_1(y)}$. If F_t is as above, we have $g(F_t) = \{t\} \times F$, because $\pi(F_t) = F$. Further, g is one-to-one because if $g(y) = g(y')$ we have $t = g_1(y) = g_1(y')$ so y and y' are in the same leaf F_t of $\mathfrak{G}$ and since $\pi \mid F_t : F_t \longrightarrow F$ is a diffeomorphism, we have $y = y'$ because $\pi(y) = \pi(y')$. Let $h : (-1,1) \times F \longrightarrow V$ be the inverse of g. We have then that $h(t,x)$ is the unique point of $F_t \cap \pi^{-1}(x)$. One easily verifies that h and g are C^r and hence are diffeomorphisms. ■

Corollary. *Let F be a leaf of $\mathfrak{G}$ such that $\#(\text{Hol}(F, x_0)) = 1$ and let $V(F)$ be a neighborhood of F as in Lemma 6. If ℓ is a segment transverse to $\mathfrak{G}$ such that $\ell \subset V(F)$ then for every leaf $F' \subset V(F)$ we have $\#(\ell \cap F') \leq 1$. If the endpoints of ℓ are in ∂V then $\#(\ell \cap F') = 1$ for every leaf $F' \subset V(F)$.*

Proof. This follows immediately from the preceding lemma.

Proof of Theorem 4. Let U be the set of points x in M such that the leaf F_x of $\mathfrak{F}$ through x is compact and has finite fundamental group. By hypothesis $U \neq \varnothing$ and by Lemma 6, U is open. Let $\tilde{U}$ be a connected component of U. We will show that $\partial \tilde{U} = \varnothing$ and, since M is connected, we will have $\tilde{U} = = U = M$.

Since U is saturated by $\mathfrak{F}$, $\tilde{U}$ and $\partial\tilde{U}$ are also saturated by $\mathfrak{F}$. Suppose by contradiction that $\partial\tilde{U} \neq \emptyset$. We prove first that all leaves contained in $\partial\tilde{U}$ are compact.

In fact, suppose by contradiction that $\tilde{F} \subset \partial\tilde{U}$ is a non compact leaf. Since M is compact, $\tilde{F}$ has an accumulation point $p \in M - \tilde{F}$. Let J be a compact segment transversal to $\mathfrak{F}$ with $p \in J$. Clearly $\tilde{F} \cap J \neq \emptyset$ and since $p \notin \tilde{F}$, it follows that $\tilde{F} \cap J$ is infinite. We can suppose that the extremities of J are p and a point of $\tilde{F}$. Moreover, since $\tilde{F} \subset \partial\tilde{U}$, from the transverse uniformity of $\mathfrak{F}$ (Theorem 3, Chapter III), it follows that any transverse segment which intersects $\tilde{F}$, intersects $\tilde{U}$. In particular $\tilde{U} \cap J = \cup_{n\in B} J_n$, where for each $n \in B$, J_n is an open segment in J and $J_n \cap J_m = \emptyset$ if $n \neq m$. Let us prove that B is infinite. Since J is a segment we can order J and consider a monotone sequence $\{x_n\}_{n\in\mathbb{N}}$ in $\tilde{F} \cap J$, such that $\lim_{n\to\infty} x_n = p$. For each $m \in B$, clearly $J_m \subset (x_n, x_{n+1})$ for some $n \in \mathbb{N}$, where (x_n, x_{n+1}) is the segment of J between x_n and x_{n+1}. Now, $(x_{n-1}, x_{n+1}) \cap \tilde{U} \neq \emptyset$, which implies that there exists $m \in B$ such that $J_m \subset (x_{n-1}, x_{n+1})$. Since there is an infinite number of such segments, it follows that B is infinite.

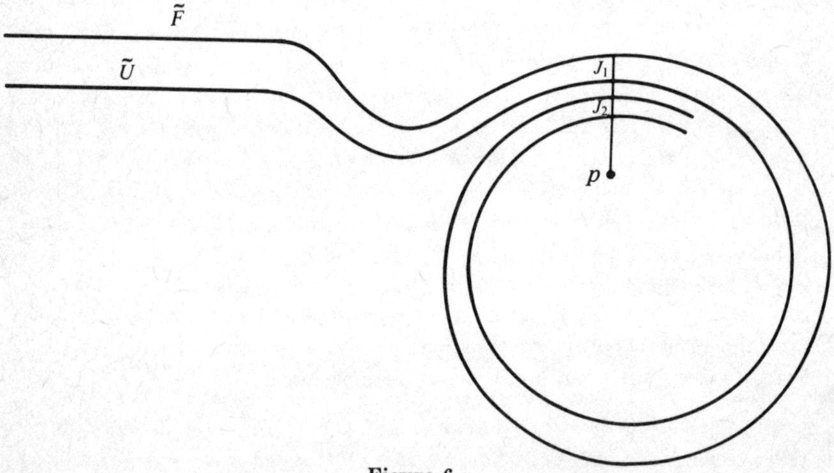

Figure 6

Now we prove that for each $n \in B$, $\text{sat}(J_n) = \tilde{U}$. Clearly $\text{sat}(J_n) \subset \tilde{U}$ and from Lemma 6, it follows that $\text{sat}(J_n)$ is open in $\tilde{U}$. Since $\tilde{U}$ is connected it is sufficient to prove that $\text{sat}(J_n)$ is closed in $\tilde{U}$. Let $\{p_j\}_{j\in\mathbb{N}}$ be a sequence in $\text{sat}(J_n)$ such that $\lim_{n\to\infty} p_n = p_\infty \in \tilde{U}$. Since $F_{p_\infty} \subset \tilde{U}$, by Lemma 6, there exists a neighborhood $V(F_{p_\infty}) = V \subset \tilde{U}$, saturated by $\mathfrak{F}$ and such that $\mathfrak{F}/V$ is equivalent to a product foliation $(-1,1) \times F_{p_\infty}$. Since $p_j \in V$ for j sufficiently big, it follows that $J_n \cap V \neq \emptyset$. Now, the extremeties of J_n can not be in V, therefore J_n intersects all the leaves of V, hence $p_\infty \in \text{sat}(J_n)$. This implies that $\text{sat}(J_n) = \tilde{U}$ for each $n \in B$.

Now, let $F' \subset \tilde{U}$ be a leaf of $\mathfrak{F}$. Since $\text{sat}(J_n) = \tilde{U}$ it follows that for each $n \in B, F' \cap J_n \neq \emptyset$. On the other hand, F' and J are compact and so $F' \cap J$ is finite, which implies that B is finite, a contradiction. Therefore $\tilde{F}$ is compact. The same argument implies that for any compact segment I which is transverse to $\mathfrak{F}$, then $I \cap \tilde{U}$ contains a finite number of open intervals.

Now, let us fix a tubular neighborhood W of $\tilde{F} \subset \partial\tilde{U}$, such that ∂W is compact, and a retraction $\pi : W \longrightarrow \tilde{F}$ such that for any $p \in \tilde{F}$, $\pi^{-1}(p)$ is a segment transverse to $\mathfrak{F}$. Let us prove that there exists a leaf F' of $\mathfrak{F}$ such that $F' \subset W \cap \tilde{U}$. Suppose by contradiction that for any leaf $F' \subset \tilde{U}$ such that $F' \cap W \neq \emptyset$ then $F' \cap \partial W \neq \emptyset$. Let $J = \pi^{-1}(x)$, where $x \in \tilde{F}$ is fixed. Since $\tilde{F} \subset \partial\tilde{U}$, x is accumulated by points of $J \cap \tilde{U}$ and so $J \cap \tilde{U}$ contains a monotone sequence $\{x_n\}_{n\in\mathbb{N}}$ such that $\lim_{n\to\infty} x_n = x$. For each $n \in N$, let $y_n \in F_{x_n} \cap \partial W$. Since ∂W is compact, by considering a subsequence if necessary, we can suppose that $\lim_{n\to\infty} y_n = y \in \partial W$. Set $S_n = \overline{\text{sat}(x_n,x)}$ and $S = \bigcap_{n\geq 1} S_n$, where (x_n,x) is the segment of J between x_n and x. Since S_n is compact, connected, saturated and $S_{n+1} \subset S_n$ for each $n \in N$, we get that S is also compact, connected and saturated. Moreover for all $n \in \mathbb{N}$, $\{x_n,y_n\} \in S_n$, which implies that $\{x,y\} \in S$. Let us consider the leaf F_y. Clearly F_y intersects the segment $[x_n,x)$ for all $n \geq 1$ and since $F_y \neq F_x$, this implies that $F_y \cap [x_1,x]$ is infinite. On the other hand, $y = \lim_{n\to\infty} y_n$, $y_n \in \tilde{U}$, and so $y \in \partial\tilde{U}$, which implies that F_y is compact. Therefore $F_y \cap [x_1,x]$ is finite, a contradiction. It follows that there exists a leaf $F' \subset \tilde{U} \cap W$.

Now F' is compact and $\pi/F' : F' \longrightarrow \tilde{F}$ is a local diffeomorphism, which implies that $\pi/F' : F' \longrightarrow \tilde{F}$ is a covering map. Since F' is compact and has finite fundamental group the same is true for $\tilde{F}$, hence $\tilde{F} \subset \tilde{U}$ and $\partial\tilde{U} = \emptyset$ as we wish.

We will now prove that all the leaves of $\mathfrak{F}$ are diffeomorphic. Given F', a leaf of $\mathfrak{F}$, let

$$U_{F'} = \{x \in M \mid F_x \text{ is diffeomorphic to } F'\} .$$

By Lemma 6, $U_{F'}$ is open in M for every F'. For the same reason $M - U_{F'}$ is also open in M. Since M is connected, we have $M = U_{F'}$ as desired.

To conclude the proof we prove that there is a submersion $f : M \longrightarrow S^1$ such that the leaves of $\mathfrak{F}$ are the surfaces $f^{-1}(\theta)$, $\theta \in S^1$. For this it is enough to show that there exists a closed curve $\gamma : S^1 \longrightarrow M$ transverse to $\mathfrak{F}$ and such that $\gamma(S^1)$ meets each leaf of $\mathfrak{F}$ in exactly one point. In fact, this being the case, it is enough to define $f(p) = \gamma^{-1}(F_p \cap \gamma(S^1))$. The existence of γ follows from the following two lemmas.

Lemma 7. *Let $\mathfrak{F}$ be a codimension one foliation defined on a compact manifold M. Then*

(a) *There is a curve* $\gamma : S^1 \longrightarrow M$ *transverse to* $\mathfrak{F}$,

(b) *If* $\mathfrak{F}$ *is transversely orientable,* $\tilde{\gamma} : S^1 \longrightarrow M$ *is transverse to* $\mathfrak{F}$ *and* $\tilde{\gamma}(S^1) \cap F_0 \neq \varnothing$ *where* F_0 *is a compact leaf of* $\mathfrak{F}$, *then there is a* $\gamma : S^1 \longrightarrow M$ *transverse to* $\mathfrak{F}$ *such that* $\gamma(S^1) \cap F_0$ *contains only one point.*

Proof. (a) Suppose initially that $\mathfrak{F}$ is transversely orientable. Let X be a vector field transverse to $\mathfrak{F}$. It suffices to get an integral curve of X which cuts a leaf F of $\mathfrak{F}$ twice. In fact, suppose that $\alpha : \mathbb{R} \longrightarrow M$ is an integral curve of X such that $\alpha(t_1)$ and $\alpha(t_2) \in F$ where $t_1 < t_2$. If $\alpha(t_1) = \alpha(t_2)$ the curve α is a closed path in X and we are done. If $\alpha(t_1) = p_1 \neq p_2 = \alpha(t_2)$ let $\delta : [0,1] \longrightarrow F$ be a simple curve such that $\delta(0) = p_1$ and $\delta(1) = p_2$. By the global trivialization lemma there is a neighborhood V of $\delta(I)$ and a diffeomorphism $h : D^{m-1} \times (-\epsilon,\epsilon) \longrightarrow V$ such that $h^*(\mathfrak{F}) = \mathfrak{F}^*$ is the foliation whose leaves are of the form $D^{m-1} \times \{t\}$, $t \in (-\epsilon,\epsilon)$ where $p_1,p_2 \in h(D^{m-1} \times \{0\}) \subset F$. Let $X^* = h^*(X)$ be the vector field $X^*(q) = (Dh(q))^{-1} \cdot (X(h(q)))$. This vector field is transverse to $\mathfrak{F}^*$ so we can assume that its last component in $D^{m-1} \times (-\epsilon,\epsilon)$ is positive. Let δ_1 and δ_2 be the segments of $\alpha[t_1,t_2] \cap V$ which contain p_1 and p_2 respectively. Since $t_1 < t_2$ we have that $h^{-1}(\delta_1) \subset D^{m-1} \times [0,\epsilon)$ and $h^{-1}(\delta_2) \subset D^{m-1} \times (-\epsilon,0]$. It suffices now to define $\gamma : [t_1,t_2] \longrightarrow M$ by modifying the curve $\alpha \mid [t_1,t_2]$ as in the figure below.

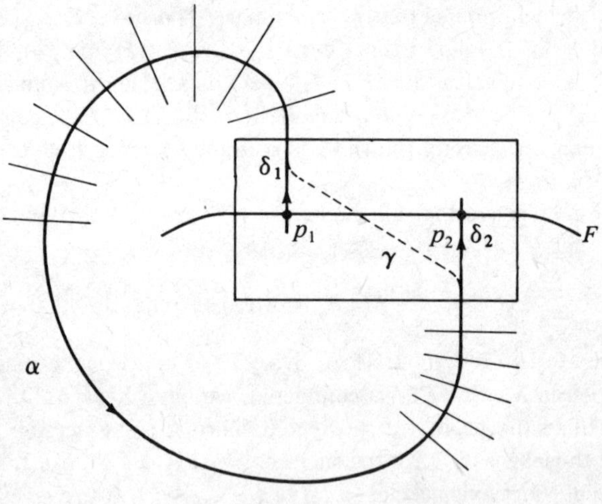

Figure 7

In the figure, α is modified in intervals of the form $[t_1,t_1+\delta]$ and $[t_2-\delta,t_2]$ so that $\gamma \mid [t_1+\delta,t_2-\delta]$, $\gamma \mid [t_1,t_1+\delta]$ and $\gamma \mid [t_2-\delta,t_2]$ are transverse to $\mathfrak{F}$, $\gamma(t_1) = \gamma(t_2)$ and $\gamma'(t_1) = \gamma'(t_2)$.

We prove now that given an integral curve α of X, not closed, then α meets a certain leaf of $\mathfrak{F}$, an infinite number of times. Let $\alpha : \mathbb{R} \longrightarrow M$ be such a curve and p an accumulation point of the sequence $\{\alpha(n)\}_{n\in\mathbb{N}}$. Let U be a foliation box of $\mathfrak{F}$ containing p. Then there exists a sequence $n_k \longrightarrow \infty$ such that $\alpha(n_k) \in U$. The segments of $\alpha(\mathbb{R}) \cap U$ which contain $\alpha(n_k)$ necessarily meet the plaque of U which contains p for k sufficiently large. This proves our claim.

If $\mathfrak{F}$ is not transversely orientable consider a line field L transverse to $\mathfrak{F}$ and take the double covering $\pi : \tilde{M} \longrightarrow M$ of L. Since $\tilde{M}$ is compact the above case applies and we obtain a curve $\tilde{\gamma} : S^1 \longrightarrow \tilde{M}$ transverse to $\pi^*(\mathfrak{F})$. The curve $\pi \circ \tilde{\gamma} = \gamma$ is transverse to $\mathfrak{F}$.

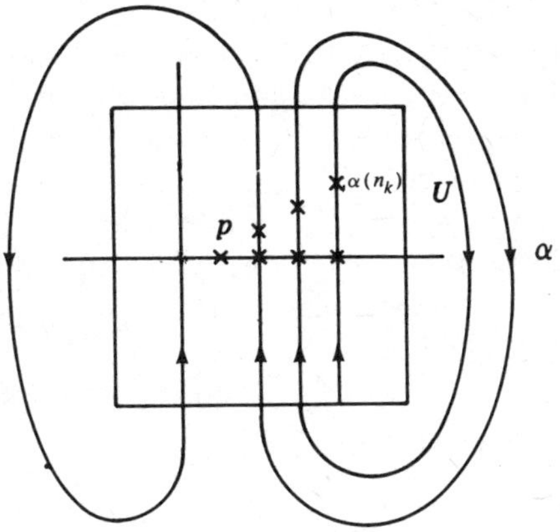

Figure 8

(b) Suppose $\mathfrak{F}$ is transversely orientable, $\tilde{\gamma} : S^1 \longrightarrow M$ is transverse to $\mathfrak{F}$ and $\tilde{\gamma}(S^1) \cap F_0 \neq \varnothing$ where F_0 is compact. Then $\tilde{\gamma}(S^1) \cap F_0$ is finite. In the case that $\tilde{\gamma}(S^1) \cap F_0$ contains more than one point we choose $\theta_1, \theta_2 \in S^1$ such that $\tilde{\gamma}(\theta_1), \tilde{\gamma}(\theta_2) \in F_0$ and $\tilde{\gamma}(\theta_1, \theta_2) \cap F_0 = \varnothing$ where (θ_1, θ_2) is one of the segments of $S^1 - \{\theta_1, \theta_2\}$. One can now repeat the first argument, since $\mathfrak{F}$ is transversely orientable. ■

Lemma 8. *Let $\mathfrak{F}$ be a transversely orientable, codimension one foliation of a compact, connected manifold M. Let $\gamma : S^1 \longrightarrow M$ be a closed curve transverse to $\mathfrak{F}$ such that every leaf in* sat$(\gamma(S^1))$ *is compact and has finite fundamental group. Then* sat$(\gamma(S^1)) = M$. *Moreover, the number of points of intersection of the leaves of $\mathfrak{F}$ with $\gamma(S^1)$ is constant.*

Proof. We will show that $W = \text{sat}(\gamma(S^1))$ is open and closed in M and hence $W = M$.

W is open. Given $x \in W$, $F_x \cap \gamma(S^1) \neq \varnothing$. If $y \in F_x \cap \gamma(S^1)$ we consider a simple curve $\mu : [0,1] \longrightarrow F_x$ such that $\mu(0) = x$ and $\mu(1) = y$. Using the global trivialization lemma it is easy to show that there is a neighborhood A of $\mu[0,1]$ such that for every $z \in A$, $F_z \cap \gamma(S^1) \neq \emptyset$ as desired.

W is closed. Given $x \in \gamma(S^1)$, by Lemma 6 there exist a neighborhood $V(F_x)$ of F_x, saturated by $\mathfrak{F}$ and a diffeomorphism $h : (-1,1) \times F_x \longrightarrow V(F_x)$ such that the leaves of $\mathfrak{F}$ in $V(F_x)$ are the sets of the form $h(\{t\} \times F_x)$, $t \in (-1,1)$. Denote by V_x the compact neighborhood of F_x, $h([-1/2,1/2] \times F_x)$. It is clear that $V_x \cap \gamma(S^1)$ contains an open interval which contains x, which we will call I_x. We have then $\cup_x I_x = \gamma(S^1)$ and since $\gamma(S^1)$ is compact there is a finite cover of $\gamma(S^1)$ by intervals $\{I_{x_i}\}_{i=1}^{\ell}$. It is easily verified that W is the finite union $V_{x_1} \cup \ldots \cup V_{x_\ell}$ of compact sets, so it is compact hence closed, from which it follows that $W = M$.

Now let $k(x) = \#(F_x \cap \gamma(S^1))$, $x \in \gamma(S^1)$. We want to prove that $k(x)$ is constant. For this we consider the set $\gamma_j = \{x \in \gamma(S^1) \mid \#(\gamma(S^1) \cap F_x) = = j\}$. We claim that γ_j is open in $\gamma(S^1)$ for every $j \in N$. In fact, given $x \in \gamma_j$, let $V(F_x)$ be a neighborhood as in Lemma 6. Then $\gamma(S^1) \cap V(F_x)$ has a finite number of components $\ell_1, \ldots, \ell_r$. For all $i = 1, \ldots, r$ the endpoints of ℓ_i are contained in $\partial V(F_x)$ so by the corollary of Lemma 6, $\#(\ell_i \cap F) = 1$ for every leaf $F \subset V(F_x)$. In particular $\#(\ell_i \cap F_x) = 1$, $i = 1, \ldots, r$ so $r = j$. Moreover for every leaf $F \subset V(F_x)$, $\#(\ell(S^1) \cap F) = j$ so γ_j is open in $\gamma(S^1)$. We have then $\gamma(S^1) = \cup_{j=1}^{\infty} \gamma_j$ so $\gamma(S^1) = \gamma_k$ for some $k \in \mathbb{N}$, since $\gamma(S^1)$ is connected.

§6. Global stability theorem. General case

Theorem 5. *Let $\mathfrak{F}$ be a C^1, codimension one foliation of a compact connected manifold M and F a compact leaf of $\mathfrak{F}$ with finite fundamental group. Then all the leaves are compact with finite fundamental group.*

Proof. By Theorem 4 we can suppose $\mathfrak{F}$ is not transversely orientable. Let P be a line field transverse to the plane field of codimension one which is tangent to the leaves of $\mathfrak{F}$. Let $\pi : \tilde{M} \longrightarrow M$ be the orientable double covering of P. Since P is not orientable, $\tilde{M}$ is connected. Consider on $\tilde{M}$ the foliation $\mathfrak{F}^* = \pi^*(\mathfrak{F})$. Since π takes leaves of $\mathfrak{F}^*$ to leaves of $\mathfrak{F}$, $\mathfrak{F}^*$ has two types of leaves:

(1) leaves F^* such that $\pi \mid F^* : F^* \longrightarrow \pi(F^*)$ is a diffeomorphism and
(2) leaves F^* such that $\pi \mid F^* : F^* \longrightarrow \pi(F^*)$ is a two-sheeted covering.

In either case, the leaf $\pi(F^*)$ is compact and has finite fundamental group. ■

Example 2. We will now construct a foliation $\mathfrak{F}$ on a three-dimensional manifold which has two leaves, F_1 and F_2, diffeomorphic to the projective plane. The other leaves of $\mathfrak{F}$ are diffeomorphic to S^2.

First consider $\tilde{M} = S^2 \times [1,2]$, foliated trivially by $S^2 \times \{t\}$, $t \in [1,2]$.

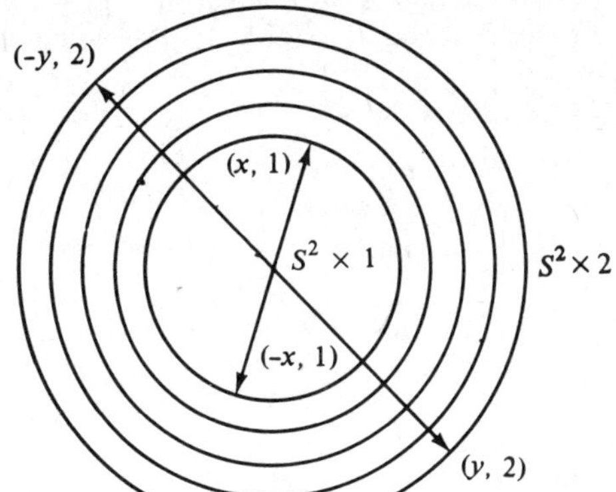

Figure 9

On M consider the equivalence relation $\sim$ such that $(x,t) \sim (x',t')$ if and only if $x = x'$, $t = t'$ or $x = -x'$ when $t = t' = 1$ or when $t = t' = 2$. Taking $M = \tilde{M}/\sim$, we have that $S^2 \times \{1\}$ and $S^2 \times \{2\}$ are transformed to two projective planes F_1 and F_2. The other leaves $S^2 \times \{t\}$ are taken to spheres; with this we get a foliation $\mathfrak{F}$ on M with the desired properties.

Example 3 ([46]). We will see here that the global stability theorem of Reeb does not generalize to foliations of codimension greater than one.

Let $M^n = S^{n-2} \times S^1 \times S^1$ and designate a point $p \in M^n$ by $p = (x, \varphi, \theta)$ where $x = (x_1, \ldots, x_{n-1}) \in \mathbb{R}^{n-1}$, $x_1^2 + \ldots + x_{n-1}^2 = 1$, and φ and θ are angular coordinates defined modulo 2π.

The differential forms

$$\eta_1 = d\theta, \qquad \eta_2 = ((1 - \sin\theta)^2 + x_1^2)\, d\varphi + \sin\theta\, dx_1$$

are linearly independent at every point and define on M^n a plane field π of codimension two by the following equation:

$$\pi(p) = \{v \in T_pM \mid \eta_1(v) = \eta_2(v) = 0\}.$$

Since $d\eta_1 \wedge \eta_1 \wedge \eta_2 = d\eta_2 \wedge \eta_1 \wedge \eta_2 = 0$, the plane field π is completely integrable, that is, it is tangent to a foliation $\mathfrak{F}$ of codimension two on M^n (see the appendix).

It is easy to verify that the surfaces given by $\sin\theta = 0$ and $\varphi =$ constant are leaves of this foliation, homeomorphic to S^{n-2}, and consequently simply connected if $n > 3$. Nevertheless the surfaces defined by $\sin\theta = 1$ and $\varphi = \varphi_0 + 1/x_1$ are non-compact leaves of $\mathfrak{F}$.

Remark 3. Reeb's stability theorem was generalized by W. Thurston [57] in the following way. Let F be a compact leaf of a transversely oriented codimension one, C^1 foliation $\mathfrak{F}$. If $H^1(F, \mathbb{R}) = 0$ then the holonomy of $\mathfrak{F}$ is trivial. Applying Lemma 6 we can conclude that $\mathfrak{F} \mid V$ is a product foliation where V is a small neighborhood of F.

Remark 4. Reeb's stability theorems (§ 4,5,6) hold even when the foliation is C^0. In the C^1 case the proof simplifies and illustrates better the techniques of the theory of foliations. For these reasons we restricted ourselves to the C^1 case.

Notes to Chapter IV

(1) Foliations induced by closed 1-forms.

Every C^r, closed 1-form $\omega\,(r \geq 1)$ defined on a manifold M induces a codimension one, C^r foliation $\mathfrak{F}(\omega)$ on $M - \text{sing}(\omega)$ where $\text{sing}(\omega) = \{p \in M \mid \omega_p = 0\}$. A system of distinguished maps for $\mathfrak{F}(\omega)$ can be defined in the following manner.

Let $p_0 \in M - \text{sing}(\omega)$. Consider a neighborhood D of p_0, homeomorphic to a disk and such that $\omega_p \neq 0$ for every $p \in D$. Since D is simply connected, by Poincaré's lemma (see [52]), there is a C^r function $f : D \longrightarrow \mathbb{R}$ such that $df = \omega$. This function is a submersion, since $df(p) = \omega_p \neq 0$ for every $p \in D$. One easily verifies that the set of all pairs (D,f), constructed as above, constitutes a system of distinguished maps and hence a foliation $\mathfrak{F}(\omega)$.

One particularly interesting fact about such foliations is that the holonomy of any leaf is trivial. This follows from Stokes' theorem, as we will see in the following. Although the result also holds for C^1 forms, we are going to suppose that ω is C^2, in order to simplify the argument. Let F be a leaf of $\mathfrak{F}(\omega)$ and $\gamma : S^1 \longrightarrow F$ be a closed curve in F. We can suppose that γ is a C^2 immersion. By Remark 2 of this chapter the holonomy of γ can be defined using a one-dimensional fibration along γ and transverse to $\mathfrak{F}(\omega)$. More precisely, there is a C^2 immersion $h : S^1 \times (-\epsilon, \epsilon) \longrightarrow M$ such that $h(\theta, 0) = \gamma(\theta)$ for every $\theta \in S^1$ and the curve $t \longmapsto h(\theta, t)$ is transverse to $\mathfrak{F}(\omega)$. The intersection of the leaves of $\mathfrak{F}(\omega)$ near F with $h(S^1 \times (-\epsilon, \epsilon))$ define the holonomy of γ. The map h can be defined, for example, by taking $h(\theta, t) = X_t(\gamma(\theta))$ where X is the C^2 vector field defined by $\omega_p(v) = \langle X(p), v \rangle_p$ for every $p \in M - \text{sing}(\omega)$ and $v \in T_pM$, letting $\langle\ ,\ \rangle$ be a Riemannian metric on M. This vector field X is also called the gradient of ω with respect to $\langle\ ,\ \rangle$ and one easily verifies that it is normal to the leaves of $\mathfrak{F}(\omega)$.

Let $\omega^* = h^*(\omega)$ be the form co-induced by ω and h (see [52]), defined by $\omega^*_{(\theta,t)}(v) = \omega_{h(\theta,t)}(dh_{(\theta,t)} \cdot v)$. Then ω^* is a closed C^1 1-form on $S^1 \times (-\epsilon, \epsilon)$ and the foliation defined by ω^*, $\mathfrak{F}^*$, is exactly the foliation $h^*(\mathfrak{F}(\omega))$. One sees also that $S^1 \times \{0\}$ is a leaf of $\mathfrak{F}^*$ and that $\mathfrak{F}^*$ is transverse to the lines $\{\theta\} \times (-\epsilon, \epsilon)$, $\theta \in S^1$. Fix a transverse line $\ell_0 = \{\theta_0\} \times (-\epsilon, \epsilon)$ and let $g : \ell_0' \subset \ell_0 \longrightarrow \ell_0$ be a representative of the holonomy of $S^1 \times \{0\}$. Then g is conjugate to the holonomy map of γ, $f_\gamma \in \text{Hol}(F)$. This follows from the fact that $\mathfrak{F}^* = h^*(\mathfrak{F}(\omega))$. It suffices to show that $g = \text{identity}$.

Suppose on the contrary that there is a $q = (\theta_0, t) \in \ell_0'$ such that $g(q) = (\theta_0, t_1) = q_1 \neq q$. In this case the leaf L_q of $\mathfrak{F}^*$ which passes through q meets ℓ_0 for the first time at q_1. Let α be the simple closed curve of $S^1 \times (-\epsilon, \epsilon)$ formed by the segments qq_1 of ℓ_0 and $q_1 q$ of L_q as in Figure 9.

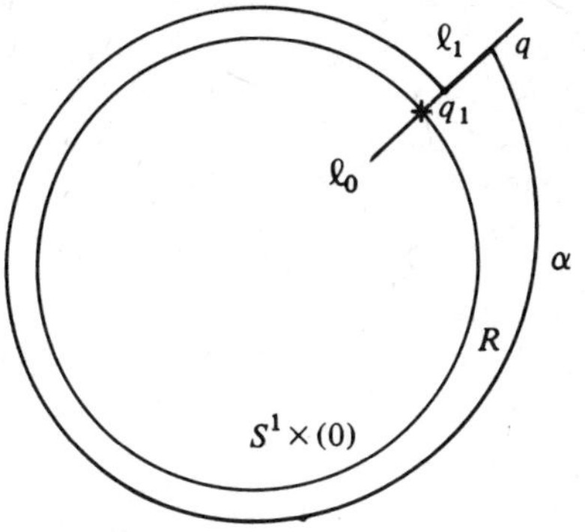

Figure 10

The curves α and $S^1 \times \{0\}$ define a region $R \subset S^1 \times (-\epsilon, \epsilon)$. Applying Stokes' theorem to the form ω^* in the region R, we have

$$\int_{S^1 \times \{0\} \cup \alpha} \omega^* = \int_R d\omega^* = 0$$

and hence $\int_\alpha \omega^* = \int_{S^1 \times \{0\}} \omega^*$. Since $S^1 \times \{0\}$ is an integral curve of ω^*, we have

$$\int_\alpha \omega^* = 0 \, .$$

Therefore

$$0 = \int_\alpha \omega^* = \int_{qq_1} \omega^* + \int_{\ell_1} \omega^* = \int_{\ell_1} \omega^* \, .$$

Since ℓ_1 is transverse to $\mathfrak{F}$, it is easy to see that $\int_{\ell_1} \omega^* \neq 0$, which is a contradiction. Hence we must have $q = q_1$, i.e., the holonomy of γ is trivial.

The following converse of the above-mentioned fact was proved by Sacksteder in [50].

Theorem. *Let $\mathfrak{F}$ be a C^2 codimension one foliation defined on a manifold M^n. Suppose the holonomy of every leaf of $\mathfrak{F}$ is trivial. Then $\mathfrak{F}$ is topologically equivalent to a foliation defined by a closed 1-form. We say that two foliations $\mathfrak{F}$ and $\mathfrak{F}'$ of M are topologically equivalent when there is a homeomorphism of M which takes leaves of $\mathfrak{F}$ to leaves of $\mathfrak{F}'$.*

(2) Singular points of completely integrable 1-forms.

A differentiable 1-form ω on $\mathbb{R}^n$ of class C^r is written

$$\omega = \sum_{i=1}^{n} a_i(x)\,dx_i$$

where the $a_i(x)$ are C^r functions. The form ω is called completely integrable when $\omega \wedge d\omega = 0$ (see the appendix).

A singularity of ω is a point p where all the $a_i(x)$ vanish simultaneously. In what follows we will show that in certain cases it is possible to determine the topological structure of the leaves of ω near a singularity using the stability theorems of Reeb.

For simplicity we suppose $a_i(0) = 0$ for $i = 1, \ldots, n$. Let $b_{ij} = (\partial a_i)(0)/(\partial x_j)$, $1 \le i, j \le n$. Then we can write

$$a_i(x) = \sum_{j=1}^{n} b_{ij} x_j + R_i(x), \qquad \lim_{x \to 0} \frac{R_i(x)}{\|x\|} = 0 .$$

Then,

$$\omega = \sum_{i,j=1}^{n} b_{ij} x_j dx_i + R, \qquad R = \sum_{i=1}^{n} R_i(x)\,dx_i .$$

The form $\omega_1 = \sum_{i,j=1}^{n} b_{ij} x_j dx_i$ is called the *linear part of* ω *at* 0.

Lemma. *The form ω_1 is completely integrable.*

Proof. From the relation $\omega \wedge d\omega = 0$ we have that

$$0 = \omega \wedge d\omega = \omega_1 \wedge d\omega_1 + \omega_1 \wedge dR + R \wedge d\omega_1 + R \wedge dR .$$

From the above expression we see that $\omega_1 \wedge d\omega_1$ is the linear part of $\omega \wedge d\omega$. So $\omega_1 \wedge d\omega_1 = 0$. ■

Proposition. *Let ω be C^1 and completely integrable and $0 \in \mathbb{R}^n$ a singularity of ω. We have two possibilities:*

(1) $d\omega \big|_{x=0} = 0$. *In this case $B = (b_{ij})_{1 \le i,j \le n}$ is symmetric.*

(2) $d\omega \big|_{x=0} \neq 0$. *In this case B has rank ≤ 2.*

Proof. Since $dR\big|_{x=0} = 0$, one has that $d\omega\big|_{x=0} = d\omega_1$. So, $d\omega\big|_{x=0} = \Sigma_{i<j} (b_{ji} - b_{ij})\,dx_i \wedge dx_j$. Clearly, if $d\omega\big|_{x=0} = 0$ then B is symmetric.

Suppose now that $d\omega_1 \neq 0$. We claim that $d\omega_1 = \theta_1 \wedge \theta_2$ where θ_1 and θ_2 are linearly independent 1-forms which do not vary with the point (i.e. $\theta_{i,x} = \theta_{i,0}$, $x \in \mathbb{R}^n$). In fact, set $d\omega_1 = \sum_{i<j} c_{ij} dx_i \wedge dx_j$, $c_{ij} = b_{ji} - b_{ij}$. Since $d\omega_1 \neq 0$, we can suppose that $c_{12} \neq$ $\neq 0$, for example. Set $\theta_1 = 1/c_{12} \sum_{j=}^{n} c_{1j} dx_j$ and $\theta_2 = \sum_{j=1}^{n} c_{2j} dx_j$ $(c_{11} = c_{22} = 0)$. Taking the product we get

(i) $$\theta_1 \wedge \theta_2 = \frac{1}{c_{12}} \sum_{i<j} (c_{1i} c_{2j} - c_{1j} c_{2i})\,dx_i \wedge dx_j .$$

On the other hand the relation $\omega_1 \wedge d\omega_1 = 0$ implies that $d\omega_1 \wedge d\omega_1 = 0$. The coefficient of $dx_i \wedge dx_j \wedge dx_k \wedge dx_\ell$ in $d\omega_1 \wedge d\omega_1$ is $2(c_{ij}c_{k\ell} + c_{ki}c_{j\ell} + c_{jk}c_{i\ell})$ so taking $k = 1$ and $\ell = 2$ we get

(ii) $$c_{12}c_{ij} + c_{1i}c_{j2} + c_{j1}c_{i2} = 0.$$

Comparing (ii) with (i), we see that $\theta_1 \wedge \theta_2 = d\omega_1$, which proves the claim.

Take now $f_1(x) = \dfrac{1}{c_{12}} \sum_{j=1}^{n} c_{1j}x_j$ and $f_2(x) = \sum_{j=1}^{n} c_{2j}x_j$.

We have then that $d\omega_1 = df_1 \wedge df_2$ so $d(\omega_1 - f_1\, df_2) = 0$. By Poincaré's lemma ([52]), there is a $g : \mathbb{R}^n \longrightarrow \mathbb{R}$ of degree 2 such that $\omega_1 = f_1\, df_2 + dg$. Take a linear change of variable $y = A(x)$ defined by

$$y_1 = f_1(x), \quad y_2 = f_2(x), \; y_k = x_k \quad \text{for} \quad k \geq 3.$$

In this system of coordinates ω_1 is written as $A_*(\omega_1) = y_1\, dy_2 + dh$, $h = g \circ A^{-1}$, $A_* = (A^{-1})^*$.

Using now the relation $\omega_1 \wedge d\omega_1 = 0$ we get $dy_1 \wedge dy_2 \wedge dh = 0$, a relation that is equivalent to $\partial h / \partial y_k = 0$ for $k \geq 3$.

Therefore h depends only on y_1 and y_2, i.e.,

$$A_*(\omega_1) = \frac{\partial h}{\partial y_1} dy_1 + (y_1 + \frac{\partial h}{\partial y_2})\, dy_2 = (e_{11}y_1 + e_{12}y_2)\, dy_1 + (e_{21}y_1 + e_{22}y_2)\, dy_2$$

Then the matrix B_* of $A_*(\omega_1)$ has rank ≤ 2. Thus it is possible to deduce that B has rank ≤ 2. We leave the final details to the reader. ■

Examples

(1) Let $\omega = x_1\, dx_1 + x_2\, dx_2 - x_3\, dx_3 = df$ where $f = 1/2\,(x_1^2 + x_2^2 - x_3^2)$.

The leaves of ω are the connected components of the level surfaces of $f \mid \mathbb{R}^3 - \{0\}$ and are given in figure 11 below.

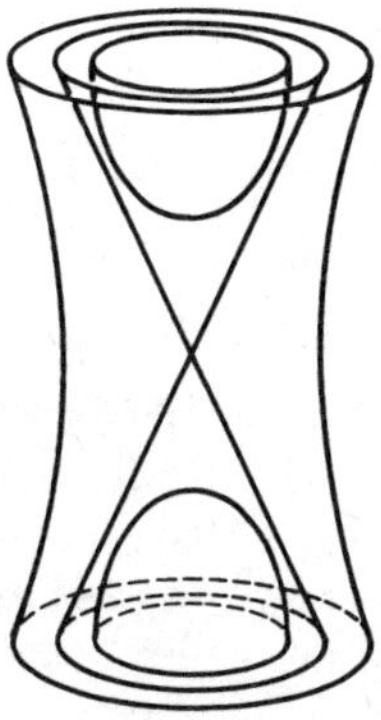

Figure 11

(2) The form $\eta = x_2\,dx_1 - 2x_1\,dx_2$ in $\mathbb{R}^3$ satisfies $\eta \wedge d\eta = 0$. Further $d\eta = -3\,dx_1 \wedge dx_2 \neq 0$. The set of singularities of η is the x_3-axis. The leaves of η are products of the x_3-axis with the orbits of the vector field $X = 2x_1(\partial/\partial x_1) + x_2(\partial/\partial x_2)$ (see figure 12).

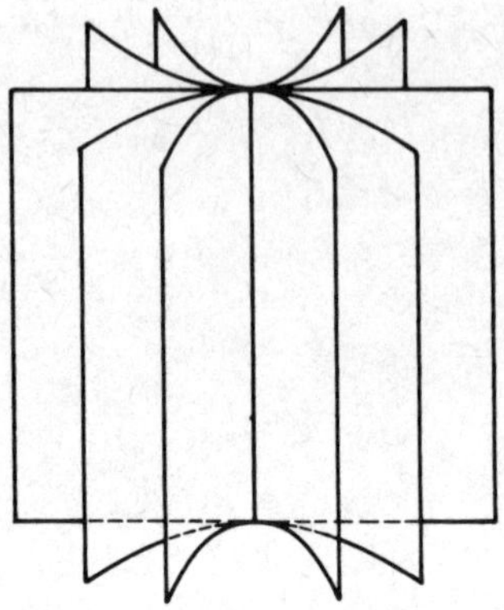

Figure 12

Study of centers in $\mathbb{R}^n$, $n \geq 3$.

Let ω be a completely integrable 1-form in $\mathbb{R}^n$ with a singularity at 0. Let $\omega_1 = \sum_{i,j} b_{ij} x_j dx_i$ be the linear part of ω at 0.

Definition. We say that 0 is a *simple* or *nondegenerate* singularity for ω when $\det(b_{ij}) \neq 0$.

This definition does not depend on the coordinate system chosen. Further, the simple singularities are isolated.

By the previous proposition if $n \geq 3$ and $0 \in \mathbb{R}^n$ is a simple singularity then necessarily $d\omega\,|_{x=0} = d\omega_1 = 0$. Consequently, $\omega_1 = df$ where f is a quadratic function. By means of a linear change of variable we can suppose f is written as the sum of squares. Thus

$$\omega_1 = d(x_1^2 + \ldots + x_k^2 - x_{k+1}^2 - \ldots - x_n^2)\ .$$

Definition. The singularity $0 \in \mathbb{R}^n$ of ω is called a *center* when it is positive definite or negative definite, that is, when ω_1 is written $\omega_1 = \pm d(x_1^2 + \ldots + x_n^2)$ in some linear system of coordinates.

In this case the leaves of ω_1 are the spheres $x_1^2 + \ldots + x_n^2 = \epsilon^2$. Our objective here is to show that if ω is C^2 then its leaves are also diffeomorphic to spheres.

Theorem (Reeb [46]). *Let ω be a completely integrable, C^2 form defined in a neighborhood of $0 \in \mathbb{R}^n$, $n \geq 3$. Suppose that 0 is a center of ω. Then there is a neighborhood of 0 in which all the nonsingular leaves of ω are diffeomorphic to the sphere S^{n-1}.*

Proof. Since 0 is a center of ω, by a linear change of coordinates we can suppose that $\omega_1 = df, f(x) = 1/2 \sum_{i=1}^{n} x_i^2$.

Consider the map $\psi : \mathbb{R} \times S^{n-1} \longrightarrow \mathbb{R}^n$ defined by $\psi(\rho,u) = \rho \cdot u$ where $u = (u_1 ,..., u_n) \in S^{n-1}$ and $\rho \in \mathbb{R}$.

Observe that the restrictions $\psi \mid (0,\infty) \times S^{n-1}$ and $\psi \mid (-\infty,0) \times S^{n-1}$ are diffeomorphisms on $R^n - \{0\}$. Therefore if F is a leaf of $\psi^*(\omega)$ contained in $(0,\infty) \times S^{n-1}$, $\psi(F)$ will be a leaf of ω in $\mathbb{R}^n - \{0\}$. We will now prove that all the leaves of $\psi^*(\omega)$ in a neighborhood of $\{0\} \times S^{n-1}$ are diffeomorphic to S^{n-1} and this implies that all the nonsingular leaves of ω in a neighborhood of $0 \in R^n$ are diffeomorphic to S^{n-1}. To do this, set

$$\omega = df + \sum_{i=1}^{n} R_i(x)\,dx_i, \qquad dR_i(0) = 0, \qquad R_i(0) = 0 .$$

We have then

$$(*) \qquad \psi^*(\omega) = d(f \circ \psi) + \sum_{i=1}^{n} R_i(\rho \cdot u)\,d(\rho \cdot u_i) .$$

Since $f \circ \psi(\rho \cdot u) = 1/2 \sum_{i=1}^{n} \rho^2 u_i^2 = 1/2\,\rho^2$, one sees that $d(f \circ \psi) = \rho d\rho$. As R_i is of class C^2, $R_i(0) = 0$ and $dR_i(0) = 0$ one obtains, $R_i(\rho \cdot u) = \rho S_i(\rho,u)$ where S_i is C^1 on $\mathbb{R} \times S^{m-1}$ and $\lim_{\rho \to 0} S_i(\rho,u) = 0$. From $(*)$ above we have

$$\psi^*(\omega) = \rho(d\rho + \eta), \qquad \eta = \Big(\sum_{i=1}^{n} u_i S_i\Big)\,d\rho + \rho \sum_{i=1}^{n} S_i du_i .$$

Consequently the form $\tilde{\omega}$ defined on $\mathbb{R} \times S^{n-1}$ by $\tilde{\omega} = 1/\rho\,\psi^*(\omega)$ for $\rho \neq 0$ and $\tilde{\omega} = d\rho$ for $\rho = 0$ is C^1, integrable and has the same integral surfaces as $\psi^*(\omega)$ on $(\mathbb{R} - \{0\}) \times S^{n-1}$.

Observe now that $\{0\} \times S^{n-1}$ is an integral surface of $\tilde{\omega}$, so, since $n \geq 3$, by Reeb's stability theorem, there is a neighborhood V of $\{0\} \times S^{n-1}$ such that all the integral surfaces of $\tilde{\omega}$ in V are diffeomorphic to S^{n-1}. This concludes the proof. ■

In the case that ω is analytic, the above theorem can be strengthened.

Theorem (Reeb). *Let ω be an analytic, integrable 1-form defined in a neighborhood of $0 \in \mathbb{R}^n$, $n \geq 3$. Suppose that $\omega_0 = 0$ and that ω has nondegenerate linear part $\omega_1 = df$. Then there exists an analytic diffeomorphism h from a neighborhood U of 0 to a neighborhood V of 0 and an analytic function $g : U \longrightarrow \mathbb{R}$ such that $g(0) = 1$, $Dh(0)$ is the identity and $h^*(\omega) = gdf$.*

The proof of this theorem can be found in [46].

Other results about the existence of first integrals of integrable forms can be found in [37] and [31].

For the study of singularities of integrable forms we suggest the following references: [33], [30] and [5].

V. FIBER BUNDLES AND FOLIATIONS

The notion of holonomy of a leaf introduced in the previous chapter is essentially of local character. It is defined by a group of germs of diffeomorphisms of a transverse section to a leaf, with a fixed point. In certain circumstances, however, it is possible to associate to the foliation a group of diffeomorphisms of a global transverse section, containing in a certain well-defined sense the holonomy of each leaf. This is the case of foliations whose leaves meet transversely all the fibers of a fiber bundle E. The importance of these foliations is in the fact that they are characterized by their holonomy, in this case given by a representation $\varphi : \pi_1(B) \longrightarrow \mathrm{Diff}(F)$ of the fundamental group of the base of E to the group of diffeomorphisms of the fiber of E. In this manner properties of φ translate to properties of the foliation. For example, the action φ has exceptional minimal sets if and only if the same occurs for the foliation. Sacksteder's example, of a C^∞ codimension one foliation with an exceptional minimal set, is a typical case of what we will see in this chapter.

A necessary and sufficient condition for the existence of a foliation transverse to the fibers of a fiber bundle is that it has discrete structure group. Once showing this, we obtain the following result due to Ehresmann [11]: let N be a compact C^2 submanifold of M. A necessary and sufficient condition in order to exist a foliation of a neighborhood of N for which N is one of the leaves is that the normal bundle to N be equivalent to one that has discrete structure group.

As an elementary application of this theory we will prove a classic result about the solutions of the differential equation of Ricatti (see note 1).

§1. Fiber bundles

A fiber bundle consists of differentiable manifolds E, B, F and a differentiable map $\pi : E \longrightarrow B$. Further E has a local product structure defined by a cover

$(U_i)_{i\in J}$ of B and by diffeomorphisms $\varphi_i : \pi^{-1}(U_i) \longrightarrow U_i \times F$, $i \in J$, which make diagrams of the type below commutative

$$\begin{array}{ccc} \pi^{-1}(U_i) & \xrightarrow{\varphi_i} & U_i \times F \\ & \searrow^{\pi} \quad {}^{p}\swarrow & \\ & U_i & \end{array}$$

that is, $\pi = p \circ \varphi_i$, where p is the projection on the first factor. In particular π is a submersion on B and E is locally diffeomorphic to the product of an open set in B and F.

The space E is called the *total space*, B the *base* and F the *fiber* of the bundle. The map π is called the *projection*. If $x \in B$, the submanifold $F_x = \pi^{-1}(x) \simeq F$ is called the *fiber of E over x*. Usually a fiber bundle is denoted by (E,π,B,F). If there is no danger of confusion we will denote it simply by E. Moreover we will assume that E is C^∞, that is, all the manifolds and maps involved in the definition of a fiber bundle are assumed to be C^∞.

Given $i,j \in J$ such that $U_i \cap U_j \neq \varnothing$, we can define $g_{ij} : U_i \cap U_j \longrightarrow \mathrm{Diff}(F)$ in such a way that the map Φ_{ij}, which makes the following diagram commute,

$$\begin{array}{ccc} & \pi^{-1}(U_i \cap U_j) & \\ {}^{\varphi_i}\swarrow & & \searrow^{\varphi_j} \\ (U_i \cap U_j) \times F & \xrightarrow{\Phi_{ij}} & (U_i \cap U_j) \times F \end{array}$$

is written $\Phi_{ij}(x,y) = (x,h_{ij}(x,y))$. For each $x \in U_i \cap U_j$, the map $h_{ij}^x : F \longrightarrow F$ defined by $y \longrightarrow h_{ij}(x,y)$ is clearly a diffeomorphism. We will use the notation $g_{ij}(x) = h_{ij}^x$.

When F and the fibers are vector spaces and all the $g_{ij}(x)$ are linear automorphisms of F, we say that E is a *vector bundle*. We say that the fiber bundle E has *discrete structure group* if, for any i and j, the map $x \mapsto h_{ij}^x$ is locally constant.

When E has discrete structure group the foliations of $\pi^{-1}(U_i)$ given by submersions

$$\pi^{-1}(U_i) \xrightarrow{\varphi_i} U_i \times F \xrightarrow{q} F, \qquad q(x,y) = y$$

define distinguished maps of a foliation whose leaves are transverse to the fibers of E. The plaques of the foliation in $\pi^{-1}(U_i)$ are the sets $(q \circ \varphi_i)^{-1}(y)$, $y \in F$.

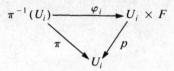

Example 1. The tangent bundle.

Let M be a C^∞, n-dimensional differentiable manifold. For each $p \in M$, let T_pM be the space of tangent vectors to M at p, and

$$TM = \{(p, v_p) \mid p \in M, \quad v_p \in T_pM\} .$$

If $\pi : TM \longrightarrow M$ is given by $\pi(p, v_p) = p$ then $(TM, \pi, M, \mathbb{R}^n)$ is a C^∞ vector bundle. In fact, given a local chart $x : U \longrightarrow \mathbb{R}^n$ of M, introduce a local chart

$$(*) \qquad \bar{x} : \pi^{-1}(U) \longrightarrow U \times \mathbb{R}^n$$

for TM in the following manner:

$$\bar{x}(p, X(p)) = \bar{x}(p, \sum_{i=1}^{n} \alpha_i \frac{\partial}{\partial x_i}) = (p, \alpha_1, \ldots, \alpha_n) = (p, Dx(p) \cdot X(p)) .$$

When $x : U \longrightarrow \mathbb{R}^n$, $y : V \longrightarrow \mathbb{R}^n$, $U \cap V \neq \varnothing$ and $X(p) = \sum_{i=1}^{n} \alpha_i \, \partial/\partial x_i = \sum_{j=1}^{n} \beta_j \, \partial/\partial y^j$, we have that

$$\bar{y} \circ \bar{x}^{-1} \mid (p, \alpha^1, \ldots, \alpha^n) = (p, \beta^1, \ldots, \beta^n)$$

with

$$(**) \qquad \beta_j = \sum_{k=1}^{n} \frac{\partial y^j}{\partial x^k} \alpha_k .$$

This permits us to define on TM both a differentiable manifold structure by $(*)$ and a vector bundle structure by $(*)$ and $(**)$.

A *section* of a fiber space is a map $\sigma : B \longrightarrow E$ such that $\pi \circ \sigma =$ identity.

A section of the tangent bundle is a vector field on M. So, a vector field X, is, in effect, a map which to each $p \in M$ associates a vector $X(p) \in T_pM$.

Example 2. The normal bundle.

Let $\langle \ , \ \rangle$ be a Riemannian metric on M^m and $N^n \subset M^m$ a submanifold of M. Given $p \in M$, let $T_pN^\perp \subset T_pM$ be the subspace of normal vectors to T_pN; we define $\nu(N) = \{(p, v_p) \mid p \in N, \ v_p \in T_pN^\perp\}$ and

$$\pi : \nu(N) \longrightarrow N, \qquad \pi(p, v_p) = p .$$

Then $(\nu(N), \pi, N, \mathbb{R}^{m-n})$ is a vector bundle.

In fact, since N is a submanifold, for each $p \in N$ there exists a local map $x : U \longrightarrow \mathbb{R}^m$ on M, $p \in U$, such that $q \in U \cap N$ if and only if $x(q)$ has its last $m - n$ coordinates zero. These coordinate functions $x^{n+1}, \ldots, x^m$ define vector fields $X_{n+1} = \operatorname{grad} x^{n+1}, \ldots, X_m = \operatorname{grad} x^m$, which form a basis for $T_q N^\perp$. The vector field grad f is defined by the relation $Df(p) \cdot u = \langle \operatorname{grad} f(p), u \rangle_p$.

So, we can define local charts $\overline{x} : \pi^{-1}(U \cap N) \longrightarrow (U \cap N) \times \mathbb{R}^{m-n}$ by

$$\overline{x}(q, X(q)) = \overline{x}(q, \sum_{i=n+r}^{m} \alpha_i X_i(q)) = (q, \alpha_{n+1}, \ldots, \alpha_m)$$

whose changes of coordinates are: if $y : V \longrightarrow \mathbb{R}^m$ is as above and $U \cap V \neq \emptyset$ then

$$(*) \quad \overline{y} \cdot \overline{x}^{-1}(q, \alpha_{n+1}, \ldots, \alpha_m) = (q, \beta_{n+1}, \ldots, \beta_m) \quad \text{for} \quad q \in U \cap V ,$$

where $X(q) = \sum_{i=n+1}^{m} \alpha_i X_i = \sum_{j=n+1}^{m} \beta_j Y_j$, $Y_j = \operatorname{grad} y^j$, $n + 1 \leq j \leq m$ and

$$\beta_j = \sum_{i=n+1}^{m} \langle X_i, \frac{\partial}{\partial y_j} \rangle \alpha_i .$$

Thus a vector bundle structure is determined on $\nu(N)$ by $(*)$.

Example 3. The unit tangent bundle.

Let $\langle \, , \rangle$ be a Riemannian metric on M^m and $T^1 M = \{(p, v_p) \mid p \in M, v_p \in T_p M, |v_p| = 1\}$.

Denote by $\pi : T^1 \mathrm{M} \longrightarrow M$ the map $\pi(p, v_p) = p$. Then $(T^1 \mathrm{M}, \pi, M, S^{n-1})$ is a fiber space whose fiber is the unit sphere $S^{n-1} \subset \mathbb{R}^n \approx T_p M$.

In fact, let $x : U \longrightarrow \mathbb{R}^n$ be a local chart of M, where $x = (x_1, \ldots, x_n)$. Let $g_{ij}^x = \langle \frac{\partial}{\partial x_i}, \frac{\partial}{\partial x_j} \rangle$. The matrix $G_x = (g_{ij}^x)_{1 \leq i,j \leq n}$ is positive definite and symmetric so it has a unique square root $A_x = (a_{ij}^x)_{1 \leq i,j \leq n}$ (i.e., $A_x^2 = G_x$) which is symmetric and positive definite. We define local charts for $T^1 M$ by setting

$$(*) \qquad \overline{x} : \pi^{-1}(U) \longrightarrow U \times S^{n-1}$$

$$\overline{x}(p, v) = (p, A_x(Dx(p) \cdot v)).$$

If $v = \sum_{i=1}^{n} v_i \, \partial / \partial x_i$ is a unit vector, we have $A_x(Dx(p) \cdot v) = \Sigma_{i,j} a_{ij}^x v_j e_i = \Sigma_i \alpha_i e_i$ where $\{e_1, \ldots, e_n\}$ is the canonical basis of $\mathbb{R}^n$, so the norm of $A_x(Dx(p) \cdot v)$ in the usual inner product is $\Sigma_j \alpha_j^2 = \Sigma_{i,j,k} v_i a_{ij}^x a_{jk}^x v_k =$

$= \Sigma_{i,k} v_i g^x_{ik} v_k = 1$, or, $A_x(Dx(p) \cdot v) \in S^{n-1}$.

If $y : V \longrightarrow \mathbb{R}^n$ is another local chart on M with $U \cap V \neq \varnothing$ then

$$\overline{y} \circ \overline{x}^{-1}(p, \alpha_1, \ldots, \alpha_n) = (p, \beta_1, \ldots, \beta_n)$$

where $\Sigma_i \alpha_i^2 = \Sigma_i \beta_i^2 = 1$ and $\beta_k = \Sigma_\ell(\Sigma_{j,k} a^y_{ij} \dfrac{\partial y_j}{\partial x_k} a^{k\ell}_x)\alpha_\ell$, letting $(a^{jk}_x)_{1 \leq j,k \leq n} = A_x^{-1}$. Since $B = (b_{i\ell})_{i \leq i, \ell \leq n}$, where $b_{i\ell} = \Sigma_{j,k} a^y_{ij} \dfrac{\partial y_j}{\partial x_k} a^{k\ell}_x$, takes unit vectors to unit vectors, B is orthogonal (that is, $B^{-1} = B^t$). We say then that $T^1 M$ is a fiber bundle with structure group $\mathcal{O}(n)$, the group of orthogonal $n \times n$ matrices.

Example 4. Let M be a compact connected manifold and $\mathfrak{F}$ a transversely orientable, codimension one foliation of M which has a compact leaf F with finite fundamental group. Then the leaves of $\mathfrak{F}$ are fibers of a fiber bundle with base S^1, fiber F and discrete structure group.

In fact, by the global stability theorem all the leaves of $\mathfrak{F}$ are diffeomorphic to F. Further, there is a submersion $f : M \longrightarrow S^1$ such that the leaves of $\mathfrak{F}$ are the sets of the form $f^{-1}(\theta)$, $\theta \in S^1$. In this case, (M, f, S^1, F) is a fiber bundle, as can easily be seen.

§2. Foliations transverse to the fibers of a fiber bundle

Let $\pi : E \longrightarrow B$ be the projection of a fiber bundle with fiber F. We say a foliation $\mathfrak{F}$ of E is transverse to the fibers when it satisfies the following properties:

(a) for every $p \in E$ the leaf L_p of $\mathfrak{F}$ which passes through p is transverse to the fiber $F_{\pi(p)}$ and $\dim(\mathfrak{F}) + \dim(F) = \dim(E)$.

(b) For every leaf L of $\mathfrak{F}$, $\pi \mid L : L \longrightarrow B$ is a covering map. It follows from the definition that for every $p \in E$ one has

$$T_p E = T_p(L_p) \oplus T_p(F_{\pi(p)}) .$$

An important observation due to Ehresmann is the following.

Proposition 1. *Suppose that the fiber F is compact. In this case* (a) *implies* (b).

Proof. Let L be a leaf of $\mathfrak{F}$. Since $\pi : E \longrightarrow B$ is a submersion, L is transverse to the fibers and $\dim(L) = \dim(B)$, it follows that $\pi \mid L : L \longrightarrow B$ is a local diffeomorphism. To prove that $\pi \mid L : L \longrightarrow B$ is a covering it suffices to show that, for every $b \in B$, there is a disk $U \subset B$ with $b \in U$ such that for each $x \in U$ the fiber $\pi^{-1}(x)$ cuts each leaf of the restricted foliation $\mathfrak{F} \mid \pi^{-1}(U)$ in exactly one point.

In fact, the connected components of $L \cap \pi^{-1}(U)$ are leaves of $\mathfrak{F} \mid \pi^{-1}(U)$. If V is one such connected component, and if $x \in U$ then $\pi^{-1}(x) \cap V$ contains exactly one point, so $\pi \mid V : V \longrightarrow U$ is one-to-one and onto and hence is a diffeomorphism, which shows that $\pi \mid L : L \longrightarrow B$ is a covering.

In order to construct U we show first that, given $z \in \pi^{-1}(b)$ there is a foliation box W_z of $\mathfrak{F}$ such that, for every $x \in \pi(W_z)$, $\pi^{-1}(x)$ cuts each plaque of W_z in exactly one point.

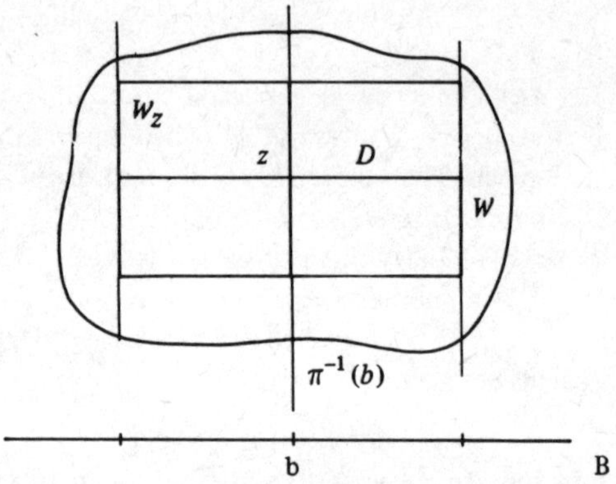

Figure 1

Let L_z be the leaf of $\mathfrak{F}$ through $z \in \pi^{-1}(b)$. Since z is an isolated point of $\pi^{-1}(b) \cap L_z$ in the intrinsic topology of L_z, there is a disk $z \in D \subset L_z$ such that for every $y \in D$, the fiber $\pi^{-1}(\pi(y))$ cuts D exactly in the point y. This implies that the map $h : \pi^{-1}(\pi(D)) \longrightarrow D$ such that $h(y) = \pi^{-1}(\pi(y)) \cap D$ is a submersion and $h(D) = D$. On the other hand, D could have been chosen so that $\overline{D}$ is contained in a plaque of a foliation box $\tilde{W}$ of $\mathfrak{F}$. It is easy to see now that it is possible to construct a foliation box W_z of $\mathfrak{F}$ such that $W_z \subset \tilde{W} \cap \pi^{-1}(\pi(D))$ and the plaques of W_z cut each fiber $\pi^{-1}(x)$ in exactly one point, if $x \in \pi(D)$. We leave the details to the reader.

Since $\pi^{-1}(b)$ is compact, we can take a finite cover $(W_i = W_{z_i})_{i=1}^{k}$ of $\pi^{-1}(b)$. For each $i \in \{1, \ldots, k\}$, $\pi(W_i)$ is a neighborhood of b in B so $\cap_{i=1}^{k} \pi(W_i)$ is a neighborhood of b in B. Let $U \subset \cap_{i=1}^{k} \pi(W_i)$ be a disk containing b. If J is a leaf of $\mathfrak{F} \mid \pi^{-1}(U)$, then J is contained in some plaque P of W_i for some $i \in \{1, \ldots, k\}$. Thus we can conclude that $\pi \mid J : J \longrightarrow U$ is injective. On the other hand, from the injectivity, it follows that $\pi(J) = \pi(P \cap \pi^{-1}(U)) = \pi(P) \cap U = U$. Therefore $\pi \mid J : J \longrightarrow U$ is a diffeomorphism, as desired.

§3. The holonomy of $\mathfrak{F}$

When $\mathfrak{F}$ is a $C^r (r \geq 1)$ foliation transverse to the fibers of E, there is a representation

$$\varphi : \pi_1 (B,b) \longrightarrow \mathrm{Diff}^r(F) \simeq \mathrm{Diff}^r(\pi^{-1}(b))$$

of the fundamental group $\pi_1(B,b)$ in the group of C^r diffeomorphisms of F, $\mathrm{Diff}^r(F)$, called the holonomy of $\mathfrak{F}$, which we define as follows.

Let $\alpha : I = [0,1] \longrightarrow B$, $\alpha(0) = \alpha(1) = b$ and $y \in \pi^{-1}(b)$. Since $\pi \mid L_y : L_y \longrightarrow B$ is a covering, there exists a unique path $\tilde{\alpha}_y : I \longrightarrow L_y$ such that $\tilde{\alpha}_y(0) = y$ and $\pi \circ \tilde{\alpha}_y = \alpha^{-1}$. If we identify $\pi^{-1}(b)$ with F, we can define a map $\varphi_\alpha : F \longrightarrow F$ by $\varphi_\alpha(y) = \tilde{\alpha}_y(1)$.

Since the endpoint of the curve $\tilde{\alpha}_y$ only depends on the homotopy class of α^{-1}, it follows that $\varphi_\alpha(y)$ only depends on the homotopy class of α^{-1} and we can write $\varphi_\alpha = \varphi_{[\alpha]}$. Further $\varphi_{[\alpha]^{-1}} = (\varphi_{[\alpha]})^{-1}$ and if $[\beta] \in \pi_1(B)$, $\varphi_{[\alpha \cdot \beta]} = \varphi_{[\alpha]} \circ \varphi_{[\beta]}$. It is easy to verify that $\varphi_{[\alpha]}$ is C^r. Since $\varphi_{[\alpha]}$ has an inverse $\varphi_{[\alpha]^{-1}}$ that is also C^r, $\varphi_{[\alpha]}$ is a C^r diffeomorphism. So, $\varphi : \pi_1(B, b_0) \longrightarrow \mathrm{Diff}^r(F)$ given by $\varphi([\alpha]) = \varphi_{[\alpha]}$ is the desired group homomorphism.

§4. Suspension of a representation

In note 1 of Chapter II we saw the simplest example of a foliation transverse to the fibers of a fiber bundle. It is that of a suspension of a diffeomorphism $f : F \longrightarrow F$ to a foliation of dimension one of the quotient space $E = F \times [0,1]/\sim$ where $\sim$ is the equivalence relation which identifies $(0,y)$ with $(1, f^{-1}(y))$. This foliation is the one induced on E by the trivial foliation of $F \times [0,1]$ with leaves $\{x\} \times [0,1]$, $x \in F$. The space E is fibered over S^1 with discrete structure group since we can take as local charts of E the quotients $\tilde{U}$, $\tilde{V}$ of $U = (\epsilon, 1-\epsilon) \times F$, $V = ([0,2\epsilon) \cup (1-2\epsilon,1]) \times F$. Then $\tilde{U} \cap \tilde{V}$ has two connected components W_{12}, W_{21}, and $g_{12} : W_{12} \longrightarrow \mathrm{Diff}(F)$, $g_{21} : W_{21} \longrightarrow \mathrm{Diff}(F)$ are given by $g_{12}(x) =$ identity and $g_{21}(x) = f$.

A natural question is the following: how does the above construction generalize to the case where we have a representation $\varphi : G \longrightarrow \mathrm{Diff}^r(F)$ where G is a group? As we have seen, G must be the fundamental group of the base B. More specifically the problem is the following: given two manifolds F, B and a representation $\varphi : \pi_1(B) \longrightarrow \mathrm{Diff}^r(F)$, to find a manifold $E(\varphi)$ fibered over B with fiber F, and a foliation $\mathfrak{F}(\varphi)$ transverse to the fibers of E such that the holonomy of $\mathfrak{F}(\varphi)$ is φ. Before giving the general construction, let us see an example.

Example 5. Let $B = T^2$ and $F = S^1$. We have that $\pi_1(T^2) = \mathbf{Z}^2$ so a representation $\varphi : \mathbf{Z}^2 \longrightarrow \text{Diff}^r(S^1)$ is determined by two diffeomorphisms of S^1, namely $f = \varphi(1,0)$ and $g = \varphi(0,1)$. These diffeomorphisms are commutative since $f \circ g = \varphi(1,0) \circ \varphi(0,1) = \varphi((1,0) + (0,1)) = \varphi(0,1) \circ \varphi(1,0) = g \circ f$. Suppose for simplicity that f preserves the orientation of S^1 but not g. We will now construct the suspension of φ. On $I^2 \times S^1$ consider the foliation $\mathfrak{F}_1$ whose leaves are $I^2 \times \{\theta\}$, $\theta \in S^1$.

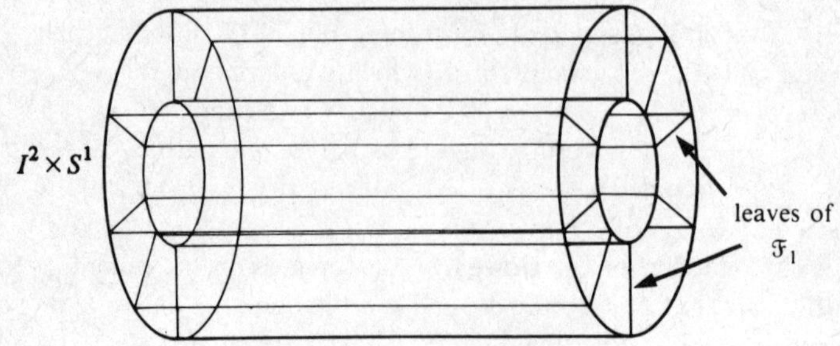

Figure 2 – The foliation $\mathfrak{F}_1$

We define $E(\varphi)$ by identifying the points of the boundary of $I^2 \times S^1$ according to the following equivalence relation: (x,y,θ) and (x',y',θ') on $I^2 \times S^1$ are equivalent if

(1) $x = 0$, $x' = 1$, $y = y'$ and $\theta = f(\theta')$ or
(2) $x = x'$, $y = 0$, $y' = 1$ and $\theta = g(\theta')$.

Making identification (1) we obtain a manifold diffeomorphic to $I \times S^1 \times S^1 = I \times T^2$ since f preserves orientation. Evidently $\mathfrak{F}_1$ induces a foliation $\mathfrak{F}_2$ transverse to the boundary whose holonomy is generated by f.

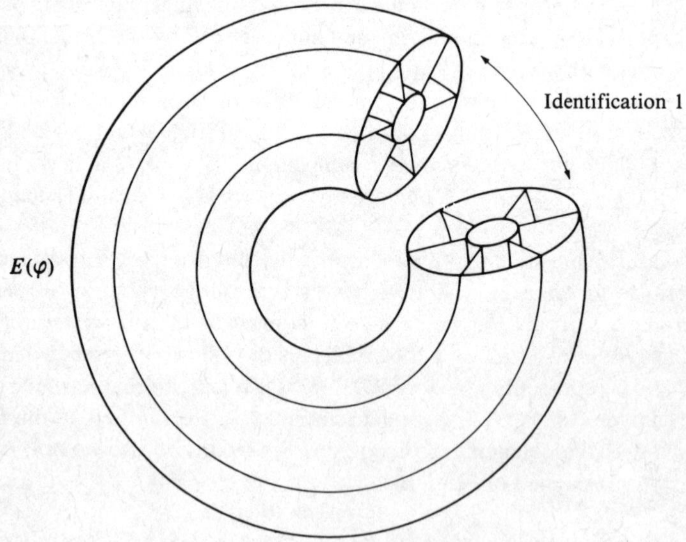

Figure 3 – The foliation $\mathfrak{F}_2$

In making identification (2) on $I \times T^2$ we obtain a nonorientable manifold $E(\varphi)$ fibered over T^2 with fiber S^1. Since g commutes with f, the leaves of $\mathcal{F}_2 \mid \{0\} \times T^2$ are taken to the leaves of $\mathcal{F}_2 \mid \{1\} \times T^2$ by identification (2). So $\mathcal{F}_2$ induces a foliation $\mathcal{F}(\varphi)$ on $E(\varphi)$. Since $\mathcal{F}_1$ is transverse to the curves $\{x\} \times \{y\} \times S^1$, $(x,y) \in I^2$, $\mathcal{F}(\varphi)$ is transverse to the fibers of $E(\varphi)$. The leaves of $E(\varphi)$ are homeomorphic to $\mathbb{R}^2$, $\mathbb{R} \times S^1$ or T^2 and in any case the projection of the fibers restricted to each leaf is a covering.

Theorem 1 (Suspension of a representation). *Let B and F be connected manifolds and $\varphi : \pi_1(B, b_0) \longrightarrow \mathrm{Diff}(F)$ a representation. Then there exist a fiber bundle $(E(\varphi), \pi, B, F)$ and a foliation $\mathcal{F}(\varphi)$ transverse to the fibers of $E(\varphi)$ whose holonomy is φ. The fiber bundle $E(\varphi)$ has discrete structure group.*

Proof. Let $P : \tilde{B} \longrightarrow B$ be the universal covering of B. The representation φ induces an action

$$\tilde{\varphi} : \pi_1(B, b_0) \longrightarrow \mathrm{Diff}^r(\tilde{B} \times F)$$

given by $\tilde{\varphi}([\alpha])(\tilde{b}, f) = ([\alpha] \cdot \tilde{b}, \varphi([\alpha])^{-1} \cdot f)$.

In the above definition, $[\alpha] \cdot \tilde{b}$ denotes the image of $\tilde{b}$ by the deck transformation associated to $[\alpha]$. Recall that a deck transformation is a diffeomorphism $g : \tilde{B} \longrightarrow \tilde{B}$ such that $P \circ g = P$. The deck transformation associated to $[\alpha]$ is defined in the following manner. Fix $\tilde{b}_0 \in \tilde{B}$ such that $P(\tilde{b}_0) = b_0$. Given $\tilde{b} \in \tilde{B}$ consider a curve $\tilde{\gamma} : [0,1] \longrightarrow \tilde{B}$ such that $\tilde{\gamma}(0) = \tilde{b}$ and $\tilde{\gamma}(1) = \tilde{b}_0$. Let $\gamma = P \circ \tilde{\gamma}$ and $\delta = \gamma * \alpha^{-1} * \gamma^{-1}$. Then $[\alpha] \cdot \tilde{b} = \tilde{\delta}(1)$ where $\tilde{\delta}$ is the unique lift of δ such that $\tilde{\delta}(0) = \tilde{b}$.

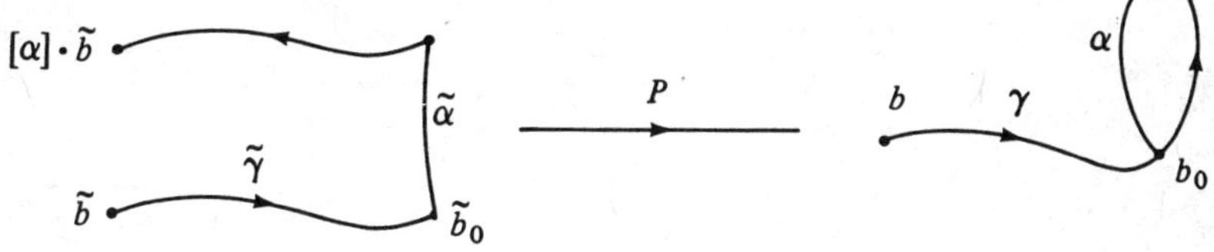

Figure 4

For the details see [29].

The set of deck transformations $\tilde{B} \xrightarrow{P} B$ with the operation of composition is a group naturally isomorphic to $\pi_1(B, b_0)$, since $\tilde{B}$ is the universal covering of B. We are identifying $[\alpha]$ with the automorphism associated to it by this isomorphism. In Example 5 the action $\tilde{\varphi}$ is defined by

$$\tilde{\varphi}(m,n) \cdot (x,y,\theta) = (x+m, y+n, f^{-m} \circ g^{\mp n}(\theta)) .$$

The action $\tilde{\varphi}$ satisfies the following property. For every $\tilde{b} \in \tilde{B}$ there is a connected neighborhood V of $\tilde{b}$ in $\tilde{B}$ such that if $g \in \pi_1(B, b_0)$ and $g \neq 1$ then $\tilde{\varphi}(g)(V \times F) \cap (V \times F) = \varnothing$. In fact, let V be a neighborhood of $\tilde{b}$ such that $P: V \longrightarrow P(V)$ is a diffeomorphism. For every $g \in \pi_1(B, b_0)$, $g \neq 1$, $g \cdot V \cap V = \varnothing$ since if there were a $y \in g(V) \cap V$, we would have that $y = g(x)$ with $x \in V$, so $P(x) = P(y)$ and P would not be injective on V. We have then that $\tilde{\varphi}(g)(V \times F) \cap (V \times F) = (g \cdot V \times \varphi(g)^{-1} \cdot F) \cap (V \times F) = \varnothing$. We will now describe the construction of $E(\varphi)$ and $\mathfrak{F}(\varphi)$.

Introduce on $\tilde{B} \times F$ the equivalence relation which identifies two points when they are in the same trajectory of $\tilde{\varphi}$, that is, $(\tilde{b}, y) \sim (\tilde{b}', y')$ if there exists a $g \in \pi_1(B, b_0)$ such that $\tilde{\varphi}(g)(\tilde{b}, y) = (\tilde{b}', y')$. In Example 5 this equivalence corresponds to the identifications (1) and (2). Using the above-mentioned property we prove that the quotient space of $\tilde{B} \times F$ by this equivalence relation is a differentiable manifold which we denote by $E(\varphi)$.

From the expression for $\tilde{\varphi}$ it follows that it preserves the fibers of $\tilde{B} \times F \xrightarrow{P_1} \tilde{B}$, $P_1(\tilde{b}, y) = \tilde{b}$. We can then define a map $\pi : E(\varphi) \longrightarrow B$ that makes the following diagram commute:

$$(*) \qquad \begin{array}{ccccc} & & \tilde{B} \times F & & \\ & \overset{Q}{\swarrow} & & \overset{P_1}{\searrow} & \\ E(\varphi) & & & & \tilde{B} \\ & \underset{\pi}{\searrow} & & \underset{P}{\swarrow} & \\ & & B & & \end{array}$$

In the above diagram Q is the projection of the equivalence relation.

Observe also that the leaves, $\tilde{B} \times \{f\}$, of the product foliation $\tilde{B} \times F$ are preserved by $\tilde{\varphi}$. So, this foliation will induce on the quotient space $E(\varphi)$ a foliation transverse to the fibers of π, which we will denote $\mathfrak{F}(\varphi)$.

We go now to the proofs.

$E(\varphi)$ is a fiber bundle with base B, fiber F and projection π: let $V \subset \tilde{B}$ be an open set such that if $g \in \pi_1(B, b_0)$, $g \neq 1$ then $g(V) \cap V = \varnothing$. From this property it follows that $Q \mid V \times F : V \times F \longrightarrow Q(V \times F) = \pi^{-1}(P(V))$ is bijective. On the other hand, $Q \mid V \times F$ is continuous and open, as is easy to see, so $Q \mid V \times F$ is a homeomorphism. It follows then that $\psi_V = (Q \mid V \times F)^{-1} : \pi^{-1}(P(V)) \longrightarrow V \times F$ is a homeomorphism. We will show that the set of such ψ_V defines on $E(\varphi)$ a C^r differentiable structure.

Let us prove first that if $U, V \subset \tilde{B}$ are two open sets as above with $P(U) \cap P(V) \neq \varnothing$ then

$$\psi_U \circ \psi_V^{-1} : (P^{-1}(P(U)) \cap V) \times F \longrightarrow (U \cap P^{-1}(P(V)) \times F$$

is given by $\psi_U \circ \psi_V^{-1} = \tilde{\varphi}(g)$ where $g \in \pi_1(B, b_0)$ is fixed in each connected component of $(P^{-1}(P(U)) \cap V) \times F$. To simplify the argument we suppose that $(P^{-1}(P(U)) \cap V) \times F$ is connected, which is the same as saying that $P(U) \cap P(V)$ is connected.

Fix $(\tilde{b}_1,f_1) \in (P^{-1}(P(U)) \cap V) \times F$. Since $\psi_V^{-1} = Q$ on $V \times F$ and $\psi_U = (Q \mid U \times F)^{-1}$ we have that $(\tilde{b}_2,f_2) = \psi_U \circ \psi_V^{-1}(\tilde{b}_1,f_1)$ is equivalent to $(\tilde{b}_1,f_1)$, so $(\tilde{b}_2,f_2) = \tilde{\varphi}(g)(\tilde{b}_1,f_1)$ for some $g \in \pi_1(B,b_0)$. Take now $(\tilde{b},f) \in (P^{-1}(P(U)) \cap V) \times F$ and set $\psi_U \circ \psi_V^{-1}(\tilde{b},f) = (\tilde{b}',f')$. In this case $(\tilde{b},f)$ is equivalent to $(\tilde{b}',f')$ and $\tilde{b}' \in U \cap P^{-1}(P(V))$. We have then that $\tilde{b}' = g' \cdot b$ and $f' = \varphi(g')^{-1} \cdot f$ for some $g' \in \pi_1(B,b_0)$. On the other hand $P^{-1}(P(V)) = \bigcup_{h \in \pi_1(B,b_0)} h(V)$, where $\{h(V) \mid h \in \pi_1(B,b_0)\}$ is the set of connected components of $P^{-1}(P(V))$. Since $\tilde{b}_2 = g \cdot b_1 \in U \cap g \cdot V \neq \emptyset$ and $U \cap P^{-1}(P(V))$ is connected, we have $U \cap P^{-1}(P(V)) = U \cap g \cdot V$, so $\tilde{b}' = g' \cdot \tilde{b} = g \cdot \tilde{c}$ with $\tilde{c} \in V$. Therefore $\tilde{c} \in (g^{-1} \cdot g' \cdot V) \cap V$ so $g^{-1} \cdot g' = 1$ and then $g = g'$. It follows therefore that $\psi_U \circ \psi_V^{-1}(\tilde{b},f) = \tilde{\varphi}(g)(\tilde{b},f)$ for every $(\tilde{b},f) \in (P^{-1}(P(U)) \cap V) \times F$, or, $\psi_U \circ \psi_V^{-1} = \tilde{\varphi}(g)$.

We can now introduce on $E(\varphi)$ a C^r manifold structure, composing each ψ_V as above with local charts $\xi : W \longrightarrow \mathbb{R}^k$ where $W \subset V \times F$ is an open set, obtaining in this manner a C^r atlas

$$\mathcal{A} = \{(\psi_V^{-1}(W), \xi \circ \psi_V / \psi_V^{-1}(W)) \mid \quad g \cdot V \cap V = \emptyset \text{ if } g \neq 1 \text{ and } (W,\xi) \text{ is a local chart on } V \times F\}.$$

The fact that $\psi_U \circ \psi_V^{-1} = \tilde{\varphi}(g)$ on each connected component of its domain implies that the change of coordinate maps are C^r.

The composition of ψ_V with $P \times id_F$ yields a diffeomorphism $\tilde{\psi}_V : \pi^{-1}(P(V)) \longrightarrow P(V) \times F$. The set $\{(V,\tilde{\psi}_V) \mid g \cdot V \cap V = \emptyset \text{ if } g \neq 1\}$ induces on $E(\varphi)$ a structure of a fiber bundle with base B, fiber F and projection π, as is easy to see from the definition of §1. Observe that $E(\varphi)$ with this structure has discrete structure group.

The foliation $\mathfrak{F}(\varphi)$ is transverse to the fibers of $E(\varphi)$. A set of foliation boxes of $\mathfrak{F}(\varphi)$ can be defined using the C^r diffeomorphisms $\psi_V : \pi^{-1}(P(V)) \longrightarrow V \times F$ defined before. The plaques of $\mathfrak{F}(\varphi)$ on $\pi^{-1}(P(V))$ are the sets of the form $\psi_V^{-1}(V \times \{f\})$, $f \in F$. Since $\psi_U \circ \psi_V^{-1} = \tilde{\varphi}(g)$ for some $g \in \pi_1(B,b_0)$ and $\tilde{\varphi}$ preserves the leaves $\tilde{B} \times \{f\}$ of the product foliation of $\tilde{B} \times F$, it follows that $\mathfrak{F}(\varphi)$ is well-defined.

Observe that from the construction it also follows that $\mathfrak{F}(\varphi)$ is transverse to the fibers of $E(\varphi)$. In fact, the charts (V,ψ_V) are both local charts for $E(\varphi)$ as well as for the foliation $\mathfrak{F}(\varphi)$, so the fibers of $E(\varphi)$ in $\psi_V^{-1}(V \times F) = \pi^{-1}(P(V))$ are represented in $V \times F$ by submanifolds of the form $\{v\} \times F$ and the plaques of $\mathfrak{F}(\varphi)$ are represented by submanifolds of the form $V \times \{f\}$; hence the condition

(a) $$T_pE(\varphi) = T_p(L_p) \oplus T_p(F_{\pi(p)})$$

from the definition of section 2 is immediate.

On the other hand the map $\pi/\pi^{-1}(P(V))$ is represented in $V \times F$ by $P \circ P_1/V \times F$, where P_1 is the first projection. Therefore the restriction of π to each plaque of $\mathfrak{F}(\varphi)/\pi^{-1}(P(V))$ is represented by $P \circ P_1 : V \times \{f\} \longrightarrow P(V)$, which is a homeomorphism. This implies that for any leaf L of $\mathfrak{F}(\varphi)$, $\pi \mid L : L \longrightarrow B$ is a covering, as we wanted.

The holonomy of $\mathfrak{F}(\varphi)$ is φ. Take b_0 as base point. The fiber $\pi^{-1}(b_0)$ is naturally diffeomorphic to $\{\tilde{b}_0\} \times F$ by the diffeomorphism

$$Q/(\{\tilde{b}_0\} \times F) : \{\tilde{b}_0\} \times F \longrightarrow \pi^{-1}(b_0) \ ,$$

where $P(\tilde{b}_0) = b_0$. Identify the points of F with those of $\pi^{-1}(b_0)$ by the diffeomorphism $\theta(f) = Q(\tilde{b}_0,f)$. Now consider $g \in \pi_1(B,b_0)$. Let $\tilde{\alpha} : [0,1] \longrightarrow \tilde{B}$ be a curve such that $\tilde{\alpha}(0) = \tilde{b}_0$ and $\tilde{\alpha}(1) = g \cdot \tilde{b}_0$. Let $\alpha^{-1} = P \circ \tilde{\alpha}$. From the construction it is clear that $[\alpha] = g$ and $\tilde{\alpha}$ is a lift of α^{-1} which satisfies $\tilde{\alpha}(0) = \tilde{b}_0$. We will see now that the image of $f \in F$ by the holonomy transformation associated to g is $\varphi(g) \cdot f$ (taking into account the identification of F with $\pi^{-1}(b_0)$ by θ). First, note that the leaf L of $\mathfrak{F}(\varphi)$ which passes through $Q(\tilde{b}_0,f) \in \pi^{-1}(b_0)$ is $Q(\tilde{B} \times \{f\})$. Further, $Q/\tilde{B} \times \{f\} : \tilde{B} \times \{f\} \longrightarrow L$ is a covering map, where on L we take the intrinsic C^r manifold structure. These facts follow from the construction of $\mathfrak{F}(\varphi)$ and from the commutativity of the diagram (*). Let $\tilde{\alpha} : [0,1] \longrightarrow L$ be the covering of α^{-1} such that $\tilde{\alpha}(0) = Q(\tilde{b}_0,f)$. The image of $(\tilde{b}_0,f)$ by the holonomy map associated with $[\alpha]$ will be $\tilde{\alpha}(1)$. On the other hand the lift of $\tilde{\alpha}$ to $\tilde{B} \times \{f\}$ by $Q/\tilde{B} \times \{f\}$ which begins at $(\tilde{b}_0,f)$ is the curve $t \longrightarrow (\tilde{\alpha}(t),f)$, as is easy to see. The endpoint of the curve is $(\tilde{\alpha}(1),f) = (g \cdot \tilde{b}_0,f)$ which is identified in $E(\varphi)$ with the point $(\tilde{b}_0,\varphi(g) \cdot f) \in \{\tilde{b}_0\} \times F$. We have then that $\tilde{\alpha}(1) = Q(\tilde{b}_0,\varphi(g) \cdot f)$ which is identified by θ with $\varphi(g) \cdot f$. This proves the theorem.

Definition. We say that two representations $\varphi : \pi_1(B,b_0) \longrightarrow \mathrm{Diff}(F)$ and $\varphi' : \pi_1(B,b_0) \longrightarrow \mathrm{Diff}(F')$ are *C^s-conjugate* if there is a C^s diffeomorphism (if $s \geq 1$) or homeomorphism (if $s = 0$) $h : F \longrightarrow F'$ such that, for every $[\alpha] \in \pi_1(B,b_0)$, we have $\varphi([\alpha]) = h^{-1} \circ \varphi'([\alpha]) \circ h$.

Theorem 2 (Uniqueness of the suspension). *Let φ and φ' be C^s-conjugate representations as above. There exists a C^s diffeomorphism $H : E(\varphi) \longrightarrow E(\varphi')$ (homeomorphism if $s = 0$) such that*

(a) *$\pi' \circ H = \pi$ and consequently H takes fibers of $E(\varphi)$ to fibers of $E(\varphi')$.*
(b) *H takes leaves of $\mathfrak{F}(\varphi)$ to leaves of $\mathfrak{F}(\varphi')$.*

Proof. Let $\tilde{H} : \tilde{B} \times F \longrightarrow \tilde{B} \times F'$ be the C^s homeomorphism defined by $\tilde{H}(\tilde{b},f) = (\tilde{b},h(f))$, where h conjugates φ and φ'. It is immediate that $\tilde{H}$ preserves the fibers of the fibration $\{\tilde{b}\} \times F$ and also takes the leaf $\tilde{B} \times \{f\}$ of the product foliation to the leaf $\tilde{B} \times \{h(f)\}$. Further, for every $g \in \pi_1(B,b_0)$ we have $\tilde{H}(\tilde{\varphi}(g)) = \tilde{\varphi}'(g) \circ \tilde{H}$, where $\tilde{\varphi}$ and $\tilde{\varphi}'$ are as in the construction of Theorem 1.

Therefore $\tilde{H}$ takes equivalent points of $\tilde{B} \times F$ to equivalent points of $\tilde{B} \times F'$. Thus it follows that there exists a $C^s(s \geq 1)$ diffeomorphism (or homeomorphism if $s = 0$) $H : E(\varphi) \longrightarrow E(\varphi')$ such that the diagram below commutes.

$$\begin{array}{ccc} \tilde{B} \times F & \xrightarrow{\tilde{H}} & \tilde{B} \times F' \\ {\scriptstyle Q}\downarrow & & \downarrow{\scriptstyle Q'} \\ E(\varphi) & \xrightarrow{H} & E(\varphi') \end{array}$$

It is immediate that H takes fibers of $E(\varphi)$ to fibers of $E(\varphi')$ and leaves of $\mathfrak{F}(\varphi)$ to leaves of $\mathfrak{F}(\varphi')$. ■

Note. Theorem 2 can be proved in a little more general context. Let B, B' and F, F' be connected manifolds. Suppose that $\varphi : \pi_1(B,b_0) \longrightarrow \mathrm{Diff}^r(F)$ and $\varphi' : \pi_1(B',b_0') \longrightarrow \mathrm{Diff}^r(F')$ are conjugate representations in the following sense: there exist diffeomorphisms $g : B \longrightarrow B'$ and $h : F \longrightarrow F'$ such that $g(b_0) = b_0'$ and for every $[\alpha] \in \pi_1(B,b_0)$ we have

$$\varphi([\alpha]) = h^{-1} \circ \varphi'(g_*([\alpha])) \circ h$$

where $g_* : \pi_1(B,b_0) \longrightarrow \pi_1(B',b_0')$ is the isomorphism induced by g between fundamental groups. Then there exists a diffeomorphism $H : E(\varphi) \longrightarrow E(\varphi')$ which takes fibers of $E(\varphi)$ to fibers of $E(\varphi')$ and leaves of $\mathfrak{F}(\varphi)$ to leaves of $\mathfrak{F}(\varphi')$.

Suppose now that (E,π,B,F) is a fiber bundle and that $\mathfrak{F}$ is a foliation on E transverse to the fibers. Let φ be the holonomy of $\mathfrak{F}$ (with respect to the fiber $\pi^{-1}(b_0)$, $b_0 \in B$). How can one relate the foliation $\mathfrak{F}(\varphi)$ on $(E(\varphi),\pi_\varphi,B,F)$ with the foliation $\mathfrak{F}$? The answer is given in the following theorem.

Theorem 3. *Let $\mathfrak{F}$ be a $C^r(r \geq 1)$ foliation transverse to the fibers of (E,π,B,F) whose holonomy in the fiber $\pi^{-1}(b_0)$, $b_0 \in B$, is $\varphi : \pi_1(B,b_0) \longrightarrow \mathrm{Diff}^r(F)$. Then there exists a C^r diffeomorphism $H : E \longrightarrow E(\varphi)$ which takes leaves of $\mathfrak{F}$ to leaves of $\mathfrak{F}(\varphi)$ and such that $\pi_\varphi \circ H = \pi$. In particular, H takes fibers of E to fibers of $E(\varphi)$.*

We leave the proof to the reader (see exercise 38).

Definition. We say that the fiber bundles (E,π,B,F) and (E',π',B,F) are

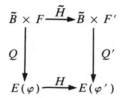

(C^r) equivalent if there exists a C^r diffeomorphism $H : E \longrightarrow E'$ such that $\pi' \circ H = \pi$. In the case of vector bundles we require also that H takes fibers to fibers linearly.

As a consequence of Theorem 3, we have the following.

Theorem 4. *Let (E,π,B,F) be a fiber bundle. Then there exists a foliation on E transverse to the fibers if and only if (E,π,B,F) is equivalent to a fiber bundle with discrete structure group.*

Proof. Suppose there is a foliation $\mathfrak{F}$ transverse to the fibers of E. Let $\varphi : \pi_1(B,b_0) \longrightarrow \mathrm{Diff}^r(F)$ be the holonomy of $\mathfrak{F}$. As we saw (E,π,B,F) is equivalent to $(E(\varphi),\pi_\varphi,B,F)$ which by Theorem 1 has discrete structure group.

Let us consider the converse. We can suppose that (E,π,B,F) has discrete structure group. In this case we define the foliation $\mathfrak{F}$ using the local trivialization for the fiber bundle. Let $\{U_i,\psi_i\}_{i\in S}$ where $\bigcup_{i\in S} U_i = B$ and for every $i \in S$, U_i is open in B and $\psi_i : \pi^{-1}(U_i) \longrightarrow U_i \times F$ is a chart of the fiber bundle. Since the fiber bundle has discrete structure group, the change of variables

$$\psi_i \circ \psi_j^{-1} : (U_i \cap U_j) \times F \longrightarrow (U_i \cap U_j) \times F$$

is written in the form

$$\psi_i \circ \psi_j^{-1}(b,f) = (b,\Phi_{ij}(f))$$

that is, the second component Φ_{ij} does not depend on b. So the charts $\psi_i : \pi^{-1}(U_i) \longrightarrow U_i \times F$ induce a foliation $\mathfrak{F}$ on E whose plaques on $\pi^{-1}(U_i)$ are of the form $\psi_i^{-1}(U_i \times \{f\})$, $f \in F$. It follows immediately from the construction that if L is a leaf of $\mathfrak{F}$ then $\pi \mid L : L \longrightarrow B$ is a covering and hence $\mathfrak{F}$ is transverse to the fibers of E. ∎

We will now see some applications.

§5. Existence of germs of foliations

In this section we will study the following problem. Let $N \subset M$ be a submanifold imbedded in M. Under what conditions does there exist a foliation $\mathfrak{F}$ defined on a neighborhood of N and such that N is a leaf of $\mathfrak{F}$?

Let $\nu(N) = \{(p,v) \in TM \mid p \in N \text{ and } v \in T_pN^\perp\}$ be the normal bundle of N with respect to a fixed Riemannian metric $\langle\ ,\ \rangle$ on M. Then the zero section of $\nu(N)$, $N_0 = \{(p,0) \in \nu(N)\}$ is diffeomorphic to N and by the tubular neighborhood theorem [27], $\nu(N)$ is diffeomorphic to a neighborhood U of N in M by a diffeomorphism $f : \nu(N) \longrightarrow U$ such that $f(N_0) = N$. We see then that the existence of a foliation defined in a neighborhood of N which has N as a leaf is equivalent to the existence of a foliation in a neighborhood V of N_0 in $\nu(N)$ which has N_0 as a leaf.

Suppose that $\nu(N)$ is equivalent to a vector bundle E which has a discrete structure group. The equivalence in this case must be a diffeomorphism $H : \nu(N) \longrightarrow E$ which takes fibers to fibers linearly; in particular taking the zero section N_0 of $\nu(N)$ to the zero section of E. By Theorem 4 there is a foliation $\overline{\mathfrak{F}}$ on E transverse to the fibers, so $\mathfrak{F} = H^*(\overline{\mathfrak{F}})$ is a foliaton on $\nu(N)$ transverse to the fibers. The local charts of $\overline{\mathfrak{F}}$, as we saw in Theorem 4 are also local charts of the fiber bundle E and since they are linear in the fibers the zero section of E must be a leaf of $\overline{\mathfrak{F}}$ which implies that N_0 is a leaf of $\overline{\mathfrak{F}}$.

Conversely, suppose there exists a foliation $\mathfrak{F}$ defined in a neighborhood of N_0 in $\nu(N)$ such that N_0 is a leaf of $\mathfrak{F}$.

Consider a cover $(U_i)_{i\in\mathcal{S}}$ of N_0 by foliation boxes of $\mathfrak{F}$, where there are defined submersions $f_i : U_i \longrightarrow \mathbb{R}^p$, $p = m - n$, such that the plaques of $\mathfrak{F} \mid U_i$ are of the form $f_i^{-1}(x)$, $x \in \mathbb{R}^p$ and $f_i(N_0 \cap U_i) = 0$. Consider also the changes of coordinates $h_{ki} : f_i(U_i \cap U_k) \longrightarrow f_k(U_i \cap U_k)$ such that $f_k = h_{ki} \circ f_i$.

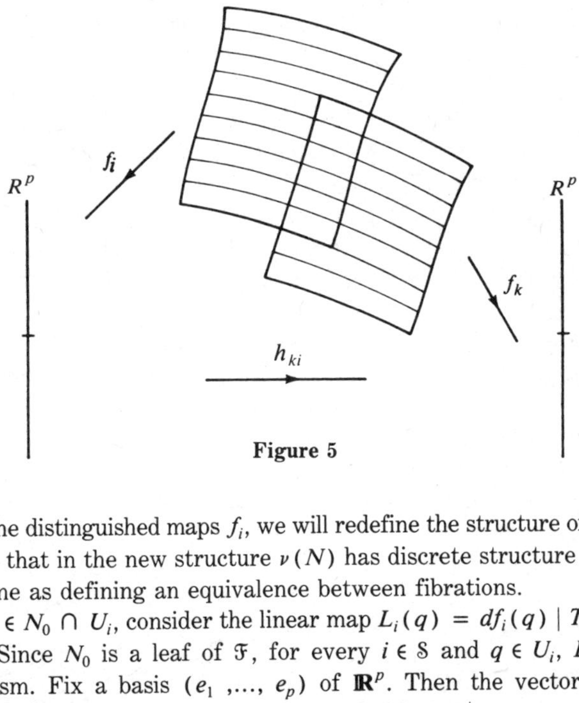

Figure 5

Using the distinguished maps f_i, we will redefine the structure of the normal bundle so that in the new structure $\nu(N)$ has discrete structure group. This is the same as defining an equivalence between fibrations.

Given $q \in N_0 \cap U_i$, consider the linear map $L_i(q) = df_i(q) \mid T_q N_0^{\perp} : T_q N_0^{\perp} \longrightarrow \mathbb{R}^p$. Since N_0 is a leaf of $\mathfrak{F}$, for every $i \in \mathcal{S}$ and $q \in U_i$, $L_i(q)$ is an isomorphism. Fix a basis $(e_1, \ldots, e_p)$ of $\mathbb{R}^p$. Then the vectors $X_i^j(q) = (L_i(q))^{-1}(e_j)$, $j = 1, \ldots, p$, form a basis of $T_q N_0^{\perp}$.

We can now define local charts $\overline{x}_i : \pi^{-1}(N_0 \cap U_i) \longrightarrow (N_0 \cap U_i) \times \mathbb{R}^p$

$$\overline{x}_i(q, X(q)) = (q, L_i(q) \cdot X(q)) = (q, \sum_{j=1}^{p} \alpha_j e_j),$$

where $X(q) = \sum_{j=1}^{p} \alpha_j X_i^j(q) \in T_q N_0^{\perp}$.

If $U_i \cap U_k \cap N_0 \neq \emptyset$, the change of coordinates will be $\overline{x}_k \circ \overline{x}_i^{-1}(q,v) = (q, g_{ki}(q) \cdot v)$ where $g_{ki}(q) = L_k(q) \circ L_i^{-1}(q) = Dh_{ki}(0)$. Therefore, the new structure group of $\nu(N)$ is discrete. We have then the following.

Theorem 5. *Let $N \subset M$ be an imbedded C^2 submanifold of M. There is a neighborhood $U \supset N$ and a foliation $\mathfrak{F}$ on U such that N is one of the leaves of $\mathfrak{F}$ if and only if the normal bundle of N has some fibered structure with discrete structure group.*

Note that the hypothesis of C^2 in the statement of Theorem 5 permits us to use the tubular neighborhood theorem of [27].

§6. Sacksteder's example

As was mentioned in note 2 of Chapter III, a codimension one C^r $(r \geq 2)$ foliation of T^2 does not have exceptional minimal sets (Denjoy's theorem). In this section we describe a codimension one C^∞ foliation of a compact three manifold which has an exceptional minimal set. The example is due to Sacksteder ([49]) and shows that the original version of Denjoy's theorem is not valid in dimensions larger than 2. Sacksteder, in 1964, proved a generalization of Denjoy's theorem for foliations of codimension one and class C^r $(r \geq 2)$. For more details see note 2 of this chapter.

Let V_2 be an orientable compact surface of genus two (two-holed torus). This surface can be represented by the region bounded by an octagon as in figure 5 Chapter I, with the sides with the same letter identified according to the arrows.

The fundamental group of V_2 is a non-abelian group generated by closed curves a, b, c, and d so that the unique nontrivial relation between these curves is $aba^{-1}b^{-1}cdc^{-1}d^{-1} = 1$ (see [29]).

Let G be the subgroup of $\pi_1(V_2)$ generated by a and c. The subgroup G is free, since all relations in $\pi_1(V_2)$ involving only a and c are trivial. Consequently, given two diffeomorphisms $f, g \in \mathrm{Diff}^\infty(S^1)$ we can define a homomorphism $\varphi : G \longrightarrow \mathrm{Diff}^\infty(S^1)$ setting $\varphi(a) = f$ and $\varphi(c) = g$. The homomorphism φ induces, in turn, a homomorphism $\varphi : \pi_1(V_2) \longrightarrow \mathrm{Diff}^\infty(S^1)$ defined on the generators of $\pi_1(V_2)$ by $\varphi(b) = \varphi(d) = 1$, $\varphi(a) = f$ and $\varphi(c) = g$. Observe that φ is well-defined, since

$$\varphi(aba^{-1}b^{-1}cdc^{-1}d^{-1}) = f \circ f^{-1} \circ g \circ g^{-1} = 1 .$$

Let $\mathfrak{F}(\varphi)$ be the suspension of φ. The fiber bundle $E(\varphi)$, where $\mathfrak{F}(\varphi)$ is defined, is homeomorphic to $V_2 \times S^1$, if we choose f and g preserving orientation. Moreover, by Theorem 7 of Chapter III, $\mathfrak{F}(\varphi)$ has an exceptional minimal set if and only if the action φ of the group G on S^1 defined by f and g has a minimal set homeomorphic to the Cantor set. In what follows we construct such an action.

We consider here S^1 as $\mathbb{R}/\mathbb{Z}$. So, a diffeomorphism f of S^1 can be thought of as a diffeomorphism of $\mathbb{R}$ satisfying the property $f(x + k) = f(x) + k$ for every $x \in \mathbb{R}$ and $k \in \mathbb{Z}$. Define $f : \mathbb{R} \longrightarrow \mathbb{R}$ by $f(x) = x + 1/3$ and $g : \mathbb{R} \longrightarrow \mathbb{R}$ such that $g(x + k) = g(x) + k$ if $x \in \mathbb{R}$ and $k \in \mathbb{Z}$ and the graph of g in the interval $[0,1]$ has the following form:

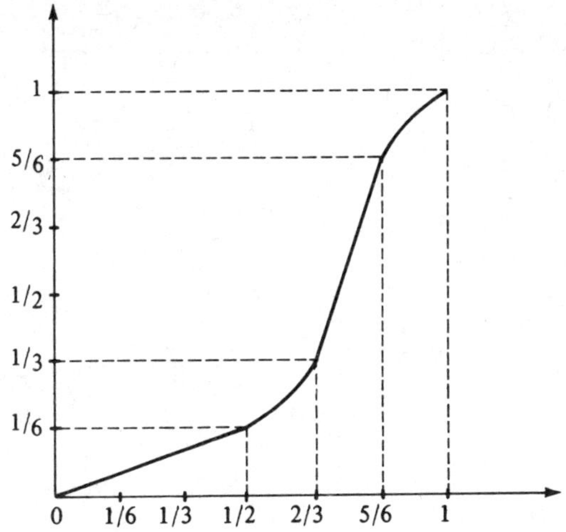

Figure 6

The diffeomorphism g is defined in such a way that $g(x) = x/3$ if $x \in [0,1/2]$, $g(x) = 3x - 5/3$ if $x \in [2/3,5/6]$, $g(1) = 1$, $g'(1) = 1/3$, $g^{(k)}(1) = 0$ if $k \geq 2$ and g is C^∞ on $[0,1]$. The diffeomorphism f corresponds to a rotation of angle $2\pi/3$ on S^1 and its expression on $\mathbb{R}/\mathbb{Z}$ is $f(x(\text{mod } 1)) = (x + 1/3)(\text{mod } 1)$.

Let K be the subset of $[0,1]$ defined by $K = \bigcap_{j=0}^{\infty} K_j$ where for every $j \in \mathbb{N}$, K_j is a union of closed intervals, defined inductively in the following manner:

$$K_0 = [0,1/6] \cup [1/3,1/2] \cup [2/3,5/6] =$$

$$[0,1] - (1/6,1/3) \cup (1/2,2/3) \cup (5/6,1) .$$

Supposing $K_{j-1} = \bigcup_{n=1}^{3\cdot 2^{j-1}} [\alpha_n^{j-1},\beta_n^{j-1}]$ then K_j is defined from K_{j-1} by removing from the interval $[\alpha_n^{j-1},\beta_n^{j-1}]$, its middle third, for every $n = 1 ,..., 3 \cdot 2^{j-1}$. Doing this we get $K_j = \bigcup_{n=1}^{3\cdot 2^j} [\alpha_n^j,\beta_n^j]$ where for every $m = 1 ,..., 3 \cdot 2^j$ and for every $n = 1 ,..., 3 \cdot 2^{j-1}$,

$$\beta_m^j - \alpha_m^j = \frac{1}{3} (\beta_n^{j-1} - \alpha_n^{j-1}) = \frac{1}{6 \cdot 3^j} .$$

It is clear that K is compact and nonempty. From the construction it is easy to verify that K is homeomorphic to the usual Cantor set, since after the first step the construction of both is the same.

Lemma: *Consider K as the subset of S^1. Then K is a minimal set of the action generated by f and g.*

Proof. First we prove that K is invariant. We will denote by G the subgroup of $\mathrm{Diff}^\infty(S^1)$ generated by f and g. From the construction of the sequence $\{K_j\}_{j\in\mathbb{N}}$, one has $f(K_j) = K_j$ for every $j \in \mathbb{N}$ so $f(K) = K$. From the definition of g and the inductive construction of K_j, we have

$$g(K_j \cap [0,1/2]) = K_{j+1} \cap [0,1/6] \qquad (j \geq 1)$$

and

$$g(K_j \cap [2/3,5/6]) = K_{j-1} \cap [1/3,5/6] \qquad (j \geq 1) .$$

Since $K_j \subset [0,1/2] \cup [2/3,5/6]$, from the above relations we get that

$$g(K_j) \subset K_{j+1} \cup K_{j-1} \subset K_{j-1} .$$

So $g(K) = g(\cap_{j=1}^{\infty} K_j) = \cap_{j=1}^{\infty} g(K_j) \subset \cap_{j=1}^{\infty} K_{j-1} = K$. Analogously

$$g^{-1}(K_j \cap [0,1/6]) = K_{j-1} \cap [0,1/2]$$

and

$$g^{-1}(K_j \cap [1/2,5/6]) = K_{j+1} \cap [2/3,5/6] ,$$

so

$$g^{-1}(K_j) \subset K_{j-1} .$$

Then $g^{-1}(K) \subset K$. Therefore $g(K) = K$. We conclude therefore that K is invariant under f and g so K is invariant under G.

Since K is compact and invariant under G, K contains a minimal subset μ. We will see next that $\mu = K$. To do this we will prove first that $1/3$ is in μ and then that the orbit of $1/3$ in G, $\mathcal{O}(1/3)$, is dense in K. We will have then that $\mu = \overline{\mathcal{O}(1/3)} = K$ since $\mathcal{O}(1/3)$ is dense in μ and μ is minimal.

$1/3 \in \mu$. Let $x \in \mu$. We can suppose, without loss of generality, that $x \in [0,1/2]$, since if $x \in (1/2,1]$ it is clear that $f^{-1}(x)$ or $f^{-2}(x) \in [0,1/2] \cap \mu$. Consider then the case that $x \in [0,1/2] \cap \mu$. Since μ is invariant under G, for every $n \geq 0$, $g^n(x) = (1/3^n) \cdot x \in \mu$, so $\lim_{n\to\infty} g^n(x) = 0 \in \mu$. Therefore $1/3 = f(0) \in \mu$.0

$\mathcal{O}(1/3)$ *is dense in* K. Let $I - K = \cup_{n\in\mathbb{N}} (\alpha_n,\beta_n)$. Since K is homeomorphic to the usual Cantor set, the set $B = \{\beta_n \mid n \in \mathbb{N}\}$ is dense in K. Observe that $1/3 \in B$, since the interval $(1/6,1/3)$ is one of the intervals of the decomposition of $I - K$. Next we will prove that $\mathcal{O}(1/3) = B$ and hence $\overline{\mathcal{O}(1/3)} =$ $= K$.

If suffices to prove that $I - K = \cup_{h \in G} h(1/6,1/3)$. Indeed, if there were the case, for every $n \in \mathbb{N}$ there would exist an $h \in G$ such that $h(1/6,1/3) = = (\alpha_n, \beta_n)$. Since $dh/dt > 0$ for every $h \in G$ we see that $h(1/3) = \beta_n$ and therefore $\mathcal{O}(1/3) = B$. We will prove that $I - K = \cup_{h \in G} h(1/6,1/3)$. In the first place, it is clear that $\cup_{h \in G} h(1/6,1/3) \subset I - K$. To see that $I - K \subset \cup_{h \in G} h(1/6,1/3)$ it is sufficient to prove that for every $j \in \mathbb{N}$, $I - K_j \subset \cup_{h \in G} h(1/6,1/3)$. We prove this by induction on j.

For $j = 0$, we have $I - K_0 = (1/6,1/3) \cup (1/2,2/3) \cup (5/6,1)$ and $(1/2,2/3) = f(1/6,1/3)$, $(5/6,1) = f^2(1/6,1/3)$. So $I - K_0 \subset \cup_{h \in G} h(1/6,1/3)$. Suppose by induction that $I - K_j \subset \cup_{h \in G} h(1/6,1/3)$ for $0 \leq j \leq \ell - 1$. We have that $I - K_\ell \supset I - K_{\ell-1}$ and $(I - K_\ell) - (I - K_{\ell-1}) = \cup_{n=1}^{3 \cdot 2^{\ell-1}} I_n^\ell$, where the intervals I_n^ℓ have length $1/6 \cdot 3^\ell$. Since $(I - K_\ell) - (I - K_{\ell-1}) \subset (0,1/6) \cup (1/3,1/2) \cup (2/3,5/6)$, $\ell \geq 1$, given $n \in \{1,2,\ldots, 3 \cdot 2^{\ell-1}\}$, there exists an $i \in \{0,-1,-2\}$ such that $f^i(I_n^\ell) \subset (0,1/6)$. On the other hand, as we saw, $g^{-1}(K_j \cap [0,1/6]) = = K_{j-1} \cap [0,1/2]$ for every $j \in \mathbb{N}$, so $g^{-1}((I - K_\ell) \cap (0,1/6)) = = (I - K_{\ell-1}) \cap (0,1/2)$ and therefore $g^{-1} \circ f^i(I_n^\ell) \subset I - K_{\ell-1}$. By the induction hypothesis $g^{-1} \circ f^i(I_n^\ell) = h(1/6,1/3)$ for some $h \in G$, so $I_n^\ell = = f^{-i} \circ g \circ h(1/6,1/3)$, and so $I_n^\ell \subset \cup_{h \in G} h(1/6,1/3)$. ■

Let us look at a simple way to see the foliation $\mathcal{F}(\varphi)$ geometrically. Let $\tilde{b}$ and $\tilde{d}$ be two closed curves in V_2 which do not meet and are homotopic to b and d respectively. Let $\tilde{V}_2 = V_2 - (\tilde{b} \cup \tilde{d})$ and $\tilde{M} = \tilde{V}_2 \times S^1$. The figure below represents $\tilde{M}$.

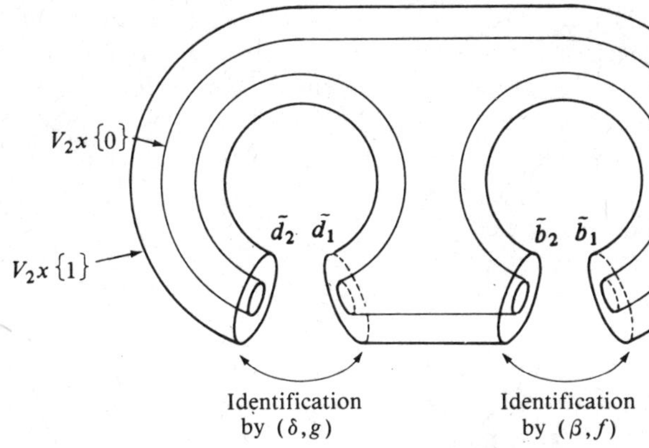

Figure 7

In this figure a manifold with boundary $\tilde{V}_2 \times [0,1]$ is drawn. The manifold $\tilde{M}$ should be thought of as $\tilde{V}_2 \times [0,1]$, where we identify the points of $\tilde{V}_2 \times 0$ and $\tilde{V}_2 \times 1$ with the same first coordinate. Observe that $\tilde{M}$ is a 3-dimensional manifold such that $\partial\tilde{M}$ is made up of four copies of T^2, $\tilde{b}_1 \times$

S^1, $\tilde{b}_2 \times S^1$, $\tilde{d}_1 \times S^1$ and $\tilde{d}_2 \times S^1$, where $\tilde{b}_1$ and $\tilde{b}_2$ (resp. $\tilde{d}_1$ and $\tilde{d}_2$) are the results of lifting $\tilde{b}$ (resp. $\tilde{d}$) from V_2. In order to obtain V_2 from $\tilde{V}_2$ we need two diffeomorphisms of identification, say $\beta : \tilde{b}_2 \longrightarrow \tilde{b}_1$ and $\delta : \tilde{d}_2 \longrightarrow \tilde{d}_1$. Let us construct M from $\tilde{M}$. Given the diffeomorphisms of S^1, f and g define $\psi : (\tilde{b}_2 \times S^1) \cup (\tilde{d}_2 \times S^1) \longrightarrow (\tilde{b}_1 \times S^1) \cup (\tilde{d}_1 \times S^1)$ by $\psi(x,y) = (\beta(x), f(y))$ if $(x,y) \in \tilde{b}_2 \times S^1$ and $\psi(x,y) = (\delta(x), g(y))$ if $(x,y) \in \tilde{d}_2 \times S^1$.

Consider on $\tilde{M}$ the equivalence relation $\sim$ defined by $(x,y) \sim (x',y')$ if $x = x'$ and $y = y'$ or $(x,y) = \psi(x',y')$, or $(x,y) = \psi^{-1}(x',y')$ in the case $(x',y') \in \partial\tilde{M}$. Then $M = \tilde{M}/\sim$ is a compact manifold of dimension 3 and we have defined a submersion $\pi : M \longrightarrow V_2$ which makes the following diagram commute:

$$\begin{array}{ccc}
\tilde{V}_2 \times S^1 = \tilde{M} & \xrightarrow{\pi_1} & \tilde{V}_2 \\
\Big\downarrow \tilde{p} & & \Big\downarrow p \\
M & \xrightarrow{\pi} & V_2
\end{array}$$

In this diagram π_1 is projection on the first factor $\pi_1(x,y) = x$, $\tilde{p}$ is the projection of the identification and p is the natural projection from $\tilde{V}_2$ onto V_2. By construction $\pi : M \longrightarrow V_2$ defines on M a fibered space structure with base V_2 and fiber S^1.

Let $\tilde{\mathfrak{F}}$ be the product foliation on $\tilde{M}$, whose leaves are of the form $\tilde{V}_2 \times \{x\}$, $x \in S^1$. A leaf $\tilde{V}_2 \times \{x\}$ meets $\partial\tilde{M}$ in four circles $\tilde{b}_1 \times \{x\}$, $\tilde{b}_2 \times \{x\}$, $\tilde{d}_1 \times \{x\}$, and $\tilde{d}_2 \times \{x\}$. Observe that the diffeomorphism of identification ψ takes the circle $\tilde{b}_2 \times \{x\}$ to the circle $\tilde{b}_1 \times \{f(x)\}$ and the circle $\tilde{d}_2 \times \{x\}$ to the circle $\tilde{d}_1 \times \{g(x)\}$, or, as we identify the points of $\partial\tilde{M}$ by ψ we automatically glue the boundary of a leaf of $\tilde{\mathfrak{F}}$. In this way $\tilde{\mathfrak{F}}$ induces on M a foliation $\mathfrak{F}$ equivalent to $\mathfrak{F}(\varphi)$.

Notes to Chapter V

(1) An application to the study of Ricatti's equation.

The equation of Ricatti in the complex domain is a differential equation of the following type:

$$\frac{dy}{dx} = a(x)y^2 + b(x)y + c(x), \qquad x \in \mathbb{C} ,$$

where a, b and c are entire functions, i.e. holomorphic in $\mathbb{C}$.

In this note we will prove that the solutions of the Ricatti equation are meromorphic functions from $\mathbb{C}$ to $\mathbb{C}$. This result was originally proved by Painlevé [41]. Observe that this fact is not true for example for the equation $dy/dx = y^3, x, y \in \mathbb{C}$. In fact, this equation can be integrated by separation of variables and the solution such that $g(0) = y_0 \neq 0$ is $y(x) = y_0/\pm\sqrt{1-2xy_0^2}$, which is *not* a function in the usual sense of the word, since it is multi-valued, that is, for each $x \neq 1/2y_0^2$, $y(x)$ assumes two distinct values.

Before we prove this result we will remember some facts about the theory of meromorphic functions. Consider the Riemann sphere of radius $1/2$, $S^2 \subset \mathbb{R}^3$ with a parametrization defined by the stereographic projections

$$\varphi_1 : \mathbb{C} \longrightarrow S^2 - \{(0,0,-1/2)\} \text{ and } \varphi_2 : \mathbb{C} \longrightarrow S^2 - \{(0,0,1/2)\}$$

$$\varphi_1(z = z_1 + iz_2) = \frac{1}{1+|z|^2}\left(z_1, z_2, \frac{1}{2}(1-|z|^2)\right)$$

$$\varphi_2(z = z_1 + iz_2) = \frac{1}{1+|z|^2}\left(z_1, -z_2, \frac{1}{2}(|z|^2-1)\right).$$

The changes of parametrizations are given by

$$\varphi_2^{-1} \circ \varphi_1(z) = \varphi_1^{-1} \circ \varphi_2(z) = \frac{1}{z}, \quad \varphi_2^{-1} \circ \varphi_1 : \mathbb{C} - \{0\} \longrightarrow \mathbb{C} - \{0\}.$$

A meromorphic function $f : \mathbb{C} \longrightarrow \mathbb{C}$ can be considered a holomorphic function $\overline{f} : \mathbb{C} \longrightarrow S^2$. In fact, if $z \in \mathbb{C}$ is not a singularity of f, set $\overline{f}(z) = \varphi_1 \circ f(z)$ and if z is a singularity of f, set $\overline{f}(z) = (0,0,-1/2)$. We claim that $\overline{f} : \mathbb{C} \longrightarrow S^2$ is holomorphic. Proof: if z_0 is not a singularity of f then $\varphi_1^{-1} \circ \overline{f}$ is holomorphic in a neighborhood of z_0 and if $f(z_0) \neq 0$, $\varphi_2^{-1} \circ \overline{f}(z) = 1/f(z)$ is also holomorphic in a neighborhood of z_0. In case z_0 is a singularity of f then z_0 is a pole of f and therefore $f(z) = (z-z_0)^{-k} g(z)$, where $k \geq 1$ and g is holomorphic and nonvanishing in a neighborhood of z_0. We have then that $\varphi_2^{-1} \circ \overline{f}(z) = (z-z_0)^k \cdot (g(z))^{-1}$, which is holomorphic.

Conversely, if $\overline{f} : \mathbb{C} \longrightarrow S^2$ is a holomorphic function then the function $f(z) = \varphi_1^{-1} \circ \overline{f}(z)$, $z \in \mathbb{C}$, is meromorphic.

Consider now the Ricatti equation

$$\frac{dy}{dx} = a(x)y^2 + b(x)y + c(x), \qquad y \in \mathbb{C}.$$

The local solutions of this equation are (complex) integral curves of the complex 1-form on $\mathbb{C}^2$

$$\omega = dy - (a(x)y^2 + b(x)y + c(x))\,dx.$$

From the real point of view, $\omega = \omega_1 + i\omega_2$, where

$$\omega_1 = dy_1 - \mathrm{Re}(F(x,y))\,dx_1 + \mathrm{Im}(F(x,y))\,dx_2$$

$$\omega_2 = dy_2 - \operatorname{Im}(F(x,y))\,dx_1 - \operatorname{Re}(F(x,y))\,dx_2$$

where

$$F(x,y) = a(x)y^2 + b(x)y + c(x) .$$

Therefore the plane field defined by $\omega = 0$ on $\mathbb{C}^2$ has real codimension 2 and complex 1. This plane field is integrable and can be considered as a plane field on $\mathbb{C} \times S^2$, transverse to the fibers $\{z\} \times S^2$. Indeed, consider the parametrizations of $\mathbb{C} \times S^2$, $\psi_1 : \mathbb{C} \times \mathbb{C} \longrightarrow \mathbb{C} \times (S^2 - \{(0,0,-1/2)\})$ and $\psi_2 : \mathbb{C} \times \mathbb{C} \longrightarrow \mathbb{C} \times (S^2 - \{(0,0,1/2)\})$ defined by $\psi_1(x,y) = (x, \varphi_1(y))$ and $\psi_2(x,y) = (x, \varphi_2(y))$.

On $\mathbb{C} \times (S^2 - \{(0,0,-1/2)\})$ the plane field is defined by the complex 1-form $(\psi_1^{-1})^*(\omega)$, and on $\mathbb{C} \times (S^2 - \{(0,0,1/2)\})$ the plane field is defined by the complex 1-form $(\psi_2^{-1})^*(\eta)$ where $\eta = -dy - (a(x) + b(x)y + c(x)y^2)\,dx$. We must verify that this plane field is well-defined. It is enough to show that they coincide on $\mathbb{C} \times (S^2 - \{(0,0,1/2),(0,0,-1/2)\})$. In this region, the plane field defined by $(\psi_2^{-1})^*(\eta)$ is equivalent to the plane field defined by the form $\psi_1^*(\psi_2^{-1})^*(\eta) = (\psi_2^{-1} \circ \psi_1)^*(\eta)$ on $\mathbb{C} \times (\mathbb{C} - \{0\})$, which can be expressed

$$(\psi_2^{-1} \circ \psi_1)^*(\eta) = -d(y^{-1}) - (a(x) + b(x)y^{-1} + c(x)y^{-2})\,dx =$$

$$y^{-2}(dy - (a(x)y^2 + b(x)y + c(x))\,dx) =$$

$$= y^{-2} \cdot \omega = y^{-2}\psi_1^*(\psi_1^{-1})^*(\omega) .$$

Hence the two complex 1-forms ω and $\psi_1^*(\psi_2^{-1})^*(\eta)$ differ by the factor y^{-2} which does not vanish on $\mathbb{C} \times (\mathbb{C} - \{0\})$, and this proves that the plane field is well-defined on $\mathbb{C} \times S^2$.

On the other hand, since $\omega_{(x,y)}(0,\dot{y}_1 + i\dot{y}_2) = \dot{y}_1 + i\dot{y}_2$ and $\eta_{(x,y)}(0,\dot{y}_1 + i\dot{y}_2) = -\dot{y}_1 - i\dot{y}_2$, we have that the plane field is transverse to the fibers $\{z\} \times S^2$ of the fibration $(\mathbb{C} \times S^2, \pi_1, \mathbb{C}, S^2)$ where π_1 is projection on the first factor. Let $\mathfrak{F}$ be the foliation of complex codimension 1 defined by the plane field on $\mathbb{C} \times S^2$. Since the fiber is compact, by Proposition 1, given a leaf L of $\mathfrak{F}$, the map $\pi_1 \mid L : L \longrightarrow \mathbb{C}$ is a covering. Since $\mathbb{C}$ is simply connected, we can conclude that $\pi_1 \mid L : L \longrightarrow \mathbb{C}$ is a diffeomorphism, so $L \subset \mathbb{C} \times S^2$ is the graph of a function $\overline{y} : \mathbb{C} \longrightarrow S^2$. The function $y(x) = \varphi_1^{-1}(\overline{y}(x))$ is single-valued and meromorphic, being also a solution to the Ricatti equation. We can then state the following

Theorem. *Let $a,b,c : \mathbb{C} \longrightarrow \mathbb{C}$ be holomorphic functions. The solutions of the Ricatti equation*

$$\frac{dy}{dx} = a(x)y^2 + b(x)y + c(x), \qquad x,y \in \mathbb{C}$$

are meromorphic functions from $\mathbb{C}$ to $\mathbb{C}$.

(2) Sacksteder's Theorem ([50]).

In 1965, Sacksteder proved the following theorem.

Theorem. *Let $\mathfrak{F}$ be a codimension one, C^2 foliation of a compact manifold M^m $(m \geq 2)$. Suppose that $\mathfrak{F}$ has an exceptional minimal set μ. Then there is a leaf $L \subset \mu$ and a closed curve $\gamma : I \longrightarrow L$ such that, if f is the germ of holonomy of γ on a transverse segment Σ passing through $p = \gamma(0) = \gamma(1)$ then $f'(p) < 1$. In particular, the holonomy of L is not trivial.*

Denjoy's theorem (see note 2 of Chapter III) is a consequence of the above theorem. In fact, if μ is an exceptional minimal set of a foliation $\mathfrak{F}$ of codimension 1 defined on a compact manifold M^2 then all the leaves are homeomorphic to $\mathbb{R}$ and hence simply connected; so the holonomy of such leaves is trivial. By Sacksteder's theorem such a foliation cannot be C^2. In fact, the calculations made by Sacksteder for the proof of the above theorem are similar to those made by Denjoy.

In the example of Sacksteder described in the text, two leaves which satisfy the conclusion of the theorem are those corresponding to the fixed points 0 and 5/6 of the diffeomorphism g defined in §6.

(3) Hector's example ([19]).

We will see here an example of a codimension one foliation on $\mathbb{R}^3 = \mathbb{R}^2 \times \mathbb{R}$ whose leaves are dense and transverse to the fibers $\{x\} \times \mathbb{R}$, $x \in \mathbb{R}^2$. This foliation satisfies condition (a) of the definition of §2, but not condition (b), that is, $\pi \mid L : L \longrightarrow \mathbb{R}^2$ is not a covering, where L is a leaf and π the projection on the first factor.

Consider the solid cylinder $D^2 \times \mathbb{R} \subset \mathbb{R}^3$ where $D^2 = \{(x_1,x_2) \in \mathbb{R}^2 \mid x_1^2 + x_2^2 < 1\}$ and let $S^1 \times \mathbb{R} = \partial(D^2 \times \mathbb{R}^2)$. Given a diffeomorphism $f : \mathbb{R} \longrightarrow \mathbb{R}$ such that $f(x) = x$ for every $x \in (a,b)$, we are going to define a foliation $\mathfrak{F}(f)$ on $D^2 \times \mathbb{R} - (\{0\} \times (-\infty,a] \cup \{0\} \times [b,\infty))$, as follows.

Consider on $S^1 \times \mathbb{R}$ a foliation $\mathcal{G}(f)$, the suspension of f, such that for every $z \in (a,b)$ the circle $S^1 \times \{z\}$ is a leaf of $\mathcal{G}(f)$ and the leaves of $\mathcal{G}(f)$ are transverse to the lines $\{q\} \times \mathbb{R}$, $q \in S^1$.

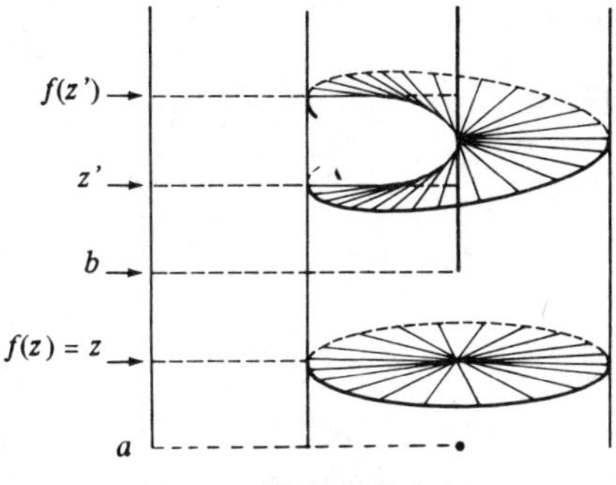

Figure 8

Define now the leaves of $\mathfrak{F}(f)$ saturating the leaves of $\mathcal{G}(f)$ by the flow of the radial field $X(x,y,z) = (-x,-y,0)$ inside $D^2 \times S^1$, as in figure 8.

So, if $t \longrightarrow (\cos t, \sin t, z(t)) = \gamma(t)$ is the parametrization of a leaf $\mathcal{G}(f)$ which passes through a point $(x,y,z) \in S^1 \times D^2$ such that $z > b$ or $z < a$, then $\gamma(r,t) =$ $= (r\cos t, r\sin t, z(t))$, $t \in \mathbb{R}$, $0 < r < 1$, is the parametrization of a leaf of $\mathfrak{F}(f)$. The leaves of $\mathfrak{F}(f)$ in the region $D^2 \times (a,b)$ are the disks $D^2 \times \{z\}$, $z \in (a,b)$.

Now consider a C^∞ diffeomorphism $h : D^2 \times \mathbb{R} \longrightarrow D^2 \times \mathbb{R} - (\{0\} \times (-\infty, a] \cup \{0\} \times [b, \infty))$ which satisfies the following properties:

(a) $h(\{0\} \times (a,b)) = \{0\} \times \mathbb{R}$,
(b) for every $x \in D^2 - \{0\}$, $h(\{x\} \times \mathbb{R}) = \{x\} \times \mathbb{R}$,
(c) h is the identity in a product neighborhood of $S^1 \times \mathbb{R}$.

We can define a diffeomorphism h as above taking $h(x_1,x_2,z) = (x_1,x_2,\alpha(r^2)\beta(z) + (1 - \alpha(r^2))z)$, where $r^2 = x_1^2 + x_2^2$ and the graphs of α and β are as in figure 9.

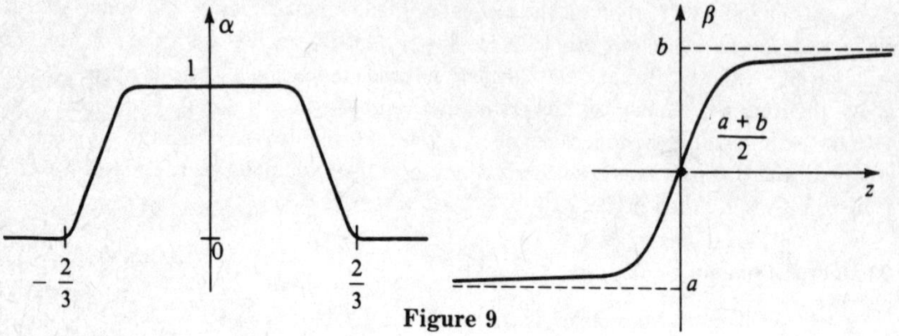

Figure 9

It is clear that in the above construction we can have $a = -\infty$ or $b = +\infty$. Let $\mathfrak{F}^*(f) = h^*(\mathfrak{F}(f))$. It is clear that $\mathfrak{F}^*(f)$ is a foliation on $D^2 \times \mathbb{R}$ whose leaves are transverse to the fibers $\{x\} \times \mathbb{R}$, $x \in D^2$. Further, $\mathfrak{F}^*(f)$ coincides with $\mathfrak{F}(f)$ on a neighborhood of $S^1 \times \mathbb{R}$. The idea now is to glue together various foliations like this defined initially on solid, disjoint cylinders, identifying the boundaries of these cylinders in the following manner.

Suppose that $f_1 : \mathbb{R} \longrightarrow \mathbb{R}$ and $f_2 : \mathbb{R} \longrightarrow \mathbb{R}$ are diffeomorphisms with $f_i(x) = x$ if $x \in (a_i, b_i)$, $i = 1,2$. We can suppose that the leaves of $\mathcal{G}(f_i)$, $i = 1,2$, are horizontal in the region V_i of $S^1 \times \mathbb{R}$ defined by $V_i = \{(1,\theta,z) \in S^1 \times \mathbb{R} \mid \theta_i - \pi/2 < \theta < \theta_i + \pi/2\}$ (where we are writing the point $(x_1,x_2) \in S^1$ in polar coordinates and $\theta_i \in S^1$ is fixed). Figure 11 illustrates what we want.

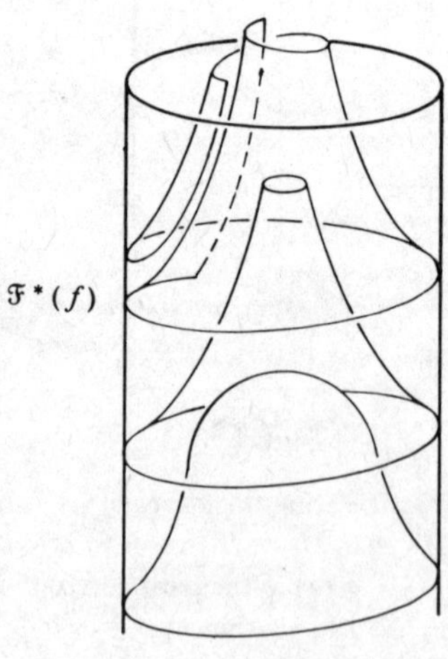

Figure 10

With this restriction the leaves of $\mathfrak{F}(f_i)$ are horizontal in the region $\tilde{V}_i = \{(r,\theta,z) \in D^2 \times \mathbb{R} \mid 0 < r < 1, \theta_i - \pi/2 < \theta < \theta_i + \pi/2\}$.

Since $\mathfrak{F}^*(f_i)$ coincides with $\mathfrak{F}(f_i)$ in a neighborhood $A_i \times \mathbb{R}$ of $S^1 \times \mathbb{R}$, where $A_i = \{(r,\theta) \mid 1 - \epsilon < r < 1\}$, the leaves of $\mathfrak{F}^*(f_i) \mid ((A_i \cap V_i) \times \mathbb{R})$ are horizontal.

Glue together now the two cylinders by a diffeomorphism $g : (A_1 \cap V_1) \times \mathbb{R} \longrightarrow (A_2 \cap V_2) \times \mathbb{R}$ of the form $g(x,z) = (\alpha(x),z)$ where $\alpha : A_1 \cap V_1 \longrightarrow A_2 \cap V_2$ is a diffeomorphism. Thus we obtain a foliation $\mathfrak{F}(f_1,f_2)$ as in Figure 11.

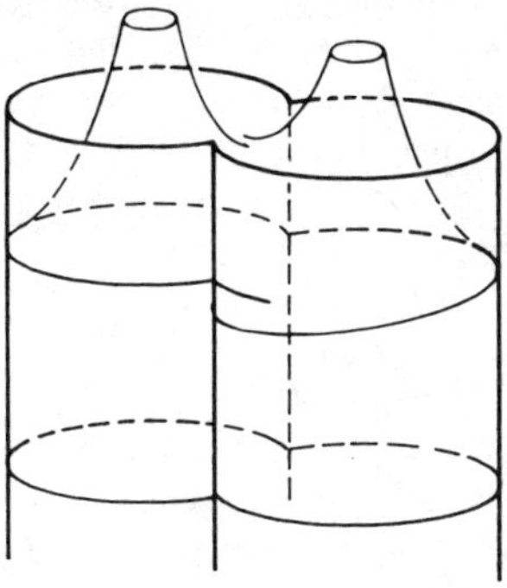

Figure 11

The region where $\mathfrak{F}(f_1, f_2)$ is defined is a product of the form $U \times \mathbb{R}$ where U is diffeomorphic to the unit disk D^2. So we can consider $\mathfrak{F}(f_1,f_2)$ as being a foliation of $D^2 \times \mathbb{R}$ transverse to the fibers $\{x\} \times \mathbb{R}$, $x \in D^2$.

Let us see now how a leaf F of $\mathfrak{F}(f_1,f_2)$ meets a fiber $\{x_0\} \times \mathbb{R}$, $x_0 \in D_2$. Observe first that in every step of the construction vertical fibers were taken to vertical fibers. If ℓ_1 and ℓ_2 are the axis of the initial cylinders, there are $p \neq q \in D^2$ such that ℓ_1 is taken to $\{p\} \times \mathbb{R}$ and ℓ_2 to $\{q\} \times \mathbb{R}$, at the end of the process. If we consider $\mathfrak{F}(f_1,f_2)$ restricted to $(D^2 - \{p,q\}) \times \mathbb{R}$, it is easy to see that, for every leaf F of the new foliation, $\pi \mid F : F \longrightarrow D^2 - \{p,q\}$ is a covering, where π is the projection $\pi(x,z) = x$, $x \in D^2$, $z \in \mathbb{R}$. We have then that the new foliation satisfies the hypothesis (a) and (b) of the definition of §2.

Given $x_0 \in D^2 - \{p,q\}$, if $\gamma : I \longrightarrow D^2 - \{p,q\}$ is a continuous curve such that $\gamma(0) = \gamma(1) = x_0$, we can get a holonomy transformation $g_{[\gamma]} : \{x_0\} \times \mathbb{R} \longrightarrow \{x_0\} \times \mathbb{R}$ as in Theorem 1. Observe that the fundamental group of $D^2 - \{p,q\}$ is the free group of two generators, α and β, which can be taken as in the figure below:

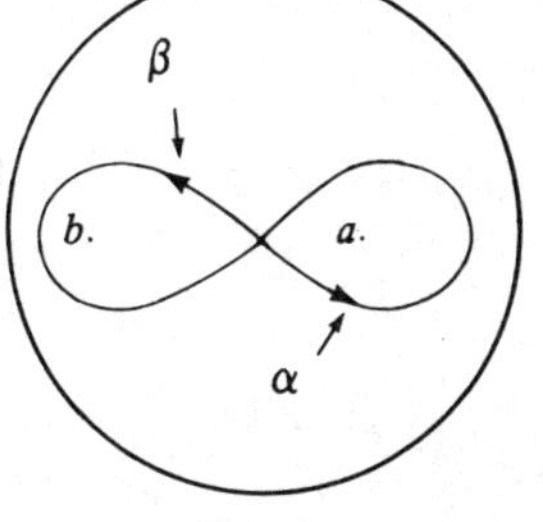

Figure 12

It is easy to see that the holonomy transformations $g_{[\alpha]}$ (or $g_{[\alpha]^{-1}}$) and $g_{[\beta]}$ (or $g_{[\beta]^{-1}}$) are conjugate to f_1 and f_2 respectively, by the same diffeomorphism. We have then that the intersection of the leaf F of $\mathfrak{F}(f_1,f_2)$ which passes through (x_0,z) with $\{x_0\} \times \mathbb{R}$ can be represented by the set

$$\mathcal{O}(z) = \{f_1^{k_1} \circ f_2^{\ell_1} \circ f_1^{k_2} \circ f_2^{\ell_2} \circ \dots \circ f_2^{\ell_j}(z) \mid k_1 ,\dots, k_j,\ell_1 ,\dots, \ell_j \in Z\} .$$

We now consider diffeomorphisms $f_1 ,\dots, f_k : \mathbb{R} \longrightarrow \mathbb{R}$ as before. Using the same process as before, we can inductively define a foliation $\mathfrak{F}(f_1 ,\dots, f_k)$ on $D^2 \times \mathbb{R}$ transverse to the fibers $\{x\} \times \mathbb{R}$, $x \in D^2$. As before, the intersection of the leaf of $\mathfrak{F}(f_1 ,\dots, f_k)$ which passes through (x_0,z) with $\{x_0\} \times \mathbb{R}$ can be represented by the set

$$\mathcal{O}(x) = \{ \prod_{i=1}^{n} f_{j_i}^{\ell_i}(\tau) \mid n \in \mathbf{Z}, \ell_i \in \mathbf{Z} \text{ and } j_i \in \{1 ,\dots, k\}\}$$

where $\prod_{i=1}^{n} f_{j_i}^{\ell_i}(z) = f_{j_1}^{\ell_1} \circ \dots \circ f_{j_n}^{\ell_n}(z)$.

Finally observe that since D^2 is diffeomorphic to $\mathbb{R}^2$, we can consider $\mathfrak{F}(f_1 ,\dots, f_k)$ as being a foliation on $\mathbb{R}^3 = \mathbb{R}^2 \times \mathbb{R}$ transverse to the fibers $\{x\} \times \mathbb{R}$, $x \in \mathbb{R}^2$.

Now we will see that there exist diffeomorphisms $(f_i)_{i=1}^{4}$ such that all the leaves of $\mathfrak{F}(f_1 ,\dots, f_4)$ are dense in $\mathbb{R}^3$.

First fix $\varphi : \mathbb{R} \xrightarrow{C^\infty} \mathbb{R}$ of class C^∞ such that $\varphi(x) = 0$ if $x \leq 0$, $\varphi(x) = 1$ if $x \geq 1$ and $\varphi'(x) \geq 0$ for every $x \in \mathbb{R}$. Define $f_1(x) = x + \alpha\varphi(x)$, where $\alpha > 0$ is irrational. It is clear that f_1 is a C^∞ diffeomorphism and that $f_1(x) = x$ if $x \leq 0$ and $f_1(x) = x + \alpha$ if $x \geq 1$. Set $f_2(x) = f_1^{-1}(x + \alpha)$. It is clear that f_2 is a C^∞ diffeomorphism and that $f_2(x) = x$ if $x \geq 1$. Analogously define $f_3(x) = x + \varphi(x)$ and $f_4(x) = f_3^{-1}(x + 1)$. We claim that if α is irrational, all the leaves of $\mathfrak{F}(f_1 ,\dots, f_4)$ are dense in $\mathbb{R}^3$. In fact, let $g_1(x) = f_1 \circ f_2(x) = x + \alpha$ and $g_2(x) = f_3 \circ f_4(x) = x + 1$. It suffices to prove that, for every $z \in \mathbb{R}$, the set $B = \{g_1^m g_2^n(z) \mid m,n \in \mathbf{Z}\}$ is dense in $\mathbb{R}$. Indeed, $g_1^m g_2^n(z) = z + m\alpha + n$ and, since α is irrational the set $\{m\alpha + n \mid m,n \in \mathbf{Z}\}$ is dense in $\mathbb{R}$, so B is dense in $\mathbb{R}$. ■

(4) Foliations on open manifolds.

As was already mentioned not every compact n-dimensional manifold has a codimension k foliation where $0 < k < n$. However when the manifold is open (i.e., not compact) the situation changes. The following theorem in this direction is known.

Theorem. *Every codimension one plane field defined on an open manifold is homotopic to a plane field tangent to a codimension one foliation.*

This theorem is contained in a more general result due to Gromov [15] and Phillips [44] about homotopies of plane fields of arbitrary codimension, of which the following is a sample.

Theorem. *Let σ be a codimension k plane field on an open Riemannian manifold. If the normal fibration $\sigma^\perp$ of σ has discrete structure group then σ is homotopic to a completely integrable plane field.*

(5) Foliations on $\mathbb{R}^n$.

A problem which has been studied recently is that of the characterization of a foliation when one imposes strong conditions on the topology of the leaves. A typical result in this direction is the following, due to C.F. Palmeira ([42]).

Theorem. *Let $\mathfrak{F}$ be a C^r $(r \geq 0)$, codimension one foliation defined on $\mathbb{R}^n$ $(n \geq 3)$. Suppose that all the leaves of $\mathfrak{F}$ are closed subsets of $\mathbb{R}^n$ and are homeomorphic to $\mathbb{R}^{n-1}$. Then there is a foliation $\tilde{\mathfrak{F}}$ on $\mathbb{R}^2$ and a C^r diffeomorphism (homeomorphism if $r = 0$) $f : \mathbb{R}^n \longrightarrow \mathbb{R}^2 \times \mathbb{R}^{n-2}$ such that the leaves of $\mathfrak{F}$ are of the form $f^{-1}(F \times \mathbb{R}^{n-2})$, where F is a leaf of $\tilde{\mathfrak{F}}$. In other words $\mathfrak{F}$ is topologically equivalent to a product foliation of a foliation on $\mathbb{R}^2$ with fibers $\mathbb{R}^{n-2}$.*

This theorem can in fact be strengthened, to the following statement ([42]): Let $\mathfrak{F}$ and $\mathcal{G}$ be two foliations on $\mathbb{R}^n$ $(n \geq 3)$ whose leaves are closed and homeomorphic to $\mathbb{R}^{n-1}$. Then $\mathfrak{F}$ and $\mathcal{G}$ are topologically equivalent if and only if the leaf spaces of $\mathfrak{F}$ and $\mathcal{G}$ are homeomorphic. This fact is not true for $n = 2$.

The above result admits other generalizations which can be found in [42].

VI. ANALYTIC FOLIATIONS OF CODIMENSION ONE

A codimension n foliation $\mathfrak{F}$ of an m-dimensional manifold is analytic when the change of coordinate maps which define $\mathfrak{F}$ are analytic local diffeomorphisms of $\mathbb{R}^m$. Under these conditions any element of the holonomy of a leaf of $\mathfrak{F}$ has a representation which is an analytic local diffeomorphism of $\mathbb{R}^n$.

For example, the Reeb foliation of S^3 is not analytic. Indeed, the Reeb foliation has a compact leaf whose holonomy is represented by the germs of two commuting diffeomorphisms of the line, f and g, with the following characteristics: $f(x) = x$, $g(x) < x$ if $x \in (0,\infty)$ and $f(x) < x$, $g(x) = x$ if $x \in (-\infty,0)$. These diffeomorphisms are not analytic since an analytic map which coincides with the identity on an interval is the identity. It is natural then to ask if there exists analytic foliations of S^3. In this chapter we will see that the answer is no. More generally, we will show that no compact manifold with finite fundamental group has an analytic foliation of codimension one. The results of this chapter are due to A. Haefliger ([17]) and [18]). The obstruction to the existence of analytic foliations of codimension one is contained in the following theorem.

Theorem 1. *Let $\mathfrak{F}$ be a codimension one C^2 foliation of a manifold M. Suppose there exists a closed curve transverse to $\mathfrak{F}$ homotopic to a point. Then there exist a leaf F of $\mathfrak{F}$ and a closed curve $\Gamma \subset F$ whose germ of holonomy in a segment J transverse to $\mathfrak{F}$, with $x_0 = J \cap \Gamma$, is the identity on one of the components of $J - \{x_0\}$ but differ from the identity on any neighborhood of x_0 in J.*

Corollary. *Let M be a real analytic manifold on which is defined an analytic foliation $\mathfrak{F}$ of codimension one. Every closed curve transverse to $\mathfrak{F}$ represents an element of infinite order in $\pi_1(M)$.*

Proof of the Corollary. Let $\gamma : S^1 \longrightarrow M$ be a curve transverse to $\mathfrak{F}$ and $[\gamma]$ its homotopy class. Suppose, by contradiction, that $[\gamma]$ has finite order, $[\gamma]^n = 1$. Then the path $\alpha : S^1 \longrightarrow M$ defined by $\alpha(e^{i\theta}) = \gamma(e^{in\theta})$ is homotopic to a point. This yields a contradiction, since by Theorem 1, there is a path in a leaf of $\mathfrak{F}$ whose holonomy germ cannot be analytic.

§1. An outline of the proof of Theorem 1

In this section we will present a sketch of the proof of Theorem 1. In subsequent sections this proof will be developed in detail.

Let $\alpha : S^1 \longrightarrow M$ be a curve transverse to $\mathfrak{F}$ homotopic to a point. This means that there is an extension of α to a map $A : D^2 \longrightarrow M$. This map is continuous, but by the Stone-Weierstrass approximation theorem, we can assume that A is C^∞ (see [26]).

The map A is next approximated by a map $g : D^2 \longrightarrow M$ transverse to $\mathfrak{F}$, except at a finite number of points $\{p_1, \ldots, p_\ell\}$, where the tangency of g with the leaves of $\mathfrak{F}$ is nondegenerate, that is, of one of the following types:

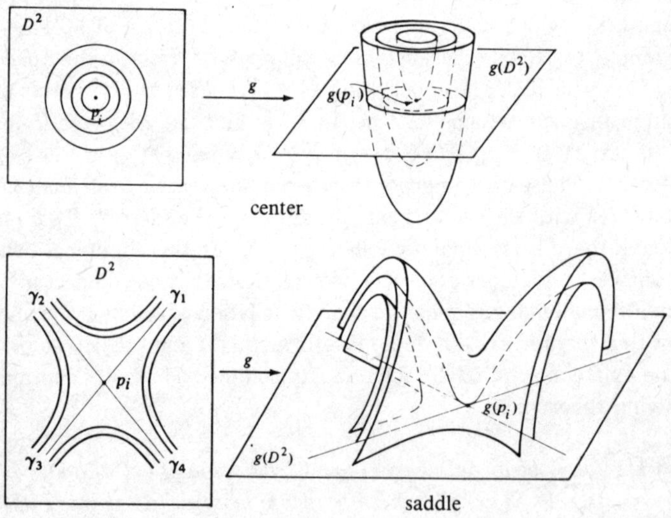

Figure 1

From this we obtain a foliation with singularities on D^2, $\mathfrak{F}^* = g^*(\mathfrak{F})$, transverse to the boundary and whose singular set $\{p_1, \ldots, p_\ell\}$ is made up of centers and saddles.

When p_i is a saddle of $\mathfrak{F}^*$, if we restrict to a small neighborhood V of p_i as in figure 1, $\mathfrak{F}^* \mid V$ has four integral manifolds $\gamma_1, \gamma_2, \gamma_3, \gamma_4$ which accumulate on p_i. These leaves are called local separatrices of p_i. If γ is a leaf of $\mathfrak{F}^*$ such that $\gamma \cap V$ contains a local separatrix of p_i, we say that γ is a separatrix of p_i, and if $\gamma \cap V$ contains two local separatrices of p_i, γ is a *self connection of a saddle.* When γ is a separatrix of two distinct saddles, we say that γ is a *saddle connection* and that the two saddles are connected. In the final part of the proof of the theorem we will use the fact that it is possible to obtain g so that no two distinct saddles of $\mathfrak{F}^*$ are connected.

Observe that if p_i and p_j are two connected saddles of $\mathfrak{F}^*$, then $g(p_i)$ and $g(p_j)$ are contained in the same leaf of $\mathfrak{F}$. When this occurs, we can modify g to obtain a function $\tilde{g} : D^2 \longrightarrow M$, near g, such that $\tilde{g}(\tilde{p}_i)$ and $\tilde{g}(\tilde{p}_j)$ are in distinct leaves of $\mathfrak{F}$ whenever $\tilde{p}_i \neq \tilde{p}_j$ are saddles of $\tilde{g}^*(\mathfrak{F})$. We can then suppose, without loss of generality, that $\mathfrak{F}^*$ does not have distinct connected saddles, possibly having self-connections (which are indestructible under small perturbations). In the figure below we illustrate an example of what can happen:

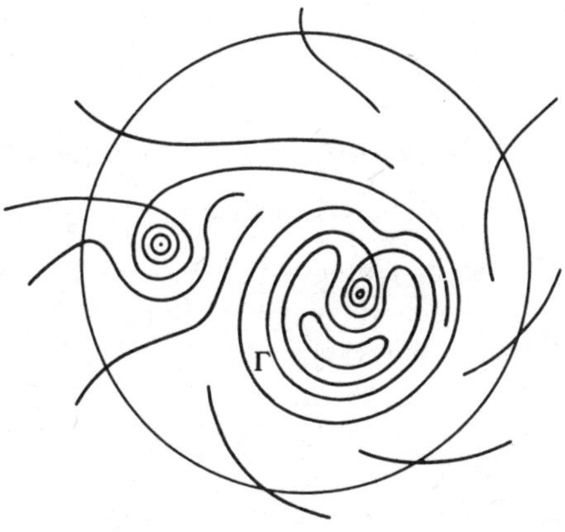

Figure 2

Observe that the foliation $\mathfrak{F}^*$ is locally orientable (even in a neighborhood of a singularity). Since D^2 is simply connected there is a vector field Y on D^2, with singularities $\{p_1, \ldots, p_\ell\}$, whose regular orbits are the leaves of $\mathfrak{F}^*$. Applying Poincaré-Bendixson theory to the vector field Y, one shows that there is a closed curve Γ, invariant under Y, and a segment J, transverse to Y, such that it is possible to define a first-return map f, in a neighborhood of $x_0 = \Gamma \cap J$ in J, following the positive orbits of Y, which is the identity on one of the components of $J - \{x_0\}$, but is not the identity on any neighborhood of x_0 in J. The image of Γ under g defines a closed curve $g(\Gamma)$ in a leaf of $\mathfrak{F}$ whose holonomy is conjugate to f. This concludes the proof.

§2. Singularities of maps $f : \mathbb{R}^n \longrightarrow \mathbb{R}$

A *singularity* of a differentiable function $f : \mathbb{R}^n \longrightarrow \mathbb{R}$ is a point $p \in \mathbb{R}^n$ where $df(p) = 0$. If $(x_1, \ldots, x_n)$ is a coordinate system about p, this means that

$$\frac{\partial f}{\partial x_1}(p) = \ldots = \frac{\partial f}{\partial x_n}(p) = 0.$$

A singularity p of f is *nondegenerate* if the symmetric quadratic form

$$H(p) \cdot u = D^2 f(p)(u,u) = \sum_{i,j=1}^{n} \frac{\partial^2 (f \circ x^{-1})}{\partial x_i \partial x_j}(x(p)) u_i \cdot u_j$$

is nondegenerate, where $(u_1, \ldots, u_n) = dx(p) \cdot u$. Since p is a singularity of f the form $H(p)$ is independent of the system of coordinates chosen. So, if $y = (y_1, \ldots, y_n) : U \longrightarrow \mathbb{R}^n$, $p \in U$, is another system of coordinates and $dy(p) \cdot u = (v_1, \ldots, v_n)$, one easily verifies that

$$(*) \sum_{i,j=1}^{n} \frac{\partial^2 (f \circ y^{-1})}{\partial y_i \partial y_j}(y(p)) v_i \cdot v_j = \sum_{i,j=1}^{n} \frac{\partial^2 (f \cdot x^{-1})}{\partial x_i \partial x_j}(x(p)) u_i \cdot u_j .$$

Associated to the quadratic form $H(p)$ and to the system of coordinates $x : V \longrightarrow \mathbb{R}^n$, we have the symmetric matrix $\left(\frac{\partial^2 (f \circ x^{-1})}{\partial x_i \partial x_j}(x(p)) \right)_{1 \le i,j \le n}$ which is called the Hessian of f at p with respect to x.

The form $H(p)$ is nondegenerate if and only if the Hessian matrix of f, in some system of coordinates, is nondegenerate.

These notions extend naturally to differentiable functions defined on manifolds. In fact, if $f : M \longrightarrow \mathbb{R}$ is a differentiable function, a singular point of f is a point $p \in M$ such that $df(p) = 0$. A singular point of f is nondegenerate if for some system of coordinates $x : U \subset M \longrightarrow \mathbb{R}^n$, $p \in U$, the function $f \circ x^{-1} : x(U) \longrightarrow \mathbb{R}$ has a nondegenerate singularity at $x(p)$. In virtue of $(*)$, this definition is independent of the system of coordinates. Moreover the Hessian quadratic form of f at p remains well-defined by

$$H(p) \cdot u = \sum_{i,j=1}^{n} \frac{\partial^2 (f \circ x^{-1})}{\partial x_i \partial x_j}(x(p)) \cdot u_i \cdot u_j$$

where $(u_1, \ldots, u_n) = dx(p) \cdot u$.

Let $p \in M$ be a singularity of $f : M \longrightarrow \mathbb{R}$. If E is a subspace of $T_p M$, we say that $H(p)$ is positive (respectively negative) definite on E if $H(p) \cdot u > 0$ (resp. $H(p) \cdot u < 0$) for every $u \in E - \{0\}$. The *index* of f at p is the largest possible dimension for a subspace $E \subset T_p M$ where $H(p)$ is negative definite:

$$\operatorname{ind}(f,p) = \max\{\dim(E) \mid E \text{ is a subspace of } T_pM$$
$$\text{and } H(p)\cdot u < 0 \text{ if } u \in E - \{0\}\} .$$

The proof of the result below can be found in [36].

Theorem (Morse Lemma). *Let p be a nondegenerate singular point of f with* $\operatorname{ind}(f,p) = k$. *There is a system of coordinates $x = (x_1, \ldots, x_n)$ on a neighborhood U of p such that*

$$f(q) = f(p) - x_1^2(q) - \ldots - x_k^2(q) + x_{k+1}^2(q) + \ldots + x_n^2(q) .$$

In other words, the function f is locally equivalent to the function

$$x \longmapsto f(p) - x_1^2 - \ldots - x_k^2 + x_{k+1}^2 + \ldots + x_n^2 .$$

In the case $n = 2$ one has three possible canonical forms:

(a) $f(x) = f(p) - x_1^2 - x_2^2$
(b) $f(x) = f(p) + x_1^2 + x_2^2$
(c) $f(x) = f(p) - x_1^2 + x_2^2$.

In the first two cases the level curves of f in a neighborhood of p are diffeomorphic to circles and p is a local maximum or minimum, respectively. In the third case p is a saddle point of f and the equation $f(x) = f(p)$ defines the four local separatrices of p (see figure 1).

We say that $f : M \longrightarrow \mathbb{R}$ is a Morse function when all its singularities are nondegenerate.

Next we will see how one defines the C^r uniform convergence topology on the space of C^r functions of M to $\mathbb{R}$, $C^r(M,\mathbb{R})$, in the case when M is compact. For $r = 0$ and $f \in C^0(M,\mathbb{R})$, define

$$\|f\|_0 = \sup_{x \in M} |f(x)| .$$

One easily verifies that $\| \ \|_0$ defines a norm on $C^0(M,\mathbb{R})$. With this norm $C^0(M,\mathbb{R})$ is a Banach space. For $r \geq 1$, the C^r norm can be defined by means of finite covers of M by systems of coordinates. Since M is compact, consider covers $(U_i)_{i=1}^k$ and $(V_i)_{i=1}^k$ of M such that for every $i \in \{1, \ldots, k\}$, one has $\overline{U_i} \subset V_i$ and V_i is the domain of a local chart $\varphi_i : V_i \longrightarrow \mathbb{R}^n$. Given $f \in C^r(M,\mathbb{R})$, define the C^r norm of f on $\overline{U_i}$ by

$$\|f\|_{r,i} = \max_{0 \leq j \leq r} \left(\sup_{p \in \overline{U_i}} \{\|D^j(f \circ \varphi_i^{-1})(\varphi_i(p))\|\}\right) .$$

For each $j \geq 1$ and each $p \in \overline{U_i}$, $D^j(f \circ \varphi_i^{-1})(\varphi_i(p)) = L_j$ is a symmetric j-linear map from $\mathbb{R}^n \times \ldots \times \mathbb{R}^n = (\mathbb{R}^n)^j$ to $\mathbb{R}$, where $\|L_j\| = \sup\{\|L_j(u_1, \ldots, u_j)\| \mid \|u_1\| = \ldots = \|u_j\| = 1\}$.

The C^r norm of f with respect to the cover $(U_i)_{i=1}^k$ and $(V_i)_{i=1}^k$ is defined by $\|f\|_r = \max_{i=1,\ldots,k} \|f\|_{r,i}$. One verifies that $\| \; \|_r$ is a norm on $C^r(M, \mathbb{R})$. With this norm this space is Banach.

Let $(U_i')_{i=1}^{\ell}$, $(V_i')_{i=1}^{\ell}$ be other covers of M such that for every $i \in \{1, \ldots, \ell\}$, $\overline{U_i'} \subset V_i'$ is the domain of a local chart $\psi_i : V_i' \longrightarrow \mathbb{R}^n$. Let $\| \; \|_r'$ be the C^r norm induced on $C^r(M, \mathbb{R})$ by the covers $(U_i')_{i=1}^{\ell}$ and $(V_i')_{i=1}^{\ell}$. One verifies easily that there exist constants $K_2 > K_1 > 0$ such that

$$K_1 \| \; \|_r' \leq \| \; \|_r \leq K_2 \| \; \|_r' .$$

Therefore, the C^r uniform convergence topology is well-defined. For more details see [27]. The result stated below proves, in particular, the existence of Morse functions.

Theorem. *Let M be a compact manifold, with or without boundary. If $r \geq 2$ the set of C^r Morse functions is open and dense in $C^r(M, \mathbb{R})$.*

The proof can be found in [36].

The following lemma will be used in the next section:

Lemma 1. *Let U, V and W be open sets of $\mathbb{R}^n$ where $\overline{U} \subset V \subset \overline{V} \subset W$ and W is an open ball. Let $h : \overline{W} \longrightarrow \mathbb{R}$ be C^r, $r \geq 2$. For every $\epsilon > 0$, there is $f : \overline{W} \longrightarrow \mathbb{R}$ which is C^r and such that $\|f - h\|_r < \epsilon$, $f \mid (W - V) = h \mid (W - V)$ and $f \mid U$ is a Morse function.*

Proof. By the preceding theorem, given $\delta > 0$ there is an $f_0 : \overline{W} \longrightarrow \mathbb{R}$ such that f_0 is Morse and $\|f_0 - h\|_r < \delta$. Let $\varphi : \mathbb{R}^n \longrightarrow \mathbb{R}$ be C^∞ and such that $\varphi \mid U \equiv 1$, $\varphi \mid (\mathbb{R}^n - V) \equiv 0$ and $\varphi \geq 0$. Let $f = \varphi f_0 + (1 - \varphi) h$. We have $(f - h) \equiv \varphi(f_0 - h) = \varphi \cdot \alpha$. Differentiating $\varphi \cdot \alpha$ k times we get $D^k(\varphi \cdot \alpha) = \sum_{j=0}^{k} C_j^k D^j(\varphi) D^{k-j}(\alpha)$, where for every $j = 0, \ldots, k$, C_j^k is a constant independent of α and of φ. We have then

$$\|D^k(\varphi \cdot \alpha)\|_0 \leq \sum_{j=0}^{k} |C_j^k| \, \|D^j \varphi\|_0 \cdot \|D^{k-j} \alpha\|_0$$

$$\leq \delta \sum_{j=0}^{k} |C_j^k| \, \|D^j \varphi\|_0 .$$

Letting

$$\mu = \max_{0 \leq k \leq r} \{ \sum_{j=0}^{k} |C_j^k| \, \|D^j \varphi\|_0 \} ,$$

we get $\|D^k(\varphi \cdot \alpha)\|_0 \leq \mu \cdot \delta$ for $0 \leq k \leq r$. So, if δ is sufficiently small, we will have $\|f - h\|_r < \epsilon$. ■

In an analogous manner one defines the C^r uniform convergence topology for maps between manifolds (see [27]).

§3. Haefliger's construction

In this section $\mathfrak{F}$ will denote a $C^r (r \geq 2)$ codimension-one foliation defined on a manifold M of dimension ≥ 3. If $h : D^2 \longrightarrow M$ is a C^∞ map, we say that $p \in D^2$ is a point of tangency of h with $\mathfrak{F}$ if $Dh(p) \cdot \mathbb{R}^2 \subset T_{h(p)}\mathfrak{F}$.

Proposition 1. *Let $A : D^2 \longrightarrow M$ be a C^∞ map such that the restriction $A \mid \partial D^2$ is transverse to $\mathfrak{F}$. Then for every $\epsilon > 0$ and every $r \geq 2$, there exists $g : D^2 \longrightarrow M$, C^∞, ϵ near A in the C^r topology and satisfying the following properties:*

(a) *$g \mid \partial D^2$ is transverse to $\mathfrak{F}$.*
(b) *For every point of tangency $p \in D^2$ of g with $\mathfrak{F}$, there exist a foliation box U of $\mathfrak{F}$ with $g(p) \in U$ and a distinguished map $\pi : U \longrightarrow \mathbb{R}$ such that p is a nondegenerate singularity of $\pi \circ g : g^{-1}(U) \longrightarrow \mathbb{R}$. In particular there are only a finite number of tangency points, since they are isolated, and they are contained in the open disk $\mathring{D}^2 = \{z \in \mathbb{R}^2 \mid \|z\| < 1\}$.*
(c) *If $T = \{p_1, \ldots, p_\ell\}$ is the set of points of tangency of g with $\mathfrak{F}$, then $g(p_i)$ and $g(p_j)$ are contained in distinct leaves of $\mathfrak{F}$, for every $i \neq j$. In particular the singular foliation $\mathfrak{F}^* = g^*(\mathfrak{F})$ has no distinct connected saddles.*

Proof. Let $Q_1, \ldots, Q_k$ be foliation boxes for $\mathfrak{F}$ whose union covers $A(D^2)$ and $\pi_i : Q_i \longrightarrow \mathbb{R}$ distinguished maps of $\mathfrak{F}$, $i = 1, \ldots, k$. Observe that A meets $\mathfrak{F}$ transversely at $p \in A^{-1}(Q_i)$ if and only if $D(\pi_i \circ A)(p) \neq 0$. We will use the notation $W_i = A^{-1}(Q_i)$, $i = 1, \ldots, k$.

It is clear that $\cup_{i=1}^k W_i = D^2$ and since D^2 is compact, there exist covers $(V_i)_{i=1}^k$ and $(U_i)_{i=1}^k$ of D^2 by open sets such that $\overline{U}_i \subset V_i \subset \overline{V}_i \subset W_i$ for every $i = 1, \ldots, k$.

Let $\delta = \min_{1 \leq i \leq k} \inf\{d(A(x), y) \mid x \in \overline{V}_i \text{ and } y \in \partial Q_i\}$, where d is a metric on M. If $f : D^2 \longrightarrow M$ is such that $d_0(f,A) < \delta$ where $d_0(f,A) = \sup\{d(f(x),A(x)) \mid x \in D^2\}$, then $f(\overline{V}_i) \subset Q_i$ and therefore the maps $\pi_i \circ f : \overline{V}_i \longrightarrow \mathbb{R}$ are well-defined. All the maps we consider from here on are such that $d_0(f,A) < \delta$.

For each $i = 1, \ldots, k$ consider a foliation chart of $\mathfrak{F}$, $\varphi_i : Q_i \longrightarrow \mathbb{R}^n$, such that $\varphi_i = (\varphi_1^i, \ldots, \varphi_{n-1}^i, \pi_i)$. The nonempty subsets of Q_i defined by $\pi_i^{-1}(t)$ are the plaques of $\mathfrak{F}$ in Q_i. We have then $\varphi_i \circ A = (A_1^i, \ldots, A_{n-1}^i, \pi_i \circ A)$. A point $p \in V_i$ is a point of tangency of A with $\mathfrak{F}$ if and only if

$D(\pi_i \circ A)(p) = 0$. The construction of g satisfying (a) and (b) will be made by inductive modifications of $A \mid \cup_{i=1}^{j} W_i$, $j \in \{1 ,..., k\}$.

For $j = 1$. By Lemma 1 of §2 there exists $f_1 : V_1 \longrightarrow \mathbb{R}$ such that $f_1 \mid U_1$ is Morse $f_1 \mid (W_1 - V_1) = \pi_1 \circ A \mid (W_1 - V_1)$ and $\|f_1 - \pi_1 \circ A\|_r < \delta_1$. We now define $g_1 : D^2 \longrightarrow M$ setting $g_1 \mid (D^2 - V_1) \equiv A \mid (D^2 - V_1)$ and $g_1 \mid W_1 = \varphi_1^{-1} \circ (A_1^1 ,..., A_{n-1}^1, f_1)$. Since $\pi_1 \circ A$ coincides with f_1 on $W_1 - V_1$, it is clear that g_1 is well-defined and that $\pi_1 \circ g_1 \mid U_1 = f_1 \mid U_1$ is a Morse function. Further, taking δ_1 sufficiently small, we have $d_r(g_1, A) < \epsilon/k$.

Suppose by induction that a map $g_j : D^2 \longrightarrow M$, $1 \le j \le k - 1$, has been constructed so that $d_r(A, g_j) < j\epsilon/k$, $g_j \mid (D^2 - \cup_{i=1}^{j} V_i) = A \mid (D^2 - \cup_{i=1}^{j} V_i)$ and for every $i = 1 ,..., j$, $\pi_i \circ g_j : U_i \longrightarrow \mathbb{R}$ is a Morse function. By Lemma 1, there exists $f_{j+1} : W_{j+1} \longrightarrow \mathbb{R}$ such that $f_{j+1} \mid (W_{j+1} - V_{j+1}) = \pi_{j+1} \circ g_j \mid (W_{j+1} - V_{j+1})$, $f_{j+1} \mid U_{j+1}$ is a Morse function and $\|f_{j+1} - \pi_{j+1} \circ g_j\|_r < \delta_{j+1}$. Then set $g_{j+1} \mid (D^2 - V_{j+1}) = g_j \mid (D^2 - V_{j+1})$ and $g_{j+1} \mid W_{j+1} = \varphi_{j+1}^{-1} \circ (g_1^j ,..., g_{n-1}^j, f_{j+1})$ where $(g_1^j ,..., g_{n-1}^j, \pi_{j+1} \circ g_j) = \varphi_{j+1} \circ g_j$. It is clear that g_{j+1} is well-defined since $f_{j+1} \mid (W_{j+1} - V_{j+1}) = \pi_{j+1} \circ g_j \mid (W_{j+1} - V_{j+1})$. Moreover, if δ_{j+1} is sufficiently small we will have $d_r(g_{j+1}, g_j) < \epsilon/k$ and also $\pi_i \circ g_{j+1} : U_i \longrightarrow \mathbb{R}$ a Morse function, $i = 1 ,..., j + 1$, since the set of Morse functions is open and $\|\pi_i \circ g_{j+1} - \pi_i \circ g_j\|_r$ can be taken arbitrarily small (in $U_i \cap U_{j+1}$). Proceeding inductively we get finally $g_k : D^2 \longrightarrow M$ such that for every $i = 1 ,..., k$, $\pi_i \circ g_k : U_i \longrightarrow \mathbb{R}$ is a Morse function and $d_r(g_k, A) < \epsilon$. From this we obtain g_k satisfying (b). Observing that $A \mid \partial D^2$ is transverse to $\mathfrak{F}$, we see that $g_k/\partial D^2$ is transverse to $\mathfrak{F}$ if ϵ is sufficiently small.

Let $T = \{p_1 ,..., p_\ell\}$ be the set of tangencies of g_k with $\mathfrak{F}$. Next we will indicate how it is possible to modify g_k to obtain g such that the set of tangencies of g is T and g satisfies (c). Suppose for example that $g(p_1)$ and $g(p_2)$ are contained in the same leaf F of $\mathfrak{F}$ and that $p_1 \in U_i$. Set $f_i = \pi_i \circ g_k \mid V_i : V_i \longrightarrow \mathbb{R}$, $t_1 = f_i(p_1)$ and fix neighborhoods $U \subset V$ of p_1 such that $\overline{U} \subset V \subset \overline{V} \subset V_i$ and p_1 is the only singularity of f_i in V. Let $\alpha : V_i \longrightarrow \mathbb{R}$ be a C^∞ function such that $\alpha \mid U \equiv 1$, $\alpha \mid (V_i - V) \equiv 0$ and $\alpha \ge 0$. Define $\tilde{f}_i = f_i + \delta\alpha$ where $\delta > 0$. Since $d\tilde{f}_i + \delta d\alpha$, one verifies easily that if δ is sufficiently small then p_1 is the only singularity of $\tilde{f}_i$ in V, which is non-degenerate. Moreover the set of singularities of $\tilde{f}_i$ coincides with that of f_i, since $(\tilde{f}_i - f_i) \mid (V_i - V) \equiv 0$. Define then $\tilde{g}_k : D^2 \longrightarrow M$ setting $\tilde{g}_k \mid V_i \equiv \varphi_i^{-1} \circ (g_1^i ,..., g_{n-1}^i, \tilde{f}_i)$ and $\tilde{g}_k \mid (D^2 - V) \equiv g_k \mid (D^2 - V)$. One easily verifies that $\tilde{g}_k$ satisfies (a) and (b) if δ is sufficiently small and that the set of tangencies of $\tilde{g}_k$ is T.

Observe now that $F \cap Q_i$ contains at most a countable number of plaques of $\mathfrak{F}$ so there exists $\delta \ge 0$ arbitrarily small so that $\pi_i^{-1}(t_1 + \delta) = \pi_i^{-1}(\tilde{f}_i(p_1)) \not\subset F$, or $\tilde{g}_k(p_1)$ and $\tilde{g}_k(p_2)$ are in distinct leaves of $\mathfrak{F}$. Repeating the process a finite number of times it is possible to obtain g, arbitrarily near A, whose set of tangencies $\{p_1, \ldots, p_\ell\}$ is such that $g(p_i)$ and $g(p_j)$ are in distinct leaves of $\mathfrak{F}$ if $i \neq j$. ■

Corollary. *Let $\gamma : S^1 \longrightarrow M$ be a C^∞ curve transverse to $\mathfrak{F}$ homotopic to a constant. Then there is a C^∞ map $g : D^2 \longrightarrow M$ such that $g^*(\mathfrak{F}) = \mathfrak{F}^*$ is a foliation with singularities, satisfying the following properties:*

(a) *$g \mid \partial D^2$ is homotopic to γ and $\mathfrak{F}^*$ is transverse to ∂D^2,*
(b) *the singularities of $\mathfrak{F}^*$ are saddles or centers* (see §2),
(c) *$\mathfrak{F}^*$ has no distinct connected saddles.*

Proof. Since γ is homotopic to a constant in M, there is a C^∞ map $A : D^2 \longrightarrow M$ such that $A \mid \partial D^2 = \gamma$ (see [26]). It suffices now to take g as in Proposition 1, where $d_r(g, A) < \epsilon$, ϵ sufficiently small. ■

§4. Foliations with singularities on D^2

In this section we denote by $\mathfrak{F}^*$ a foliation of D^2 with a finite number of singularities and transverse to the boundary ∂D^2. The set of singularities of $\mathfrak{F}^*$ will be denoted by T. We say that $\mathfrak{F}^*$ is C^r locally orientable if for every $p \in D^2$ there exist a neighborhood U of p and a C^r vector field Y on U such that if $q \in U - T$ then $Y(q) \neq 0$ and is tangent to the leaf of $\mathfrak{F}^*$ through q, and if $q \in T$ then $Y(q) = 0$. We say that $\mathfrak{F}^*$ is C^r orientable if there is a vector field Y as above, defined on the whole disk D^2.

Proposition 2. *If $\mathfrak{F}^*$ is a foliation with singularities on D^2, C^r locally orientable, then $\mathfrak{F}^*$* is *C^r orientable.*

Proof. For every $p \in D^2$ there is a C^r vector field Y defined on a neighborhood U of p such that Y is tangent to $\mathfrak{F}^*$ and $Y(q) = 0$ if and only if q is a singularity of $\mathfrak{F}^*$. Let $U_1, \ldots, U_k$ be a cover of D^2 by such neighborhoods and $Y_1, \ldots, Y_k$ the corresponding vector fields.

The problem of determining a vector field Y as desired reduces to conveniently orienting $Y_1, \ldots, Y_k$, obtaining a new collection of vector fields $\tilde{Y}_1, \ldots, \tilde{Y}_k$ such that for every $i \in \{1, \ldots, k\}$, $\tilde{Y}_i = Y_i$ or $\tilde{Y}_i = -Y_i$ and, further, if $q \in (U_i \cap U_j) - T$ then $\tilde{Y}_i(q)$ and $\tilde{Y}_j(q)$ have the same orientation. In fact, suppose for a moment that this is possible. In that case, it suffices to consider a partition of unity $(\varphi_i)_{i=1}^k$ and define

$$Y = \sum_{i=1}^{k} \varphi_i \tilde{Y}_i.$$

Since $\varphi_i \mid (D^2 - U_i) \equiv 0$, the vector field $\varphi_i \tilde{Y}_i$ extends to a C^r vector field on D^2 setting $(\varphi_i \tilde{Y}_i)(q) = 0$ if $q \notin U_i$. So Y is well-defined and is C^r on D^2. It is tangent to $\mathfrak{F}^*$, since each $\varphi_i \tilde{Y}_i$ is. Further $Y(q) = 0$ if and only if q is a singularity of $\mathfrak{F}^*$ since if q is not a singularity of $\mathfrak{F}^*$, at least one of the vectors $\varphi_i \cdot \tilde{Y}_i(q)$ is not zero and those that are not zero coincide in direction and sense.

Now we will see how it is possible to obtain vector fields $\tilde{Y}_1, \ldots, \tilde{Y}_k$ as above. First observe that we can assume without loss of generality that the open sets $U_1, \ldots, U_k$ are connected so that if $U_i \cap U_j \neq \varnothing$ then $U_i \cap U_j$ is connected. On $(U_i \cap U_j) - T$ we can write $Y_i(q) \equiv \lambda_{ij}(q) Y_j(q)$ where $\lambda_{ij} : (U_i \cap U_j) - T \longrightarrow \mathbb{R}$ is continuous and does not vanish. So $\lambda_{ij} > 0$ or $\lambda_{ij} < 0$. If $\lambda_{ij} > 0$ we say Y_i and Y_j have the same orientation. If $\lambda_{ij} < 0$ we say they have opposite orientations.

Fix $\tilde{Y}_1 = Y_1$ and a point $p_1 \in U_1$. Given $j > 1$, fix a curve $\gamma : [0,1] \longrightarrow D^2$ such that $\gamma(0) = p_1$ and $\gamma(1) = p_j \in U_j$. Consider a cover $(V_i)_{i=1}^m$ of $\gamma(I)$ by neighborhoods so small that $V_i \cap V_j \neq \varnothing$ if and only if $j = i + 1$ and for every i we have $V_i \subset U_{\ell_i}$. Now define $\tilde{Y}_{\ell_s}$, $1 \leq s \leq n$, inductively so that $\tilde{Y}_{\ell_1} = Y_1$ and $\tilde{Y}_{\ell_{i+1}}$ has the same orientation as $\tilde{Y}_{\ell_i}$. Note that if $\ell_{i+1} = \ell_i$ we have $\tilde{Y}_{\ell_{i+1}} = \tilde{Y}_{\ell_i} = \pm Y_{\ell_i}$. Then set $\tilde{Y}_j = \tilde{Y}_{\ell_m}$. Further, if we take a curve $\delta : [0,1] \longrightarrow D^2$ such that $\delta(I) \subset \cup_{i=1}^m V_i$, the definition of $\tilde{Y}_j$ along δ will coincide with the definition of $\tilde{Y}_j$ along γ.

It follows from this fact and from D^2 being simply connected that the definition of $\tilde{Y}_j$ is independent of the curve chosen. We leave it to the reader to verify the details. ■

Corollary. *Let $\mathfrak{F}$ be a $C^r (r \geq 2)$ codimension one foliation on M and $g : D^2 \longrightarrow M$ a C^r map such that the points of tangency of g with $\mathfrak{F}$ are nondegenerate, in the sense of* (b) *of Proposition 1. Let $\mathfrak{F}^* = g^*(\mathfrak{F})$. Then $\mathfrak{F}^*$ is C^{r-1} orientable.*

Proof. Let $Q_1, \ldots, Q_k$ be foliation boxes of $\mathfrak{F}$ such that $\cup_{i=1}^k Q_i \supset g(D^2)$ and $\pi_i : Q_i \longrightarrow \mathbb{R}$ be distinguished maps of $\mathfrak{F}$ on Q_i. The nonsingular leaves of $\mathfrak{F}^*$ in Q_i are the connected components of level curves of the function $\pi_1 \circ g = g_i$. One easily verifies that for every $i \in \{1, \ldots, k\}$ the vector field $Y_i = \left(\frac{\partial g_i}{\partial x_2}, -\frac{\partial g_i}{\partial x_1}\right)$ is tangent to the level curves of g_i. This proves that $\mathfrak{F}^*$ is C^{r-1} locally orientable, hence C^{r-1}-orientable. ■

We will now see some properties of the vector field Y constructed above, in the case that $\mathfrak{F}^*$ satisfies the properties of the corollary of Proposition 1.

Consider first a singularity p of Y, where $p \in W = g^{-1}(Q)$, Q being a distinguished map of $\mathfrak{F}$. Let $\pi : Q \longrightarrow \mathbb{R}$ be a distinguished map of $\mathfrak{F}$. Then p is a nondegenerate singularity of the map $f = \pi \circ g : W \longrightarrow \mathbb{R}$ and the orbits of Y are contained in the level curves of f. By the Morse Lemma there is a diffeomorphism $\alpha : U \longrightarrow V$, where $p \in V \subset W$ and $\alpha(0) = p$, such that

$$f \circ \alpha(x,y) = f(p) + q(x,y)$$

where $q(x,y) = x^2 + y^2$ or $-(x^2 + y^2)$ or $x^2 - y^2$. In the first two cases p is a center of Y, that is, the nonsingular trajectories of Y in W are all closed. In the last case p is a saddle of Y, that is, $f^{-1}(f(p))$ is made up of two differentiable curves δ_1 and δ_2 that meet transversely at p so that $\delta_1 \cup \delta_2 - \{p\}$ contains four regular trajectories of $Y \mid V$, which we call $\alpha_1, \alpha_2, \alpha_3, \alpha_4$ where $\alpha_1 \cup \alpha_3 \cup \{p\} = \delta_1$ and $\alpha_2 \cup \alpha_4 \cup \{p\} = \delta_2$.

The trajectories α_j, $1 \leq j \leq 4$, are the local separatrices of Y. Two of them, say α_1 and α_3, are characterized by the property $\lim_{t \to \infty} Y_t(q) = p$, $q \in \alpha_i$ $(i=1,3)$ and the other two are characterized by the property $\lim_{t \to -\infty} Y_t(q) = p$, $q \in \alpha_i$ $(i = 2,4)$. The trajectories of the first type are called stable and those of the second type unstable.

The fact of $g \mid \partial D^2$ being transverse to $\mathfrak{F}$ implies that Y is transverse to $\partial D^2 = S^1$. We can suppose that Y points to the inside of D^2 along S^1, since in the other case we simply substitute Y by $-Y$.

Given a trajectory γ (that is, $\gamma(t) = Y_t(q)$, $q \in D^2$) we define $\omega(\gamma) = \{x \in D^2 \mid x = \lim_{n \to \infty} \gamma(t_n)$ where $\lim_{n \to \infty} t_n = \infty\}$ and $\alpha(\gamma) = \{x \in D^2 \mid x = \lim_{n \to \infty} \gamma(t_n)$ where $\lim_{n \to \infty} t_n = -\infty\}$.

By the Poincaré-Bendixson theorem (see [20]), the set $\omega(\gamma)$ can only be a singularity, a closed orbit or a graph of Y. A graph is a connected set Γ formed by saddles and separatrices of saddles such that if p is a saddle of Γ then at least a stable and an unstable separatrix of p are contained in Γ. Analogously, $\alpha(\gamma)$ can only be a singularity, a closed orbit, a graph or $\alpha(\gamma) = \varnothing$ in the case where γ meets ∂D^2.

Definition. We say that $\Gamma \subset D^2$ is a limit cycle of Y if Γ does not reduce to a singularity and there is an orbit γ of Y such that $\gamma \cap \Gamma = \varnothing$ and $\omega(\gamma) = \Gamma$ or $\alpha(\gamma) = \Gamma$.

From what we saw above, a limit cycle Γ of Y is a closed orbit or a graph of Y. One also can verify that Γ is always connected and that $D^2 - \Gamma$ contains at least two connected components, one of which contains $\partial D^2 (\Gamma \cap \partial D^2 = \varnothing)$. In the figures below we illustrate some possibilities for Γ in the case that it is a graph:

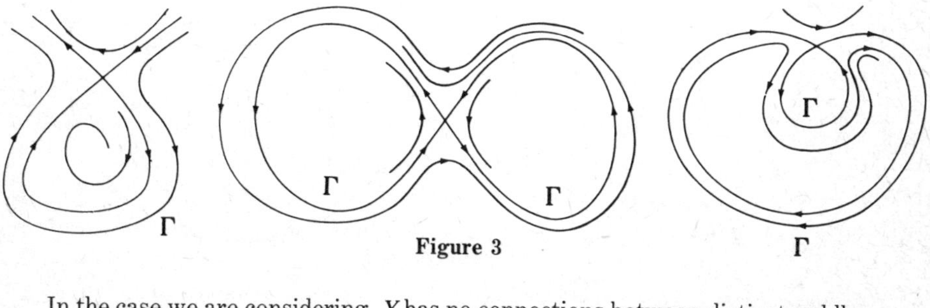

Figure 3

In the case we are considering, Y has no connections between distinct saddles and hence the above possibilities are the only ones up to homeomorphisms.

Let Σ be the set of all limit cycles of Γ. Given $\Gamma \in \Sigma$ we denote by $R(\Gamma)$ the union of the closures of the connected components of $D^2 - \Gamma$ which do not contain ∂D^2. On Σ we define a partial ordering $\leq$ by

$$\Gamma_1 \leq \Gamma_2 \Leftrightarrow R(\Gamma_1) \subset R(\Gamma_2) .$$

Observe that $R(\Gamma)$ is invariant under Y, since Γ is.

Proposition 3. *Let $(\Gamma_n)_{n \in \mathbb{N}}$ be a decreasing sequence of limit cycles and Γ_∞ the boundary of the set $\bigcap_{n \geq 1} R(\Gamma_n)$. Then Γ_∞ is a closed orbit or a graph of Y.*

Proof. Without loss of generality we can suppose that all the terms of the sequence $(\Gamma_n)_{n \in \mathbb{N}}$ are distinct. Since the number of saddles of Y is finite, Y has a finite number of graphs and hence we can also suppose that all the Γ_n are closed orbits. Under these conditions closed orbits of Y accumulate upon Γ_∞. For every $n \geq 1$, $R(\Gamma_n)$ is invariant under Y, homeomorphic to a closed disk and $R(\Gamma_{n+1}) \subset R(\Gamma_n)$. It follows therefore that $\bigcap_{n \geq 1} R(\Gamma_n)$ and Γ_∞ are invariant under Y and nonempty. Further Γ_∞ contains at least one point p such that $Y(p) \neq 0$. In fact, there are no centers in Γ_∞, since those are not accumulated upon by limit cycles. On the other hand, if Γ_∞ contains a saddle, it must also contain two separatrices.

Suppose that Γ_∞ is not a closed orbit of Y and let $p \in \Gamma_\infty$ be a regular point. We claim $\omega(p)$ (*) is a saddle contained in Γ_∞. First observe that $\omega(p)$ does not contain a center of Y. If $\omega(p)$ were not a saddle of Y, $\omega(p)$ would contain a point q such that $Y(q) \neq 0$, since $\omega(p)$ is connected.

Observe that since Γ_∞ is not a closed orbit of Y then $q \notin \mathcal{O}(p) = \{ Y_t(p) \mid t \in \mathbb{R}\}$. The argument now is analogous to that of the Poincaré-Bendixson theorem. Let δ be a small segment transverse to Y with $q \in \delta$. Then $\mathcal{O}(p)$ meets δ in a monotonic sequence $p_n = Y_{t_n}(p)$ where $\lim_{n \to \infty} t_n = \infty$ and $\lim_{n \to \infty} p_n = q$ (see figure 4.).

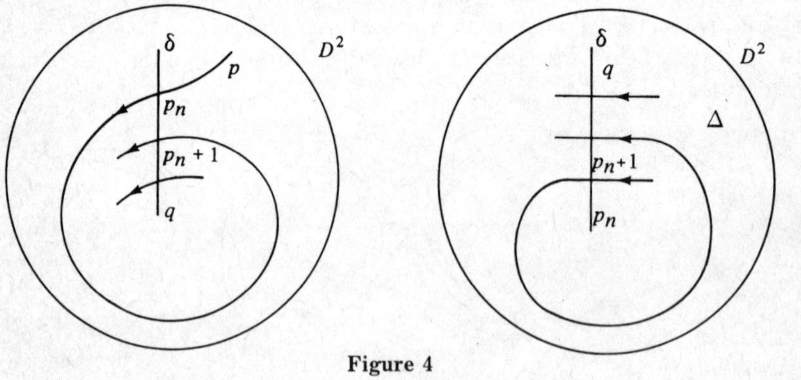

Figure 4

(*) $\omega(p) = \omega(\gamma)$ and $\alpha(p) = \alpha(\gamma)$ where γ is the orbit of Y such that $\gamma(0) = p$.

Let η be a closed curve of D^2 made up of the union of segments $[p_n, p_{n+1}] \subset \delta$ and $\mu = \{ Y_t(p) \mid t_n \leq t \leq t_{n+1} \} \subset \mathcal{O}(p)$ and let Δ be the connected component of $D^2 - \eta$ which contains q. An orbit which cuts the segment $[p_n, p_{n+1}]$ cannot be closed, since it enters the region Δ and never again leaves, so p_n cannot be an accumulation point of closed orbits of Y, a contradiction since $p_n \in \Gamma_\infty$. Hence $\omega(p)$ is a saddle. Analogously $\alpha(p)$ is a saddle.

Observe that since a saddle in Γ_∞ is an accumulation point of closed orbits of Y, then it has at least one stable and one unstable separatrix contained in Γ_∞ (see figure 5). Hence Γ_∞ is a graph.

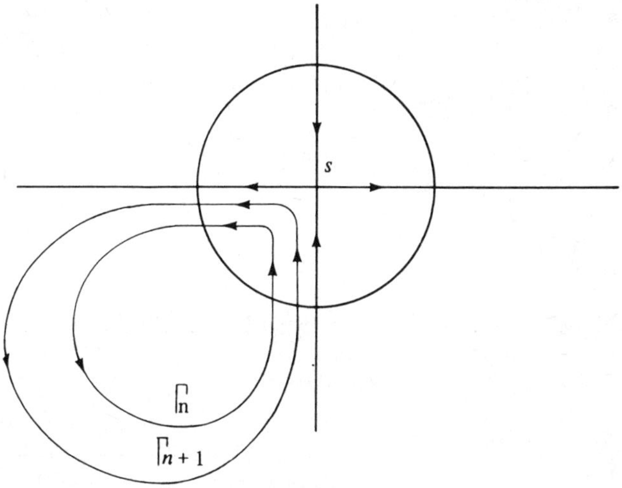

Figure 5

§5. The proof of Haefliger's theorem

Let $\mathfrak{F}$ be a C^2 codimension one foliation on M and suppose $\gamma : S^1 \longrightarrow M$ is a closed curve transverse to $\mathfrak{F}$, homotopic to a point. By the corollary to Proposition 1 there is a map C^∞ $g : D^2 \longrightarrow M$ such that $g^*(\mathfrak{F}) = \mathfrak{F}^*$ is a foliation with singularities on D^2. By the corollary to Proposition 2 there is a C^1 vector field Y, tangent to $\mathfrak{F}^*$, whose singularities are the same as those of $\mathfrak{F}^*$. This vector field Y is transverse to ∂D^2 and on ∂D^2 points inward toward D^2, its singularities are centers or saddles (finite in number) and there do not exist connections between distinct saddles.

Let $\overline{\Sigma}$ be the set of limit cycles of Y, to which we add the closed orbits and graphs obtained as in Proposition 3 of §4. Observe that $\overline{\Sigma}$ is nonempty since if $p \in \partial D^2$, $\omega(p)$ is a limit cycle. Moreover by Proposition 3, $\overline{\Sigma}$ is such that if $(\Gamma_n)_{n \in \mathbb{N}}$ is a decreasing sequence in $\overline{\Sigma}$, then there exists Γ_∞ in $\overline{\Sigma}$ such that $\Gamma_\infty \leq \Gamma_n$ for every $n \in \mathbb{N}$. The order relation on $\overline{\Sigma}$ is analogous to that already considered on Σ. By Zorn's lemma $\overline{\Sigma}$ has a minimal element Γ. Such a Γ bounds a region $R(\Gamma)$ whose interior contains no limit cycles, that is, for every $p \in R(\Gamma)$, $\mathcal{O}(p)$ is a singularity, a periodic orbit or $\mathcal{O}(p)$ is contained in a graph.

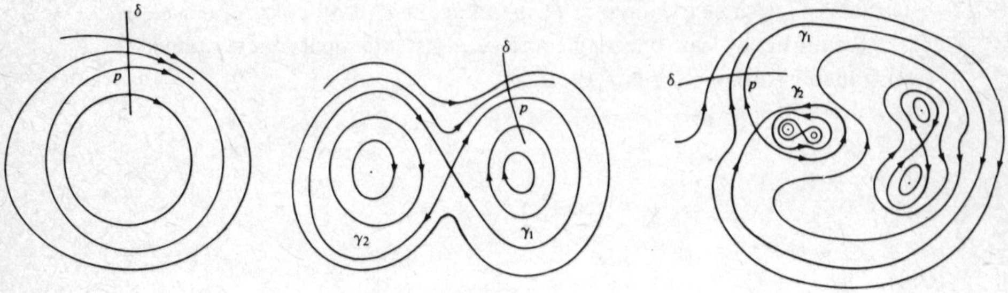

Figure 6

Consider now a segment δ, transverse to Y, cutting Γ in a point p such that $Y(p) \neq 0$. Suppose first that Γ is a closed orbit of Y. In this case it is possible to define a holonomy transformation $f: \delta_1 \longrightarrow \delta$. This transformation is the identity on $\delta_1 \cap R(T)$ but not the identity on any neighborhood of p in δ, since if it were, Γ would not be in $\overline{\Sigma}$. Let $\mu = g(\Gamma)$ and $\eta = g(\delta)$. Then μ is a closed curve in F, the leaf of $\mathfrak{F}$ which contains $g(p)$, whose holonomy h is given by $h(x) = g \circ f \circ (g/\delta_1)^{-1}(x)$. It is clear that h is the identity on one of the components of $\eta - g(p)$, but not the identity on any neighborhood of $g(p)$ in η.

Suppose now that Γ is a graph of Y. In this case the transformation f can only be defined on an open segment $\delta_1 \subset \delta - (\delta \cap R(\Gamma))$. We can extend f continuously to p so that $f(p) = p$. On $\delta \cap \operatorname{int}(R(\Gamma))$, although the orbits of Y are closed, the "holonomy transformation" in this component does not correspond to the curve Γ (see figure 6). Nevertheless the segment $g(\delta)$ is transverse to $\mathfrak{F}$ and the element h of the holonomy of $\mathfrak{F}$ corresponding to $g(\Gamma) \subset F$ is unilaterally conjugate to f, that is, if $x \in g(\delta_1)$ we have $h(x) = g \circ f \circ (g/\delta_1)^{-1}(x)$ and hence h is not the identity on any neighborhood of $g(p)$ in $g(\delta)$. It is possible to prove however that h is the identity on $g(\delta_2)$, where $\delta_2 = \delta \cap R(\Gamma)$. Indeed, let s be the saddle of Y which is contained in Γ. Any regular orbit $\gamma \subset \Gamma$ is such that $\omega(\gamma) = \alpha(\gamma) = \{s\}$, so Γ contains one or two regular orbits (see in figure 3 the last two pictures sketched). In the case that Γ contains only one regular orbit, the other two local separatrices of Y must be contained in $R(\Gamma)$, since a graph Γ as in the next figure cannot be a minimal element of $\overline{\Sigma}$.

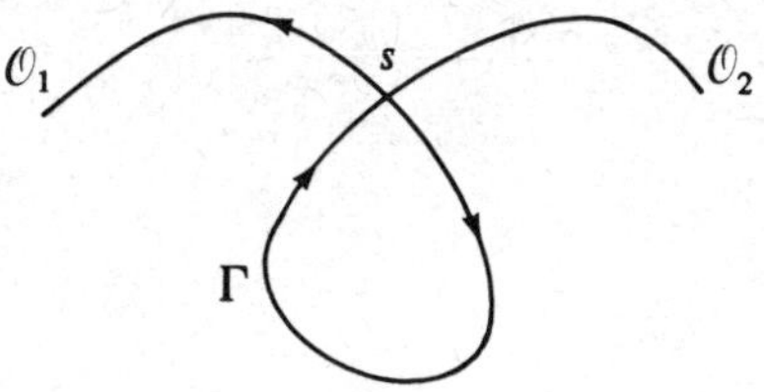

Figure 7

In figure 7, $\mathcal{O}_1$ and $\mathcal{O}_2$ are not in Γ.

We leave to the reader the verification of the above-mentioned fact. Note that this implies that the two local separatrices of s which are not in Γ must be connected, since otherwise there would be a limit cycle contained in the interior of $R(\Gamma)$.

Therefore the separatrices of s are pairwise connected, forming closed curves $\gamma_1 \cup \{s\}$ and $\gamma_2 \cup \{s\}$ as in figure 6, where $\Gamma = \gamma_1 \cup \gamma_2 \cup \{s\}$ or $\Gamma = \gamma_1 \cup \{s\}$ and $\delta \cap \gamma_1 \neq \varnothing$.

Consider for example the first case ($\Gamma = \gamma_1 \cup \gamma_2 \cup \{s\}$). Take a segment δ' transverse to Y such that $\delta' \cap \gamma_2 \neq \varnothing$. Let $\delta_2 = R(\Gamma) \cap \delta$, $\delta_2' = R(\Gamma) \cap \delta'$ and $\delta_1' = \delta' - \delta_2'$.

We can then define unilateral holonomy maps $f_1 : \delta_2 \to \delta_2$ and $f_1' : \delta_2' \to \delta_2'$. We have $f_1(x) = x$ if $x \in \delta_2$ and $f_1'(x) = x$ if $x \in \delta_2'$, since the orbits of Y in the interior of $R(\Gamma)$ are closed. Let $\mu_1 = g(\gamma_1 \cup \{s\})$ and $\mu_2 = g(\gamma_2 \cup \{s\})$ and h_1, h_1' be the holonomy maps of μ_1 and μ_2, which are defined on open segments of $g(\delta)$ and $g(\delta')$.

The restrictions $h_1 \mid g(\delta_2)$ and $h_1' \mid g(\delta_2')$ are conjugate to f_1 and f_1' so $h_1(x) = x$ if $x \in g(\delta_2)$ and $h_1'(x) = x$ if $x \in g(\delta_2')$. Since the holonomy map h of $g(\Gamma)$ is the composition of those of μ_1 and μ_2, one easily sees that $h \mid g(\delta_2) =$ identity.

The case that $\Gamma = \delta_1 \cup \{s\}$ is analogous to the above and we leave it to the reader. This ends the proof of Theorem 1.

Theorem 2. *Let M be an analytic compact manifold with finite fundamental group. Then there exists no analytic foliation of codimension one on M.*

Proof. By Lemma 7 of Chapter IV there is a closed curve $\tilde{\gamma} : S^1 \to M$ transverse to $\mathfrak{F}$. Since M has finite fundamental group, $[\tilde{\gamma}]^n = 1$ for some n, so the curve $\gamma(e^{it}) = \tilde{\gamma}(e^{int})$ is homotopic to a constant on M and is transverse to $\mathfrak{F}$. It now suffices to apply Theorem 1. ■

In particular, there are no analytic foliations of codimension one on S^3.

VII. NOVIKOV'S THEOREM

The following theorem, due to Novikov [40], is one of the deepest, most beautiful theorems in foliations.

Theorem. *Every C^2 codimension one foliation of a compact three-dimensional manifold with finite fundamental group has a compact leaf.*

In particular every C^2 codimension one foliation of S^3 has a compact leaf homeomorphic to T^2. This theorem has an elaborate proof; for this reason we will first give an informal sketch of the more important steps. Later we will see a rigorous proof.

§1. Sketch of the proof

The first step consists of showing the existence of vanishing cycles. A vanishing cycle is a map $f_0 : S^1 \longrightarrow M$ contained in a differentiable family $f_t : S^1 \longrightarrow M$, $t \in [0,\epsilon]$, $\epsilon > 0$, such that: (1) for every t, $f_t(S^1)$ is contained in a leaf A_t, (2) $f_t : S^1 \longrightarrow A_t$ is homotopic to a point (in A_t) if and only if $t \neq 0$ and (3) for every $x \in S^1$ the curve $t \mapsto f_t(x)$ is transverse to $\mathfrak{F}$.

Example 1. In the Reeb foliation of the solid torus $D^2 \times S^1$, any meridian of the boundary is a vanishing cycle.

The existence of a vanishing cycle follows, as one would expect, from the techniques of Haefliger, described in Chapter VI. The vanishing cycle is then modified to get a simple cycle in the following way: if $\pi_{A_t} : \widehat{A}_t \longrightarrow A_t$ denotes the universal covering of A_t, then the lifts $\hat{f}_t : S^1 \to \hat{A}_t$ of f_t are imbeddings for every $t \neq 0$.

The rest of the proof of Novikov's theorem consists of proving that the existence of a simple vanishing cycle implies the compactness of A_0. Having a simple vanishing cycle $(f_t)_{t \in I}$, it is possible to prove that there is a family of immersions $F_t : D^2 \to M$ such that $F_t \mid S^1 = f_t$ and $F_t(D^2) \subset A_t$ for every $t \neq 0$ and such that for every $x \in D^2$, the curve $t \mapsto F_t(x)$ is normal to $\mathfrak{F}$ (Figure 1).

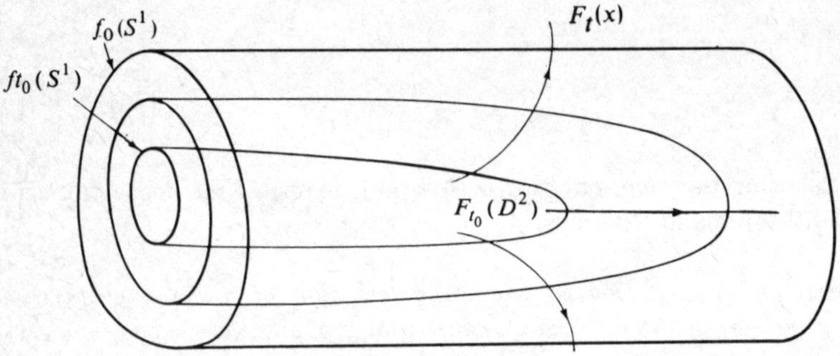

Figure 1

Next one shows that for any $\alpha > 0$ there exist t_1, t_2, $0 < t_2 < t_1 < \alpha$, such that (1) $A_{t_1} = A_{t_2}$, (2) $F_{t_1}(D^2) \subset F_{t_2}(D^2)$ and (3) for every t such that $0 < t_2 < t < t_1$ one has $F_t(D^2) \cap F_{t_1}(D^2) = \varnothing$. (Figure 2). Hence it follows that arbitrarily near A_0 there is a region R bounded by the surfaces

$$C_1 = \bigcup_{\substack{x \in S^1 \\ t_2 < t < t_1}} F_t(x) \quad \text{and} \quad C_2 = F_{t_2}(D^2) - F_{t_1}(D^2)$$

such that $A_0 \cap R = \varnothing$. This region has the following property: any simple curve transverse to $\mathfrak{F}$ that enters R is trapped to wander indefinitely without being able to leave R. Since R is arbitrarily near A_0 it follows that through the leaf A_0 there passes no closed curve transverse to $\mathfrak{F}$. On the other hand, one easily shows that if A is noncompact then there exists a closed curve transverse to $\mathfrak{F}$ passing through it. So A_0 must be compact.

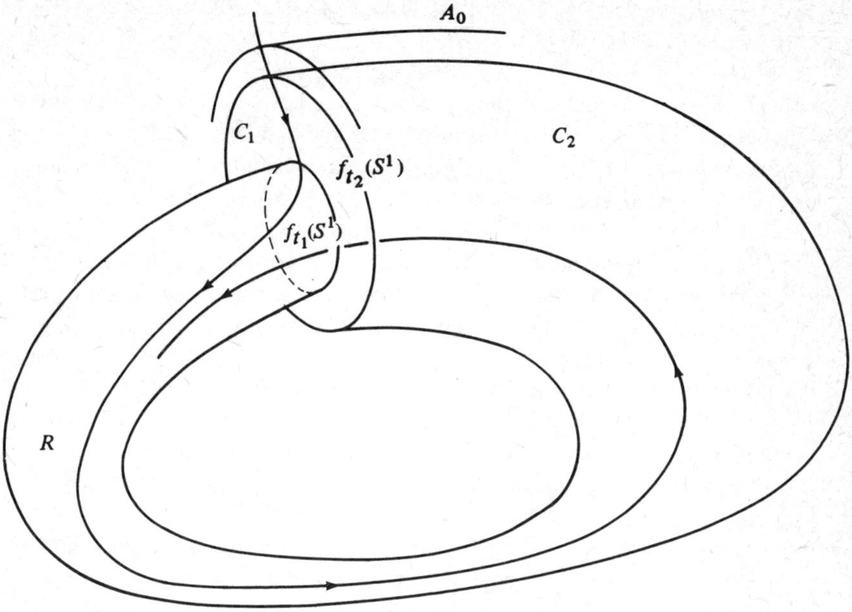

Figure 2

§2. Vanishing cycles

Consider a codimension one foliation $\mathfrak{F}$ on a manifold M.

Definition 1. Let A_0 *be a leaf of* $\mathfrak{F}$. *We say that a closed curve* $f_0 : S^1 \longrightarrow A_0$ *is a* vanishing cycle *if* f_0 *extends to a differentiable map* $F : [0,\epsilon] \times S^1 \longrightarrow M$ *satisfying the following properties:*

(a) *for every* $t \in [0,\epsilon]$, *the curve* $f_t : S^1 \longrightarrow M$, *defined by* $f_t(x) = F(t,x)$ *is contained in a leaf* A_t *of* $\mathfrak{F}$,
(b) *for every* $x \in S^1$ *the curve* $t \mapsto F(t,x)$ *is transverse to* $\mathfrak{F}$,
(c) *for* $t > 0$, *the curve* f_t *is homotopic to a constant in the leaf* A_t *and* f_0 *is* not *homotopic to a constant in* A_0.

We will call a map F *as above a coherent extension* of the vanishing cycle f_0.
In this section we will examine diverse properties of vanishing cycles. We begin with an existence theorem:

Proposition 1. *Let M be a compact manifold of dimension $n \geq 3$ with finite fundamental group and $\mathfrak{F}$ a C^2 codimension one foliation of M. Then $\mathfrak{F}$ has a vanishing cycle.*

Proof. Since M is compact and $\pi_1(M)$ is finite by Lemma 7 of Chapter IV, there is a curve $\alpha : S^1 \longrightarrow M$ transverse to $\mathfrak{F}$ and homotopic to a point. Under these circumstances, by Haefliger's construction (Chapter IV), there is a C^∞ map $g : D^2 \longrightarrow M$ such that the foliation $\mathfrak{F}^* = g^*\mathfrak{F}$ of D^2 is given by the orbits of a vector field Y on D^2 whose singularities are centers or saddles and such that the only saddle connections permitted are self-connections.

By what was seen in §5 of Chapter VI, there exists an open region R of trivial holonomy in Y such that the boundary ∂R is a curve with nontrivial holonomy (see figure 6 in Chapter VI).

The proof that $\mathfrak{F}$ has a vanishing cycle consists of finding a curve $\Gamma \subset R$ which is a periodic orbit or a graph invariant under Y such that (1) $g(\Gamma)$ is not homotopic to a constant in the leaf $A_{g(\Gamma)}$ and (2) Γ is the boundary of a band $C \subset R$, a union of periodic orbits of Y, $C = \cup_{t\in(0,1]} \gamma_t$, where $g(\gamma_t)$ is homotopic to a constant in the leaf $A_{g(\gamma_t)}$.

Since the tangency of g with $\mathfrak{F}$ at the centers is of parabolic type, all the regular trajectories γ of Y, near a center, are closed and have image $g(\gamma)$ homotopic to a point in $A_{g(\gamma)}$ (see Figure 1 of Chapter VI). It is natural then to look for Γ starting from the centers of Y in R. Observe that if the number of centers of Y is $\ell \geq 1$, then the number of saddles is $\ell - 1$. This is a consequence of the index theorem (see [35]). The same is true for the closure $\overline{R}$ of R, that is, if $c_1, \ldots, c_k$ ($k \geq 1$) are the centers of Y in R, then Y has $k - 1$ saddles in $\overline{R}$, say $s_1, \ldots, s_{k-1}$. For $i = 1, \ldots, k$ let V_i be the maximal connected subset of R that contains c_i and is a union of c_i and closed orbits of Y homotopic to a constant in their respective leaves.

The set V_i is, by construction, non empty and invariant under Y. Moreover it is open and $\partial V_i = \overline{V_i} - V_i$ is a closed connected curve in D^2 which is a union of orbits of Y.

In fact, let $\gamma_0 \subset V_i$. Since all the orbits of Y in R have trivial holonomy, it follows from Theorem 3 of Chapter IV, that there exist a cylindrical neighborhood C_0 of γ_0, $C_0 \subset R$, and a diffeomorphism $h : (-1,1) \times \gamma_0 \longrightarrow C_0$, such that for any $t \in (-1,1)$, $\gamma_t = h(\{t\} \times \gamma_0)$ is a closed orbit of Y. Denote by A_t the leaf of $\mathfrak{F}$ which contains $g(\gamma_t)$. Let us fix some Riemannian metric $\langle\ ,\ \rangle$ in M. Since $g(\gamma_0)$ is homotopic to a constant in A_0, there exists a differentiable map $f_0 : D^2 \longrightarrow A_0$ such that $f_0(\partial D^2) = g(\gamma_0)$. It follows from Lemma 4 in Chapter IV that f_0 extends to a continuous family of maps $f_t : D^2 \longrightarrow M$, $t \in (-\epsilon, \epsilon)$, such that $f_t(D^2) \subset A_t$ and for each $x \in D^2$ the curve $t \mapsto f_t(x)$ is normal to $\mathfrak{F}$. The curve $f_t(\partial D^2) = \delta_t$ is clearly homotopic to a constant in A_t. Moreover, for t small, the curves δ_t and $g(\gamma_t)$ are homotopic in A_t, because $\lim_{t\to 0} \delta_t = \lim_{t \to 0} g(\gamma_t) = g(\gamma_0)$. Therefore $g(\gamma_t)$ is homotopic to a constant in A_t, which implies that V_i is open.

Since V_i is invariant under Y, it follows that ∂V_i is either a closed orbit or a graph of Y. If ∂V_i is a closed orbit γ_0 of Y then, $g(\gamma_0)$ is not homotopic to a constant in $A_{g(\gamma_0)}$. Since $\gamma_0 \subset \overline{R}$, it follows that in the interior of γ all the orbits of Y are closed, i.e., γ_0 belongs to a family of orbits of Y, $\{\gamma_t\}_{t\in[0,\epsilon)}$, such that for $t > 0$ $g(\gamma_t)$ is homotopic to a constant in A_t, and we are done. In the other case, $\partial V_i = \Gamma_i$ is a graph of Y and we have two possibilities. The first is that there is $1 \le i \le k$ such that $g(\Gamma_i)$ is not homotopic to a constant in $A_{g(\Gamma_i)}$. In this case $g(\Gamma_i)$ is a vanishing cycle in $A_{g(\Gamma_i)}$. The second possibility is that for all $i = 1, \dots, k$, $g(\Gamma_i)$ is homotopic to a constant in $A_{g(\Gamma_i)}$. Since the difference between the number of centers and saddles in $\overline{R}$ is one, it follows that two of these graphs, let us say Γ_1 and Γ_2, contain the same saddle s_1 as in the following figures.

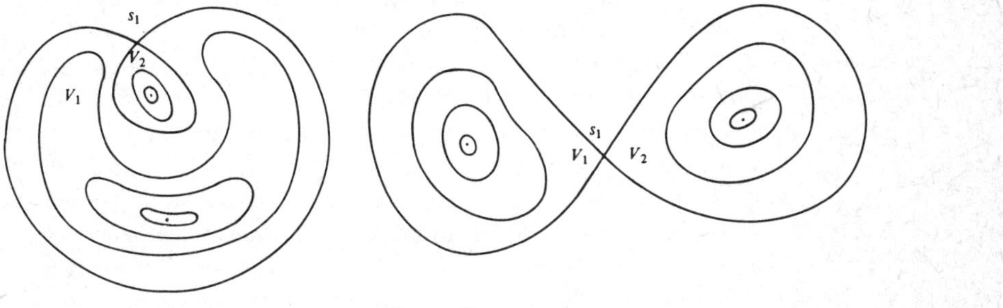

Figure 3

In this case, $\overline{V}_1 \cup \overline{V}_2$ is homeomorphic to a closed disk or to two closed disks with one point in common. Consider the graph $\Gamma^1 = \partial \overline{V}_1 \cup \partial \overline{V}_2 = \Gamma_1 \cup \Gamma_2$. Since $g(\Gamma_1)$ and $g(\Gamma_2)$ are homotopic to constants in $A_{g(\Gamma_1)} = A_{g(\Gamma_2)} = A_{g(\Gamma^1)}$, it follows that $g(\Gamma^1)$ is homotopic to a constant in $A_{g(\Gamma^1)}$.

So there is a neighborhood U of Γ^1 such that $U - \Gamma^1$ consists of closed orbits of Y, homotopic to constants in their respective leaves. This can be verified using a simple extension of the proof of Lemma 4 (Chapter IV).

Let $\Gamma^1 = \Gamma^1_1, \dots, \Gamma^1_{k_1}$, where $k_1 \le k/2$, be all the graphs of Y obtained in the same way as Γ^1. For each $i = 1, \dots, k_1$, let V^1_i be the maximal connected set which contains Γ^1_i and such that $V^1_i - \Gamma^1_i$ consists of centers and of closed orbits of Y homotopic to constants in their respective leaves. By the arguments used above V^1_i is open. Its boundary ∂V^1_i is a closed curve in R, union of orbits of Y. As before, if there exists $1 \le i \le k_1$ such that $g(\partial V^1_i)$ is not homotopic to a constant in $A_{g(\partial V^1_i)}$, then $g(\partial V^1_i)$ is a vanishing cycle in $A_{g(\partial V^1_i)}$. There remains the possibility that for every $i = 1, \dots, k_1$, $g(\partial V^1_i)$ is homotopic to a constant in $A_{g(\partial V^1_i)}$. In this case ∂V^1_i is a graph for every $i = 1, \dots, k_1$. Again at least two of the ∂V^1_i, let us say ∂V^1_1 and ∂V^1_2, contain the same saddle.

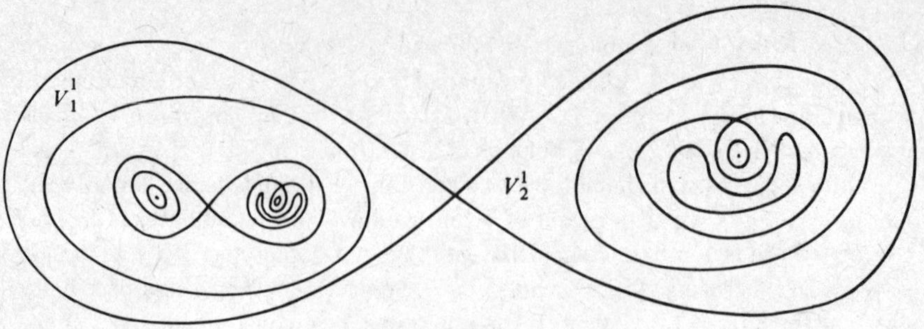

Figure 4

Let $\Gamma^2 = \partial V_1^1 \cup \partial V_2^1$. Clearly Γ^2 is a graph of Y and $g(\Gamma^2)$ is homotopic to a constant in $A_{g(\Gamma^2)}$. If k_2 is the number of graphs as above, we evidently have $1 \leq k_2 \leq (1/2)k_1$. Moreover Γ^2 has a neighborhood U such that $U - \Gamma^2$ consists of closed orbits of Y homotopic to constants in their respective leaves. Since $\partial R \cap \partial D^2 = \emptyset$ and there are only a finite number of singularities of Y in R, proceeding inductively with the above argument we get, after a finite number of steps, a closed orbit or a graph Γ of Y such that $g(\Gamma)$ is a vanishing cycle in $A_{g(\Gamma)}$. ■

Suppose now that $\mathfrak{F}$ is a transversely orientable foliation of a manifold M and X a vector field normal to $\mathfrak{F}$ with respect to a Riemannian metric $\langle\ ,\ \rangle$ on M. We say that a curve transverse to $\mathfrak{F}$, $\gamma : I \longrightarrow M$, is positive (respectively, negative) if, for every $t \in I$, $\langle \gamma'(t), X(\gamma(t))\rangle > 0$ (resp. < 0).

Let $f_0 : S^1 \longrightarrow A_0$ be a vanishing cycle and $F : [0,\epsilon] \times S^1 \longrightarrow M$ a coherent extension of f_0. We say that f_0 is a positive (resp. negative) vanishing cycle if the curves $t \longrightarrow F(t,x)$ are transverse positive (resp. negative). If further, the curves $t \longrightarrow F(t,x)$ are orthogonal to the leaves of $\mathfrak{F}$, we say F is a normal extension of f_0.

Proposition 2. *Let $f_0 : S^1 \longrightarrow A_0$ be a positive vanishing cycle. Then*

(i) *there is a normal positive coherent extension of f_0,*
(ii) *if $\overline{f}_0 : S^1 \longrightarrow A_0$ is homotopic to f_0 in A_0, then $\overline{f}_0$ is a positive vanishing cycle.*

Proof. (i). Denote by X_t the flow generated by the normal field X. The map $\tilde{F}(t,x) = X_t(f_0(x))$ is well-defined on $[0,\delta] \times S^1$ for $\delta > 0$ sufficiently small. Let us reparametrize now the curves $t \mapsto \tilde{F}(t,x)$ by $\alpha \mapsto \tilde{F}(t(\alpha,x),x)$ in such a way that the new map $\overline{F}(\alpha,x) = \tilde{F}(t(\alpha,x),x)$ is a normal positive coherent extension of f_0. Since f_0 is a positive vanishing cycle, the positive holonomy of f_0 is the identity. This implies that the integral curves of the foliation $\tilde{F}^*(\mathfrak{F})$ of $[0,\delta] \times S^1$ near $0 \times S^1$ are all simple closed curves. Fix $x_0 \in S^1$ and let S_t denote the integral curve of $\tilde{F}^*(\mathfrak{F})$ passing through $(t,x_0) \in [0,\delta] \times S^1$. Thus it follows that, for $\alpha \geq 0$ small, S_α meets the curve $t \mapsto (t,x)$ in only one point $(t(\alpha,x),x)$.

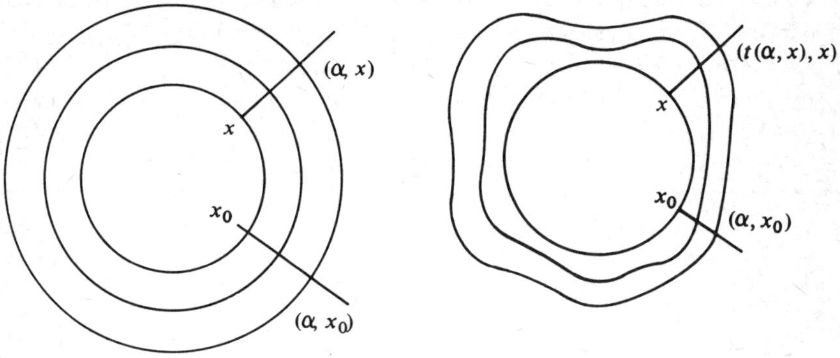

Figure 5

By construction, the map $\overline{F} : [0,\tilde{\epsilon}] \times S^1 \longrightarrow M$ given by $\overline{F}(\alpha,x) = \tilde{F}(t(\alpha,x),x)$ satisfies conditions (a) and (b) of the definition of coherent extension. We show now that by reducing more the domain of $\overline{F}$ we can guarantee that the curves $\overline{f}_\alpha(x) = \overline{F}(\alpha,x)$ are homotopic to zero for $\alpha > 0$.

Indeed, let $F : [0,\overline{\epsilon}] \times S^1 \longrightarrow M$, $f_t(x) = F(t,x)$ be a positive coherent extension for f_0. For simplicity suppose $\overline{\epsilon} \leq \tilde{\epsilon}$ and that, for $t \leq \overline{\epsilon}$, $f_t(S^1)$ and $\overline{f}_t(S^1)$ are in the same leaf. It is clear that for t small the curves f_t and $\overline{f}_t$ are close in the intrinsic topology of A_t and therefore are homotopic. Consequently, restricting $\overline{F}$ to a domain of the form $[0,\epsilon] \times S^1$, $\epsilon \leq \overline{\epsilon}$, we have that $\overline{F}$ is a normal, positive coherent extension of f_0.

(ii) Observe first that there exists $\epsilon > 0$ such that if $\overline{f}_0 : S^1 \longrightarrow A_0$ and $d_0(f_0,\overline{f}_0) < \epsilon$ then $\overline{f}_0$ is a vanishing cycle, where d_0 is the intrinsic metric on A_0. The proof of this fact can be done considering positive normal extensions $F : [0,\delta] \times S^1 \longrightarrow M$ and $\overline{F} : [0,\overline{\delta}] \times S^1 \longrightarrow M$ of f_0 and $\overline{f}_0$ respectively, where F is coherent. The proof that the restriction of $\overline{F}$ to an interval of the form $[0,\epsilon] \times S^1$ $(0 < \epsilon \leq \overline{\delta})$ is coherent reduces, as in (i), to proving that the curves f_t and $\overline{f}_t$ are close for small t, in the intrinsic metric of A_t, and therefore are homotopic.

Suppose now that f_0 and $\overline{f}_0$ are homotopic in A_0 and that f_0 is a vanishing cycle. Let $\varphi : [0,1] \times S^1 \longrightarrow A_0$ be a homotopy connecting f_0 and $\overline{f}_0$. By the preceeding observation and by compactness of the interval $[0,1]$, there exists a positive integer n_0 such that if $\varphi_{k/n_0}(x) = \varphi_k(x)$ is a positive vanishing cycle, then $\varphi_{k+1}(x)$ is a positive vanishing cycle. By transitivity, we conclude that $\varphi_{n_0}(x) = \overline{f}_0(x)$ is a positive vanishing cycle.

§3. Simple vanishing cycles

From now on we assume that M is a compact orientable three-dimensional manifold and $\mathfrak{F}$ a C^2 foliation of codimension one on M.

Let $f_0 : S^1 \longrightarrow A_0$ be a positive vanishing cycle and $F : [0,\epsilon] \times S^1 \longrightarrow M$ a positive, normal coherent extension of f_0. By transversality theory and by Proposition 2 we can assume, taking an approximation of f_0, that (i) f_0 is an immersion, (ii) the points of self-intersection of f_0 are double points and transverse, that is, the sets of the form $f_0^{-1}(p)$ contains at most two points and if $p = f_0(x_1) = f_0(x_2)$ with $x_1 \neq x_2$ in S^1 then the vectors $f_0'(x_1)$ and $f_0'(x_2)$ generate the tangent space to A_0 at p.

With this consideration, it is clear that the set $B_0 = \{p \in f_0(S^1) \mid f_0^{-1}(p)$ contains two elements$\}$ is finite. Let $B_t = \{p \in f_t(S^1) \mid f_t^{-1}(p)$ contains two elements$\}$ and $n_t = \#B_t =$ cardinality of B_t if $t \geq 0$.

Since $f_0(S^1)$ is compact and the curves $t \mapsto F(t,x)$ are reparametrizations of trajectories of the normal vector field, we can take $\epsilon > 0$ sufficiently small so that if $x_1,x_2 \in S^1$, the curves $\gamma_j = F([0,\epsilon],x_j), j = 1,2$ either do not meet or one is contained in the other. It follows therefore that $n_t \leq n_0$ for $0 \leq t \leq \epsilon$.

Given a leaf A of $\mathfrak{F}$ denote by $\hat{A}$ its universal covering and by $\pi_A : \hat{A} \longrightarrow A$ the covering projection. Since f_t, $t > 0$, is homotopic to a constant in A_t, we can lift f_t to a curve $\hat{f}_t : S^1 \longrightarrow \hat{A}_t$ such that $\pi_{A_t} \circ \hat{f}_t = f_t$ (see [29]). Our lifts always neglect base points, implicitly assuming that they are fixed.

Definition. We say F is a simple coherent extension of f_0 if the $\hat{f}_t$ are simple curves in $\hat{A}_t$ for $t > 0$.

Proposition 3. *Let $f_0 : S^1 \longrightarrow A_0$ be a positive vanishing cycle and $F : [0,\epsilon] \times S^1 \longrightarrow M$ a positive normal coherent extension of f_0. Let $f_t(x) = F(t,x)$, $x \in S^1$ and A_t be the leaf of $\mathfrak{F}$ that contains $f_t(S^1)$. Then there exists a sequence $t_n \longrightarrow 0$ such that for every $n \geq 1$, the leaf A_{t_n} contains a vanishing cycle $g_n : S^1 \longrightarrow A_{t_n}$, which has a simple, positive, normal coherent extension (it is possible that, for every $n \geq 1$, we have $t_n = 0$).*

Proof. Let $\hat{f}_t : S^1 \longrightarrow \hat{A}_t$ be a lift of f_t and $B_t = \{p \in f_t(S^1) \mid p = f_t(x_1) = f_t(x_2)$ with $x_1 \neq x_2$ in $S^1\}$. As we have already seen, we can take ϵ small enough so that $n_t = \#B_t \leq \#B_0 = k$. For $t > 0$ set $\hat{B}_t = \{p \in \hat{f}_t(S^1) \mid p = \hat{f}_t(x_1) = \hat{f}_t(x_2)$ with $x_1 \neq x_2$ in $S^1\}$. Clearly $\pi_{A_t}(\hat{B}_t) \subset B_t$ and therefore $\hat{n}_t = \#\hat{B}_t \leq n_t \leq k$. Suppose $B_0 = \{p_1 ,..., p_k\}$ and let $x_i \neq x_i'$ in S^1 be such that $f_0(x_i) = f_0(x_i') = p_i$, $i = 1 ,..., k$. By uniqueness of solutions of the normal vector field, if $p \in B_t$, then $p = f_t(x_i) = f_t(x_i')$ for some $i = 1 ,..., k$.

Let $K_1 = \{t \in [0,\epsilon] \mid f_t(x_1) = f_t(x_1')\}$ and $U_1 = \{t \in (0,\epsilon) \mid \hat{f}_t(x_1) = \hat{f}_t(x_1')\}$. Clearly K_1 is closed and $K_1 \supset U_1$. We will show that U_1 is open.

Let $S^1 - \{x_1,x_1'\} = \alpha \cup \beta$, where α and β are open segments and $\alpha \cap \beta = \emptyset$. If $t \in U_1$ we have that $\hat{f}_t(x_1) = \hat{f}_t(x_1')$ and therefore the restriction of f_t to the closure $\overline{\alpha}$ of α is a closed curve in A_t homotopic to a constant in A_t. By Lemma 4 of Chapter IV there exists a $\delta > 0$ such that for $s \in (t - \delta, t + \delta)$, $f_s \mid \overline{\alpha}$ is closed and homotopic to a constant in A_s, so $\hat{f}_s \mid \overline{\alpha}$ is a closed curve in $\hat{A}_s$, which implies that $s \in U_1$ and therefore U_1 is open. We have three possibilities:

(1) *There exists* $\delta > 0$ *such that* $[0,\delta] \cap U_1 = \emptyset$. In this case, restricting F to $[0,\delta] \times S^1$ we obtain $\hat{n}_t < n_0 = k$. We continue the argument with $U_2 = \{t \in (0,\delta] \mid \hat{f}_t(x_2) = \hat{f}_t(x_2')\}$.

(2) *There exists* $\delta > 0$ *such that* $(0,\delta] \subset U_1$. In this case $\hat{f}_t(x_1) = \hat{f}_t(x_1')$ for $t \in (0,\delta]$ so $f_t(x_1) = f_t(x_1')$ for $t \in [0,\delta]$. We can consider $g_0 = f_0 \mid \overline{\alpha}$ and $\tilde{g}_0 = f_0 \mid \overline{\beta}$ as maps of S^1 on A_0. Since f_0 is not homotopic to a constant in A_0, we can suppose, for example, that g_0 is not homotopic to a constant in A_0. Take $G = F \mid [0,\delta] \times \overline{\alpha}$ and $g_t(x) = G(t,x)$. Since $\hat{f}_t(x_1) = \hat{f}_t(x_1')$ for $t \in (0,\delta]$, g_t is homotopic to a constant in A_t for $t \in (0,\delta]$. If $\hat{n}_t^1 = \#\{p \mid \hat{g}_t(x) = \hat{g}_t(x') = p,\ x \neq x' \text{ in } \overline{\alpha}\}$, we have $\hat{n}_t^1 < \hat{n}_t \leq k$.

(3) *There exists a sequence* $\tau_m^1 \longrightarrow 0$ *such that* $\tau_m^1 \in \overline{U_1} - U_1 \subset K_1$. We can assume that for every $m \geq 1$, there exists $\epsilon_m > 0$ such that $(\tau_m^1, \tau_m^1 + \epsilon_m] \subset U_1$. As in the above case we have $f_t(x_1) = f_t(x_1')$ for $t \in [\tau_m^1, \tau_m^1 + \epsilon_m]$. Since $\tau_m^1 \in K_1 - U_1$, the restrictions $f_{\tau_m^1} \mid \overline{\alpha}$ and $f_{\tau_m^1} \mid \overline{\beta}$ are not homotopic to a constant in $A_{\tau_m^1}$ and since $(\tau_m^1, \tau_m^1 + \epsilon_m] \subset U_1$ for $\tau_m^1 < t \leq \tau_m^1 + \epsilon_m$, we have that $f_t \mid \overline{\alpha}$ and $f_t \mid \overline{\beta}$ are homotopic to constants in A_t. Therefore for every $m \geq 1$, $f_{\tau_m^1} \mid \overline{\alpha}$ is a positive vanishing cycle on $A_{\tau_m^1}$ and $F \mid [\tau_m^1, \tau_m^1 + \epsilon_m] \times \overline{\alpha}$ is a positive, normal, coherent extension of $f_{\tau_m^1} \mid \overline{\alpha}$. If $\hat{n}_t^m = \#\{p \in \hat{f}_t(\overline{\alpha}) \mid p = \hat{f}_t(x) = \hat{f}_t(x') \text{ with } \{x,x'\} \neq \{x_1,x_1'\} \text{ and } x \neq x' \text{ in } \overline{\alpha}\}$ then $\hat{n}_t^m < \hat{n}_t \leq k$.

In any case we obtained a sequence $\tau_m^1 \longrightarrow 0$ such that, for every $m \geq 1$, the leaf $A_{\tau_m^1}$ contains a vanishing cycle g_m, which has a positive, normal coherent extension $G_m^1 : [\tau_m^1, \tau_m^1 + \epsilon_m] \times S^1 \longrightarrow M$ such that the number of self-intersections of $\hat{g}_{m,t}^1$, a lift of $g_{m,t}^1 = G_m^1(t, -)$, is $\hat{n}_{m,t}^1 \leq \min\{k - 1, \hat{n}_t - 1\}$. In the first two cases $\tau_m^1 = 0$ for every $m \geq 1$. In the last case we identify the endpoints of $\overline{\alpha}$, obtaining S^1.

By an analogous argument we can obtain a sequence $\tau_{m,n}^2 \longrightarrow \tau_m^1$ such that the number of self-intersections of $\hat{g}_{m,n,t}^2$ is $n_{m,n,t}^2 \leq \min\{k - 2, \hat{n}_t - 2\}$. Extracting a subsequence $\tau_{m,n(m)}^2 = \tau_m^2 \longrightarrow 0$, we get vanishing cycles g_m^2 which have positive, normal coherent extensions $\hat{G}_m^2$ such that the number of self-intersections of $\hat{g}_{m,t}^2$ is $\hat{n}_{m,t}^2 \leq \min\{k - 2, n_t - 2\}$. Proceeding with this argument, at the end of at most $k - 2$ steps, we get a sequence $\tau_m \longrightarrow 0$ such that for every $m \geq 1$, the leaf A_{τ_m} contains a vanishing cycle satisfying the desired properties, remembering that at each stage we can always substitute for the vanishing cycle, which may not be differentiable at a point, a differentiable vanishing cycle. ■

§4. Existence of a compact leaf

The objective of this section is to prove the following theorem.

Theorem 2. *Let A_0 be a leaf of $\mathfrak{F}$ which admits a vanishing cycle with a simple, positive, normal coherent extension $F : [0,\epsilon] \times S^1 \longrightarrow M$. Then*

(a) *A_0 is a compact leaf,*
(b) *A_0 is the boundary of a submanifold $V \subset M$ such that $V = \bigcup_{t\in[0,\epsilon]} A_t$,*
(c) *for every $t \in (0,\epsilon]$, $\overline{A}_t - A_t = \lim A_t = A_0$,*
(d) *for every $t \in (0,\epsilon]$, A_t is diffeomorphic to $\mathbb{R}^2$.*

This theorem will be a consequence of the following propositions.

Proposition 4. *If A is a noncompact leaf of a foliation $\mathfrak{F}$ of codimension one of a compact manifold, then for every $p \in A$, there exists a closed curve γ, transverse to $\mathfrak{F}$, passing through p.*

Proof. Since M is compact and A is not compact, $\lim A = \overline{A} - A \neq \emptyset$. Consider a point $x_0 \in \lim A$ and let U be a foliation box of $\mathfrak{F}$, such that $x_0 \in U$. Let $\ell \subset U$ be a segment normal to $\mathfrak{F}$ such that $x_0 \in \ell$. Since $x_0 \in \lim A$, there is a sequence of distinct points $\{x_n\}_{n\in N}$ such that for every $n \in N$, $x_n \in \ell \cap A$. Consider the segment ℓ with an orientation as in the figure below.

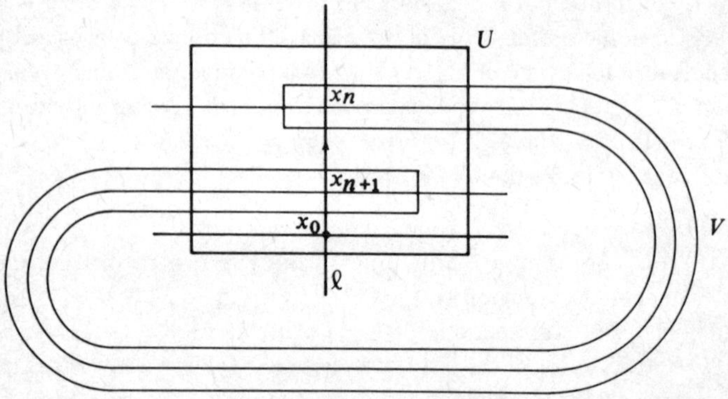

Figure 6

Let $\alpha : [0,1] \longrightarrow A$ be a differentiable curve without self-intersections, such that $\alpha(0) = x_n \neq \alpha(1) = x_{n+1}$. By the global trivialization lemma (Chapter IV), there exists a neighborhood V of $\alpha([0,1])$ such that $\mathfrak{F} \mid V$ is equivalent to the product foliation $D^{n-1} \times \{t\}$ on $D^{n-1} \times (-\rho,\rho)$. Suppose first that $\mathfrak{F}$ is transversely orientable. In this case we can assume that the orientation of ℓ is compatible with the normal orientation of $\mathfrak{F} \mid V$, so ℓ cuts V as in figure 7.

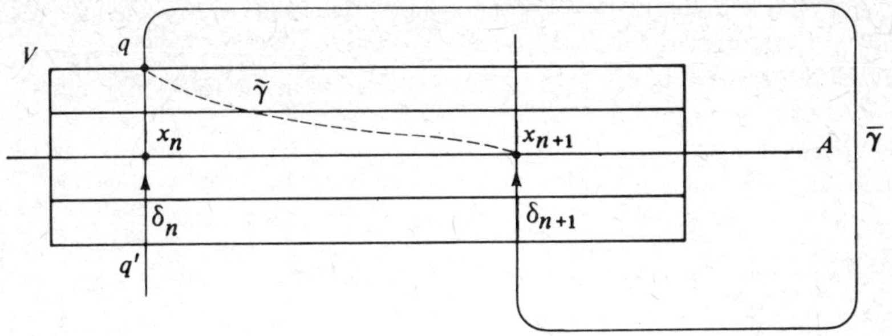

Figure 7

Let δ_n and δ_{n+1} be connected components of $\ell \cap V$ which contain x_n and x_{n+1} respectively and $\bar{\gamma} : [0,1] \longrightarrow \ell$ be the curve that begins at x_n and ends at x_{n+1}. It is clear that $\partial V \cap \delta_n$ contains two points, q and q' say, where we can assume $q > q'$ in ℓ (the orientation of ℓ induces a natural order). Since x_{n+1} is "below" q and $\mathfrak{F} \mid V$ is the horizontal foliation on V, there is a curve $\tilde{\gamma}$: $[0,1] \longrightarrow V$, transverse to $\mathfrak{F} \mid V$, compatible with the positive orientation of $\mathfrak{F} \mid V$ and such that $\tilde{\gamma}(0) = x_{n+1}$ and $\tilde{\gamma}(1) = q$. We can also assume that the tangent vectors of $\bar{\gamma}$ and $\tilde{\gamma}$ at x_{n+1} and q coincide. Hence the curve $\gamma : [0,1] \longrightarrow M$ defined by $\gamma(t) = \tilde{\gamma}(2t)$ if $t \in [0,1/2]$ and $\gamma(t) = \bar{\gamma}(2t - 1)$ if $t \in [1/2,1]$, is a C^1 curve transverse to $\mathfrak{F}$ and cuts A at x_{n+1}.

Let x be an arbitrary point of A. Let $\beta : [0,1] \longrightarrow A$ be a differentiable curve without self-intersections and such that $\beta(0) = x_{n+1}$ and $\beta(1) = x$. Let W be a neighborhood of $\beta([0,1])$ such that $\mathfrak{F} \mid W$ is equivalent to a horizontal foliation $D^{n-1} \times \{t\}$ on $D^{n-1} \times (-\epsilon,\epsilon)$. It is clear that $\gamma \cap W$ contains two points q and q' with $q > q'$, say. We can then define a curve $\tilde{\mu} : [0,1] \longrightarrow W$, transverse to $\mathfrak{F}$ and such that $\tilde{\mu}(0) = q'$, $\tilde{\mu}(1) = q$ and $\tilde{\mu}(1/2) = x$, assuming without loss of generality that the tangent vectors of γ and $\tilde{\mu}$ at q and q' coincide. Joining the curve $\tilde{\mu}$ with the segment of γ contained in the exterior of W, we get a closed curve μ, transverse to $\mathfrak{F}$ and which cuts A at x (see Figure 8).

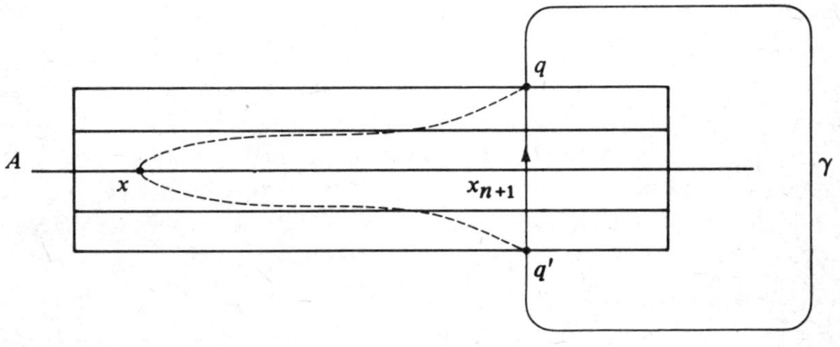

Figure 8

Suppose now that $\mathfrak{F}$ is not transversely orientable. In this case the argument is the same, except that we must consider a foliation box V that contains at least three points of the sequence $\{x_n\}_{n\in\mathbb{N}}$, as in the following figures.

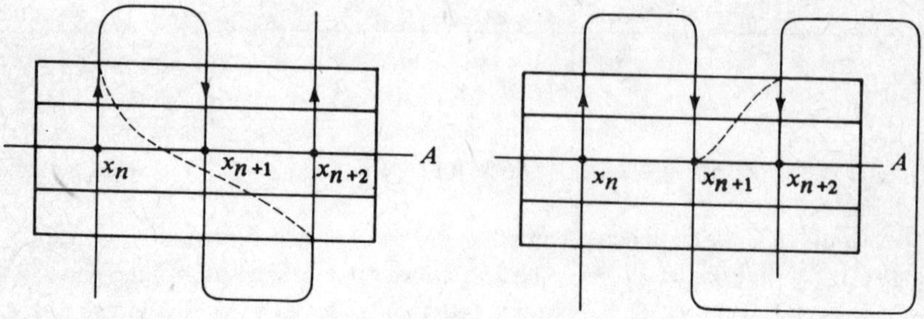

Figure 9

Proposition 5. *Let $D^2 \subset \mathbb{R}^2$ be the closed disk of radius 1. There is a differentiable*(*) *immersion $H : (0,\epsilon] \times D^2 \longrightarrow M$ satisfying the following properties:*

(a) *the curves $t \longrightarrow H(t,x)$, $x \in D^2$, are positive and normal,*
(b) *for every $t \in (0,\epsilon]$, $H(t \times D^2) \subset A_t$,*
(c) *the restriction of H to $(0,\epsilon] \times S^1$ coincides with $F \mid (0,\epsilon] \times S^1$.*

Proof. First of all, observe that for any leaf A of $\mathfrak{F}$, its universal covering $\hat{A}$ is homeomorphic to $\mathbb{R}^2$. Indeed, either $\hat{A} \simeq \mathbb{R}^2$ or $\hat{A} \simeq S^2$. If $\hat{A} \simeq S^2$ then either $A \simeq S^2$ or $A \simeq \mathbb{P}^2$. In this case, it follows from Theorem 4 of Chapter IV that $\mathfrak{F}$ has no vanishing cycle. Therefore $\hat{A} \simeq \mathbb{R}^2$.

In particular $\hat{A}_\epsilon \simeq \mathbb{R}^2$. Since F is a simple coherent extension of f_0, the curve $\hat{f}_\epsilon : S^1 \longrightarrow \hat{A}_\epsilon$ is simple and so by the Jordan curve theorem there exists $h_\epsilon : D^2 \longrightarrow A_\epsilon$ such that $\hat{h}_\epsilon \mid \partial D^2 \equiv \hat{f}_\epsilon$. Since $\hat{f}_\epsilon(S^1)$ is an embedded simple curve we can suppose that $\hat{h}_\epsilon$ is a diffeomorphism onto the disk $\hat{h}_\epsilon(D^2)$. It follows that $f_\epsilon = \pi_{A_\epsilon} \circ \hat{f}_\epsilon$ extends to an immersion $h_\epsilon = \pi_{A_\epsilon} \circ \hat{h}_\epsilon : D^2 \longrightarrow A_\epsilon$.

By Lemma 4 of Chapter IV, h_ϵ extends to an immersion $H : (\epsilon_0,\epsilon] \times D^2 \longrightarrow M$ such that (a′) the curves $t \mapsto H(t,x)$ are positive normal and (b′) $H(t \times D^2) \subset A_t$ for every $t \in (\epsilon_0,\epsilon]$. Let $\tilde{F}$ be the restriction of H to

(*) Observe that $(0,\epsilon] \times D^2$ is a manifold with boundary and corners. The notion of differentiability for mappings with these domains can be seen in [39].

$(\epsilon_0,\epsilon] \times S^1$. We have then $\tilde{F}(\epsilon,x) = h_\epsilon(x) = f_\epsilon(x) = F(\epsilon,x)$. Since the curves $t \mapsto \tilde{F}(t,x)$, $t \mapsto F(t,x)$ are reparametrizations of trajectories of the normal vector field, by considering a reparametrization of the interval $(\epsilon_0,\epsilon]$, we can suppose that $\tilde{F}(t,x) = F(t,x)$, that is: (c$'$) $H \mid (\epsilon_0,\epsilon] \times S^1 = F \mid (\epsilon_0,\epsilon] \times S^1$.

Assuming $\epsilon_0 > 0$ we are going to prove that for every $x \in D^2$ $\lim_{t \to \epsilon,\, t > \epsilon_0} H(t,x)$ exists. Indeed, the simple curve $\hat{f}_{\epsilon_0}$ is the boundary of a disk in $\hat{A}_{\epsilon_0}$ and therefore using the same argument as before, there exist $\delta > 0$ and an extension $H' : (\epsilon_0 - \delta, \epsilon_0 + \delta) \times D^2 \longrightarrow M$ satisfying (a$'$), (b$'$),(c$'$).

If $\epsilon_0 < t_0 < \epsilon_0 + \delta$, the restriction $h'_{t_0} = H' \mid t_0 \times D^2$ is homotopic to a constant in A_{t_0} and therefore h'_{t_0} lifts to $\hat{h}'_{t_0} : D^2 \longrightarrow \hat{A}_{t_0}$. By construction $\partial(\hat{h}'_{t_0}(D^2)) = \hat{f}_{t_0}(S^1) = \partial(\hat{h}_{t_0}(D^2))$, since $H' = H = F$ on $t_0 \times S^1$. Since $\hat{A}_{t_0} = \mathbb{R}^2$ and $\hat{h}_{t_0}$, $\hat{h}'_{t_0}$ are imbeddings, one sees that $\hat{h}_{t_0}(D^2) = \hat{h}'_{t_0}(D^2)$. This implies that for every $x \in D^2$ there exists $y \in D^2$ such that $\hat{h}_{t_0}(x) = \hat{h}'_{t_0}(y)$. So, by the uniqueness of trajectories of the normal field, $H(t,x) = H'(t,y)$ for every $t \in (\epsilon_0,\epsilon_0 + \delta)$ and therefore $\lim_{t \to \epsilon_0,\, t > \epsilon_0} H(t,x) = H'(\epsilon_0,y)$ exists. This argument shows that we can extend H to $(\epsilon_0 - \delta,\epsilon) \times D^2$ since f_{ϵ_0} is homotopic to a constant in A_{ϵ_0} and $\hat{f}_{\epsilon_0}$ is a simple curve in $\hat{A}_{\epsilon_0}$.

By connectivity we can extend H to $(0,\epsilon] \times D^2$.

Since the construction of H was done using orbits of the normal field, clearly H is an immersion. ■

Proposition 6. *Let $H : (0,\epsilon] \times D^2 \longrightarrow M$ be as in Proposition 5, h_t the restriction of H to $t \times D^2$ and $D_t = h_t(D^2)$. There exists a decreasing sequence $\tau_n \longrightarrow 0$, $\tau_n > 0$ such that*

(a) $A_{\tau_n} = A_{\tau_{n+1}} = A$ *for* $n \geq 1$,
(b) $D_{\tau_{n+1}} \supset D_{\tau_n}$ *for* $n \geq 1$,
(c) *for every $n \geq 1$, there exists a map $g_n : D^2 \longrightarrow D^2$ such that $g_n : D^2 \longrightarrow g_n(D^2)$ is a diffeomorphism and $h_{\tau_n} = h_{\tau_{n+1}} \circ g_n$.*

Proof. Let $U = \{x \in D^2 \mid \lim_{t \to 0,\, t > 0} H(t,x) \text{ exists}\}$. Then U is an open set in D^2 containing S^1 and $U \neq D^2$. Indeed, that U is open follows from the fact that the curves $t \mapsto H(t,x)$ are reparametrizations of orbits of the normal field and of the tubular flow theorem for vector fields. Also, $U \supset S^1$ because the restriction of H to $(0,\epsilon] \times S^1$ coincides with F. Finally if $U = D^2$, H would extend continuously to $0 \times D^2$ which would mean that f_0 is homotopic to a constant in A_0, a contradiction.

Consider $x_0 \in D^2 - U$. Since M is compact, there is a decreasing sequence $s_n \longrightarrow 0$, $s_n > 0$, such that $p_n = H(s_n,x_0)$ and $p_n \longrightarrow p_0 \in A$, a leaf of $\mathfrak{F}$. Let V be a foliation box of p_0. We can assume that $p_n \in V$ for every n.

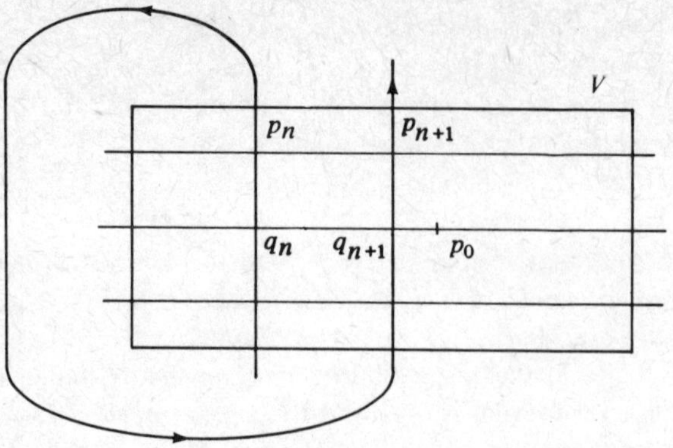

Figure 10

Since the curve $\gamma(t) = H(t,x_0)$ is normal to $\mathfrak{F}$, for every $n \geq 1$, there exists $t_n > 0$ such that $q_n = H(t_n,x_0) \in A$ and the segment of γ between s_n and t_n is contained in V (see figure 10). Clearly the sequence t_n is decreasing, $t_n \longrightarrow 0$ and $A_{t_n} = A$ for every n. This proves (a). We can assume $A \neq A_0$.

We claim that $p_0 \in D_{t_n}$ for n sufficiently large. Indeed, if not, since $q_n \in \text{int}(D_{t_n})$, there exists a sequence $b_n \in \partial D_{t_n}$ such that $b_n \longrightarrow p_0$. On the other hand, $\partial D_{t_n} \subset f_{t_n}(S^1)$, which converges uniformly to $f_0(S^1)$. So $p_0 \in f_0(S^1) \not\subset A$ which is absurd. This proves the claim.

Fix $\hat{p}_0 \in \pi_A^{-1}(p_0)$ and neighborhoods W of p_0 in A and $\hat{W}$ of $\hat{p}_0$ in $\hat{A}$ such that $\pi_A : \hat{W} \longrightarrow W$ is a diffeomorphism. Consider a sequence $\hat{q}_n \in \hat{W}$, $\hat{q}_n \longrightarrow \hat{p}_0$ with $\pi_A(\hat{q}_n) = q_n = h_{t_n}(x_0)$. Let $\hat{h}_{t_n} : D^2 \longrightarrow A$ be the unique lift of h_{t_n} such that $\hat{h}_{t_n}(x_0) = \hat{q}_n (n \geq 1)$ and set $\hat{D}_{t_n} = \hat{h}_{t_n}(D^2)$. If $m \neq n \in \mathbb{N}$, the curves $h_{t_m}(S^1) = f_{t_m}(S^1)$ and $h_{t_n}(S^1) = f_{t_n}(S^1)$ are disjoint in A and the curves $\hat{h}_{t_m}(S^1)$ and $\hat{h}_{t_n}(S^1)$ are simple and disjoint in $\hat{A}$, and therefore $\hat{D}_{t_n} \subset \hat{D}_{t_m}$ or $\hat{D}_{t_m} \subset \hat{D}_{t_n}$, since these disks have a common point p_0. Thus we conclude that $D_{t_m} \supset D_{t_n}$ or $D_{t_n} \supset D_{t_m}$.

Let $\lim(A) = \overline{A} - A$ be the adherence of A. Let d_A be the intrinsic metric on A. Since the sequence of curves $h_{t_n}(S^1)$ converge to $f_0(S^1)$, it follows that $f_0(S^1) \subset \lim(A)$; so $\lim_{n \to \infty} d_A(p_0, h_{t_n}(S^1)) = \infty$. We can then choose a decreasing subsequence $\tau_k = t_{n_k}$ of t_n such that $d_A(p_0, h_{\tau_k}(S^1))$ is increasing and tends to infinity. The sequence τ_n satisfies (a) and (b) of the lemma. Let $D_n = h_{\tau_n}(D^2)$ and $\hat{D}_n = \hat{h}_{\tau_n}(D^2)$.

Observe now that for all $n \in \mathbb{N}$, $\hat{h}_{\tau_n} : D^2 \longrightarrow \hat{D}_n$ is a diffeomorphism and that $\hat{D}_n \subset \hat{D}_{n+1}$. Define $g_n : D^2 \longrightarrow D^2$ by $g_n = (\hat{h}_{\tau_{n+1}})^{-1} \circ \hat{h}_{\tau_n}$. The diffeomorphism g_n evidently satisfies (c). ■

Remark. For every $t \in (0,\epsilon]$, there is a sequence $r_n \longrightarrow 0$ such that $D_{r_n} \subset A_t$ and $\bigcup_{n \geq 1} D_{r_n} = A_t$. This fact, which is to be proved next, implies that $\bigcup_{t \in (0,\epsilon]} D_t = \bigcup_{t \in (0,\epsilon]} A_t$. This observation will be used later.

We will verify this fact first for the leaf $A = A_{\tau_n}$. Let $\hat{d}_A$ be the metric on $\hat{A}$, co-induced by d_A. The fact that $d_A(p_0, h_{\tau_n}(S^1)) \longrightarrow \infty$ implies that $\hat{d}_A(\hat{p}_0, \hat{h}_{\tau_n}(S^1)) \longrightarrow \infty$ where $\pi_A(\hat{p}_0) = p_0$, so $\cup_{n\geq 1} \hat{D}_n = \hat{A}$. It follows therefore that $\cup_{n\geq 1} D_n = A$.

Consider now p_0 as in the above proof. Then, as we have already seen, $p_0 \in D_{\tau_n}$ for large n, or, $p_0 = H(\tau,x)$ for some $x \in D^2$ and $\tau = \tau_n$ fixed. Observe now that p_0 is in the ω-limit set of the orbit of the normal field through the point $q_n = H(\tau, x_0)$. Since the ω-limit set of an orbit is invariant under the flow ([34]), p. 13), we conclude that, if $t \in (0,\epsilon]$, then $p_t = H(t,x) = \lim_{n\to\infty} H(r_n', x_0)$ for some sequence $r_n' \longrightarrow 0$. Using now the same arguments made in the proof of Proposition 6, we can conclude that there exists a sequence $r_n \longrightarrow 0$ such that $D_{r_n} \subset D_{r_{n+1}} \subset A_t$ and $p_t \in D_{r_n}$ for $n \geq n_0$ large. Repeating what was already done for the leaf A_{τ_n}, we conclude that $\cup_{n>1} D_{r_n} = A_t$.

Consider now a manifold with boundary and corners K_n obtained from $[\tau_{n+1}, \tau_n] \times D^2$ by identifying each point of the form $(\tau_n, x) \in \tau_n \times D^2$ with $(\tau_{n+1}, g_n(x)) \in \tau_{n+1} \times D^2$ (see figure 11). Let $\pi_n : [\tau_{n+1}, \tau_n] \times D^2 \longrightarrow K_n$ be the projection of the equivalence relation.

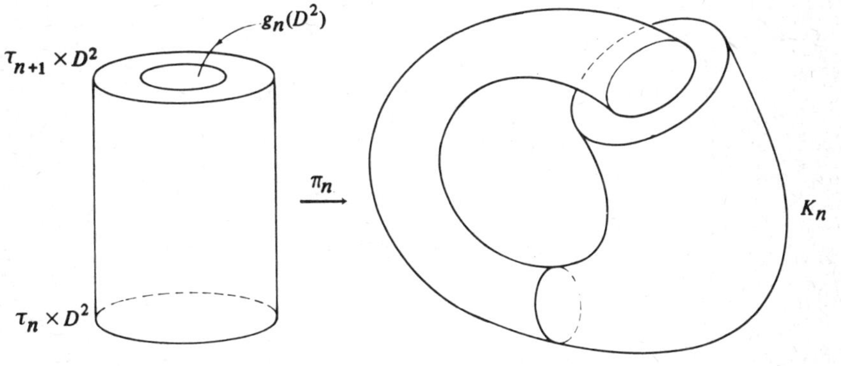

Figure 11

Observe that for every $x \in D^2$, $H(\tau_{n+1}, g_n(x)) = h_{\tau_{n+1}}(g_n(x)) = h_{\tau_n}(x) = H(\tau_n, x)$. Therefore there exists a map $\overline{H}_n : K_n \longrightarrow M$ such that $H = \overline{H}_n \circ \pi_n$. Since H and π_n are immersions, $\overline{H}_n$ is an immersion. This construction contains the main idea of the proof of Novikov's Theorem.

Proof of Theorem 2. Suppose, by contradiction, that A_0 is not compact. Fix $x_0 \in S^1$ and let $q_0 = f_0(x_0)$. By Proposition 4, there is a closed curve γ

transverse to $\mathfrak{F}$ passing through q_0. Modifying the curve γ a little we can obtain a positive closed curve $\tilde{\gamma}$ satisfying the following properties:

(i) $\tilde{\gamma}$ meets $F([0,\epsilon] \times S^1)$ along the normal segment $\{f_t(x_0) \mid t \in [0,\alpha]\}$ where $0 < \alpha \leq \epsilon$,

(ii) $\tilde{\gamma}$ does not meet the normal segment $\{f_t(x) \mid t \in [0,\alpha]\}$ if $x \neq x_0$.

Reparametrizing $\tilde{\gamma}$ we can assume without loss of generality that $\tilde{\gamma}(t) = f_t(x_0)$ for $t \in [0,\alpha]$. Let $0 < \tau_{n+1} < \tau_n < \alpha$ be as in Proposition 6 and $\overline{H}_n : K_n \longrightarrow M$ be as in the above construction.

For each $x \in D^2$ consider the curve $\beta_x(t) = \pi_n(t,x)$. By construction $\overline{H}_n(\beta_x(t)) = H(t,x)$ and hence the curve $\beta_x(t)$ is mapped by $\overline{H}_n$ to a trajectory of the normal vector field. In particular, $\tilde{\gamma}(t) = \overline{H}_n(\beta_{x_0}(t))$ for $t \in [\tau_{n+1},\tau_n]$.

We are going to lift the part of $\tilde{\gamma}$ contained in $\overline{H}_n(K_n)$ to a curve $\overline{\gamma}$ in K_n such that $\overline{H}_n \circ \overline{\gamma} = \tilde{\gamma}$. Define $\overline{\gamma}(t) = \overline{\beta}_{x_0}(t)$ if $t \in [\tau_{n+1},\tau_n]$. Since $\tilde{\gamma}(t)$ is in the interior of $\overline{H}(K_n)$ for $\tau_n < t < \tau_n + \mu$, we can lift $\tilde{\gamma}$ in the interval $[\tau_{n+1},\tau_n + \mu]$, $\mu > 0$ (see figure 12).

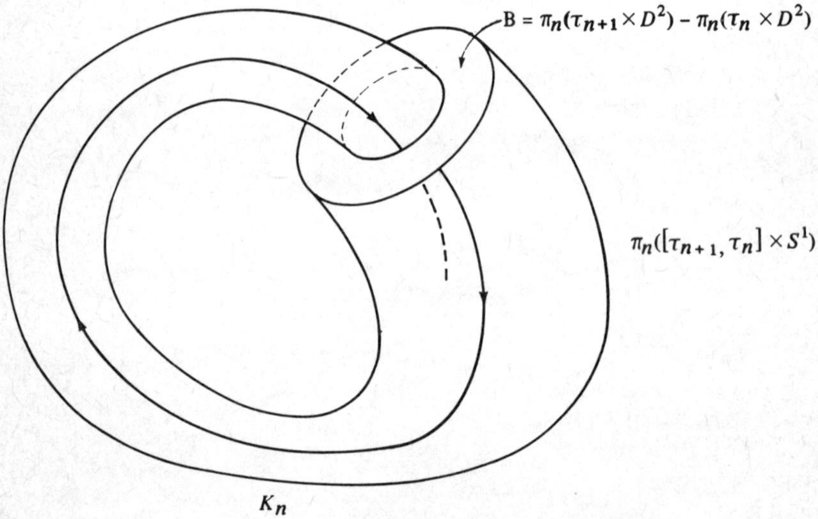

Figure 12

For, since $\overline{H}_n$ is a local diffeomorphism, we can continue to lift $\tilde{\gamma}$ while $\overline{\gamma}$ remains in the interior of K_n. We will only be obliged to stop the lift when $\overline{\gamma}$ touches ∂K_n again. On the other hand, $\overline{\gamma}$ must again meet ∂K_n, since $\tilde{\gamma}$ is a closed curve and the lift begins in ∂K_n. So, there are only two possibilities for $\overline{\gamma}$ to meet ∂K_n : (1) In $\pi_n([\tau_{n+1},\tau_n] \times S^1)$. This is impossible since $\overline{\gamma}$ cuts $H([\tau_{n+1},\tau_n) \times S^1)$ only along the segment $t \longrightarrow f_t(x_0) = \tilde{\gamma}(t)$, $t \in [\tau_{n+1},\tau_n]$.

(2) In $B = \pi_n(\tau_{n+1} \times D^2) - \pi_n(\tau_n \times D^2)$. This also is impossible. In fact, since $\overline{H}_n(B) \subset A = A_{\tau_n}$, if $\overline{\gamma}$ leaves K_n at a point $z \in B$, its tangent vector at this point would be pointing outside of K_n, so the tangent vector to $\tilde{\gamma}$ at the point $\overline{H}_n(z)$ would be pointing in the negative direction of $\mathfrak{F}$, a contradiction (see figure 12). This proves that A_0 is a compact leaf.

Consider now the compact, connected regions $V_n = \overline{H}_n(K_n) = H([\tau_{n+1},\tau_n] \times D^2) \subset M$, $n \geq 1$. Observe in the argument above that the curve $\beta_x(t)$ enters the interior of K_n and never again leaves, for every $x \in D^2$. This corresponds to the fact that the positive integral curves of the normal field, which enter V_n, never again leave. This fact will be used later and implies that $V_{n-1} \subset \bigcup_{t \in [\tau_{n+1},\epsilon]} D_t = V_n$. Let $V = \overline{\bigcup_{n \geq 1} V_n}$. It is clear that V is connected, compact and its boundary $\partial V \supset A_0$. We will show next that $\partial V = A_0$.

Observe first that $V = \bigcup_{0<t\leq\epsilon} A_t$, so V and ∂V are saturated by $\mathfrak{F}$.

Indeed, by the remark after Proposition 6, we have that $\bigcup_{0<t\leq\epsilon} D_t = \bigcup_{0<t\leq\epsilon} A_t$, so

$$\bigcup_{n\geq 1} V_n = \bigcup_{n\geq 1} \bigcup_{\tau_n \leq t \leq \epsilon} D_t = \bigcup_{0<t\leq\epsilon} D_t = \bigcup_{0<t\leq\epsilon} A_t .$$

Hence $V = \overline{\bigcup_{0<t\leq\epsilon} A_t}$.

We claim that ∂V is connected. Indeed, ∂V is the set of accumulation points of $\bigcup_{n\geq 1} \partial V_n$. This follows from the fact that $V_n \subset V_{n+1} (n \geq 1)$ and $V = \overline{\bigcup_{n\geq 1} V_n}$. Suppose by contradiction that ∂V is disconnected. In this case there exist open disjoint sets $W_1, W_2 \subset M$ such that $\partial V \subset W_1 \cup W_2$ and $\partial V \cap W_i \neq \varnothing (i = 1,2)$. This implies that there exists n_0 such that $\partial V_n \cap W_1$ and $\partial V_n \cap W_2$ are nonempty for $n \geq n_0$, so for $n \geq n_0$ there exists $p_n \in \partial V_n - (W_1 \cup W_2)$, since ∂V_n is connected. We can assume without loss of generality that $p_{n+1} \in \partial V_{n+1} - V_n$, $n \geq n_0$. Since $M - (W_1 \cup W_2)$ is compact, the sequence $\{p_n\}$ has an accumulation point p. Since $p_{n+1} \notin V_n$ it is clear that p is an accumulation point of $\bigcup_{n\geq 1} \partial V_n$, or, $p \in \partial V - (W_1 \cup W_2)$, a contradiction. So ∂V is connected. Next we will prove that A_0 is a connected component of ∂V. This will imply that $A_0 = \partial V$. Let N be a tubular neighborhood of A_0. Since $\mathfrak{F}$ is transversely orientable N is diffeomorphic to $A_0 \times (-\delta,\delta)$, where we can assume that the fibers of N are the trajectories of the normal field to $\mathfrak{F}$. Given $q \in A_0$, let $\mathcal{O}_q$ be the fiber of N which passes through q. It suffices to prove that for every $q \in A_0$, $\partial V \cap \mathcal{O}_q = \{q\}$, if the length of the fibers is sufficiently small. In fact, this will imply that $\partial V \cap N = A_0$, so A_0 will be a connected component of ∂V. Observe first that it suffices to verify the above fact for some $q \in A_0$. In fact, this follows from the fact that ∂V is invariant under $\mathfrak{F}$ and from transverse uniformity of $\mathfrak{F}$ (Chapter II). On the other hand, given $q \in f_0(S^1) \subset A_0$, let $W \subset N$ be a foliation box of $\mathfrak{F}$ which contains q. Set $\tilde{W} = H^{-1}(W)$. Since the orbits of the normal field in W meet A_0, one sees that if $(t_0,x_0) \in \tilde{W}$ then $x_0 \in U = \{x \in D^2 \mid \lim_{t\to 0} H(t,x) \text{ exists}\}$ and therefore $\lim_{t\to 0} H(t,x_0) \in A_0$. It follows therefore that the set of accumulation points of $\cup_{n\geq 1} (\partial V_n \cap W)$ is contained in A_0, or, $\partial V \cap W \subset A_0$. This proves that $\mathcal{O}_q \cap \partial V = \{q\}$, as desired.

Part (c) of Theorem 2 follows from the fact that $\partial V \supset \lim A_t \supset A_0$ for every t and that $\partial V = A_0$.

In order to prove that A_t, $t > 0$, is diffeomorphic to $\mathbb{R}^2$, we are going to use the following facts:

(a) Let N be a two-dimensional manifold without boundary, not diffeomorphic to $\mathbb{R}^2$, whose universal cover is $\mathbb{R}^2$. Then there is a deck transformation $f: \mathbb{R}^2 \longrightarrow \mathbb{R}^2$ which has infinite order (that is $f^k \neq$ identity if $k \neq 0$). Recall that a deck transformation is a diffeomorphism $f: \mathbb{R}^2 \longrightarrow \mathbb{R}^2$ such that $\pi \circ f = \pi$, where $\pi : \mathbb{R}^2 \longrightarrow N$ is the covering projection.

(b) If $f: \mathbb{R}^2 \longrightarrow \mathbb{R}^2$ is a deck transformation such that $f(x) = x$ for some $x \in \mathbb{R}^2$, then $f =$ identity.

We leave the proof of (a) and (b) as an exercise for the reader. We need the following lemma.

Lemma 1. *Let $\mathbb{R}^2 = \cup_{n\geq 1} \hat{D}_n$, where for every n, $\hat{D}_n$ is a disk, $\hat{D}_n \subset \hat{D}_{n+1}$ and $\partial\hat{D}_n \cap \partial\hat{D}_{n+1} = \emptyset$. Let $\pi : \mathbb{R}^2 \longrightarrow N$ be a covering and suppose that for every n, $\pi(\partial\hat{D}_n)$ is a closed curve in N whose self-intersections are transverse double points. Let m_k be the number of self-intersections of $\pi(\partial\hat{D}_k)$. If the sequence m_k is bounded then $\pi : \mathbb{R}^2 \longrightarrow N$ is a diffeomorphism.*

Proof. Suppose π is not a diffeomorphism. In this case, there exists a deck transformation $f: \mathbb{R}^2 \longrightarrow \mathbb{R}^2$ of infinite order. Let $z_0 \in \hat{D}_1$ and set $z_k = f^k(z_0)$. If $z_k = z_\ell$ with $k \neq \ell$, we would have $f^{k-\ell} =$ identity, contrary to the hypothesis. So the sequence $\{z_k \mid k \in Z\}$ is made up of distinct points.

On the other hand, for every $k \neq 0$ $f^k(\hat{D}_n)$ contains points outside $\hat{D}_n$, since otherwise, by the Brouwer fixed point theorem, f^k would have a fixed point in $\hat{D}_n$, a contradiction.

For n large, let k_n be such that $z_0, \ldots, z_{k_n} \in \hat{D}_n$ and $z_{k_{n+1}} \notin \hat{D}_n$. We have that $\partial\hat{D}_n \cap f^j(\partial\hat{D}_n)$ contains at least two points, $1 \leq j \leq k_n$, since $z_j \in \hat{D}_n \cap f^j(\hat{D}_n)$ and $\hat{D}_n \cap (\mathbb{R}^2 - f^j(\hat{D}_n)) \neq \emptyset$. Let $p_j, p_j' \in \partial\hat{D}_n \cap f^j(\partial\hat{D}_n)$, $1 \leq j \leq k_n$. Since $\pi \circ f^j = \pi$, $\pi(p_j)$ and $\pi(p_j')$ correspond to points of self-intersection of $\pi(\partial\hat{D}_n)$ and since these points are all double points, for $1 \leq i \leq k_n$ and $i \neq j$ we have $p_i \neq p_j, p_j'$, so if m_n is the number of self-intersections of $\pi(\partial\hat{D}_n)$, then $m_n \geq k_n$. Since $\cup_{n\geq 1} \hat{D}_n = \mathbb{R}^2$, it follows that $\lim_{n\to\infty} k_n = \infty$ and therefore m_n is not bounded, contrary to hypothesis.

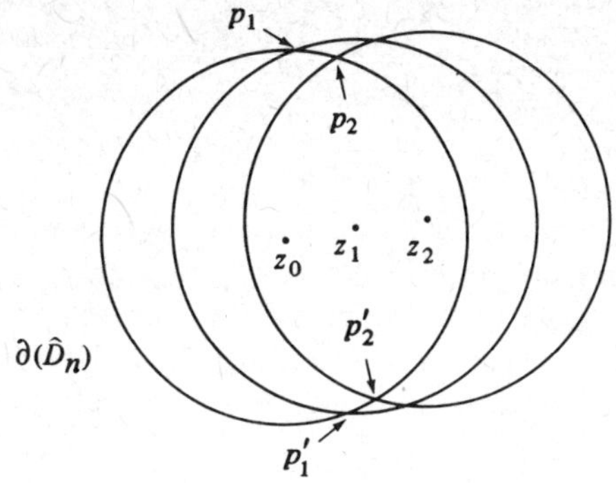

Figure 13

We now prove that $A_{\tau_n} = A \simeq \mathbb{R}^2$. The proof that $A_t \simeq \mathbb{R}^2$, $t \neq \tau_n$ is similar.

By the considerations preceding Proposition 3, if n_t is the number of self-intersections of the curve $f_t(S^1) \subset A_t$, then $n_t \leq n_0$. Since n_{τ_k} is the number of self-intersections of $\pi_A(\partial\hat{D}_k) = f_{\tau_k}(S^1)$, it follows from the above lemma that $A \simeq \mathbb{R}^2$ and $\pi_A : \hat{A} \longrightarrow A$ is a diffeomorphism. In particular $n_{\tau_k} = 0$ for every k. This proves Theorem 2. ■

As a consequence, we have the following.

Theorem 1. *Let $\mathfrak{F}$ be a transversely orientable, C^2, codimension one foliation of a compact, three-dimensional manifold with finite fundamental group. Then $\mathfrak{F}$ has a compact leaf.*

Proof. By Proposition 1 there is a vanishing cycle of $\mathfrak{F}$. By Proposition 3, $\mathfrak{F}$ has a leaf A_0 containing a vanishing cycle, which admits a simple, positive, normal coherent extension. By Theorem 2, the leaf A_0 is compact. ■

§5. Existence of a Reeb component

In this section we will present a sketch of the proof that the compact leaf A_0, found in Theorem 2, in fact bounds a Reeb component. More precisely, with the notation of Theorem 2:

Theorem 3. *The manifold V bounded by A_0 is diffeomorphic to $D^2 \times S^1$ and $\mathfrak{F} \mid V$ is topologically equivalent to the Reeb foliation on $D^2 \times S^1$.*

Proof. By Propositions 5 and 6 there is an immersion $H : (0,\epsilon] \times D^2 \longrightarrow M$, $h_t(x) = H(t,x)$, and a sequence $\tau_n \longrightarrow 0$ such that (1) $h_t(D^2) \subset A_t$ for every t, (2) the curves $t \mapsto h_t(x)$ are positive normal, (3) $A_{\tau_n} = A_{\tau_{n+1}}$ for every $n \geq 1$, (4) $h_{\tau_{n+1}}(D^2) \supset h_{\tau_n}(D^2)$. Further for every $n \geq 1$ there is a map $g_n : D^2 \longrightarrow D^2$ that is a diffeomorphism onto its image and such that $h_{\tau_n} = h_{\tau_{n+1}} \circ g_n$.

As before, let K_n be the manifold with boundary obtained from $[\tau_{n+1},\tau_n] \times D^2$ by identifying the points $(\tau_n,x) \in \tau_n \times D^2$ with $(\tau_{n+1},g_n(x)) \in \tau_{n+1} \times D^2$ and $\pi_n : [\tau_{n+1},\tau_n] \times D^2 \longrightarrow K_n$ be the quotient projection. Also let $\overline{H}_n : K_n \longrightarrow M$ be the immersion induced by H and $V_n = \overline{H}_n(K)$. The normal field to $\mathfrak{F}$ will be denoted by X. Since the set $B_n = \{t \in [\tau_{n+1},\tau_n] \mid A_t = A_{\tau_n}\}$ is discrete (since two distinct curves $f_t(S^1)$ and $f_s(S^1)$ contained in A_{τ_n} are far apart in the intrinsic metric of A_{τ_n}), we can assume that if $t \in (\tau_{n+1},\tau_n)$ then $A_t \neq A_{\tau_n}$, $n \geq 1$. We will next prove that, with this hypothesis, $\overline{H}_n : K_n \longrightarrow V_n$ is a diffeomorphism. As a consequence of the lemma of §4 we have that for every $t \in [0,\epsilon]$, the curve $f_t(S^1)$ is simple and this implies that the restriction of H to $t \times D^2$ can be taken to be injective (review the proof of Proposition 5). Since $A_{t'} \neq A_t$ if $\tau_{n+1} < t' < t \leq \tau_n$ it follows that H is injective and therefore a diffeomorphism.

On the other hand, since M is orientable, for every $n \geq 1$, V_n is orientable, so K_n is also. It follows therefore that the identification diffeomorphism $g_n : D^2 \longrightarrow D^2$, from which we obtain K_n, preserves the orientation of D^2. We conclude then that K_n and V_n are topological solid tori for every $n \geq 1$.

We are now going to construct a sequence of two-dimensional tori $\{T_n\}$ such that

(a) T_n meets each leaf of $\mathfrak{F}$ transversely in a closed curve which is the boundary of a disk in this leaf,
(b) T_n is the boundary of a solid torus $R_n \subset V = \overline{\bigcup_n V_n}$,
(c) the trajectories of the normal field cut T_n in at most one point,
(d) $T_n \subset V_{n+1} - V_{n-1}$ and, therefore, if n is sufficiently large, every positive trajectory of the normal field through a point of A_0 cuts T_n.

Consider a cylinder $C \subset D_{\tau_{n+1}} - D_{\tau_n}$ such that $\partial C = f_{\tau_{n+1}}(S^1) \cup \gamma$, where γ is a closed curve such that $\gamma \cap f_{\tau_n}(S^1) = \emptyset$. Saturating C by X in an interval $[0,\mu]$, $\mu > 0$ we obtain a region U of V_n diffeomorphic to $C \times [0,\mu]$ where the normal field is represented in $C \times [0,\mu]$ by the field $\partial/\partial t = (0,1)$ (see Figure 14). We take μ such that $f_{\tau_n}(S^1) \subset U$. Construct now T_n' by attaching the cylinder

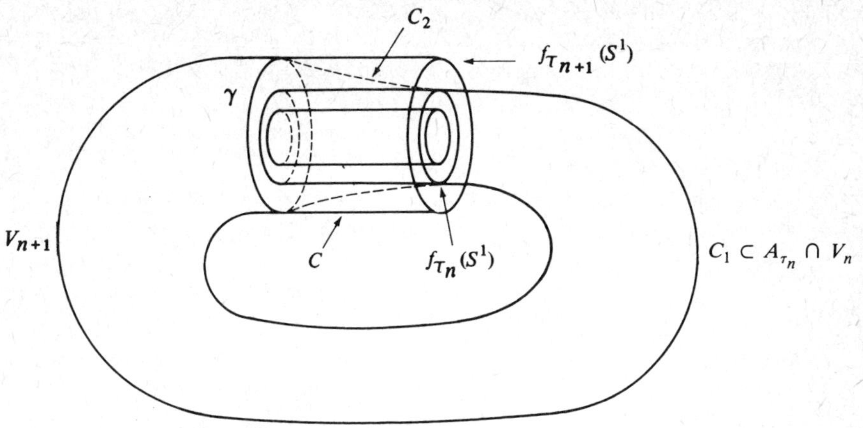

Figure 14

$C_1 = (D_{\tau_{n+1}} - D_{\tau_n}) - C$ such that $\partial C_1 = \gamma \cup f_{\tau_n}(S^1)$, to a cylinder C_2 transverse to X and to $\mathfrak{F}$, $C_2 \subset C \times [0,\mu]$ and $\partial C_2 = f_{\tau_n}(S^1) \cup \gamma$. (See Fig. 14). It is clear from the construction that $T_n' = C_1 \cup C_2$ satisfies properties (b), (c) and (d), as desired. Modifying the torus T_n' slightly, we can get a torus T_n transverse to the foliation $\mathfrak{F}$ (see Fig. 15).

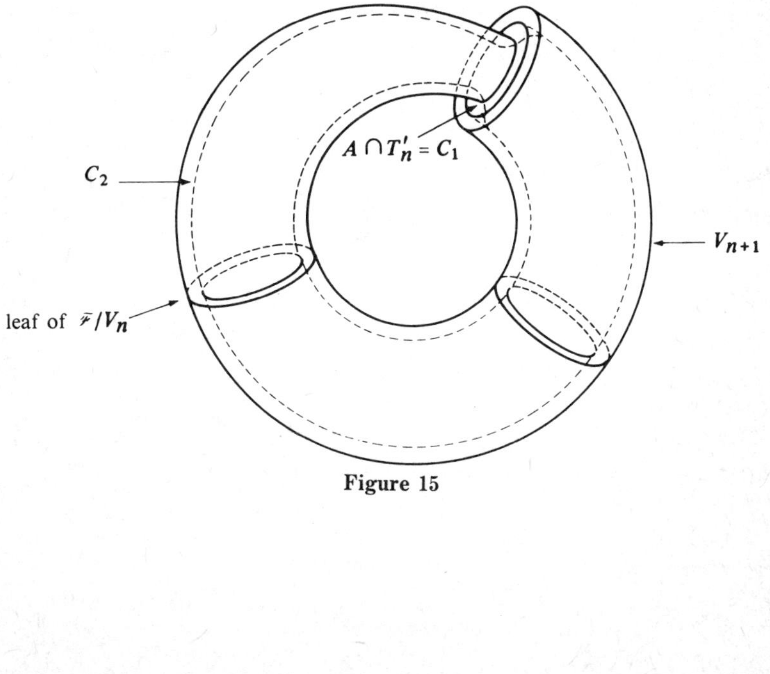

Figure 15

The above figure is a type of "dual" to figure 14. In that figure the elongated part of ∂V_{n+1} is contained in the leaf A_{τ_n} and the short part is transverse to $\mathfrak{F}$ while in figure 15 the elongated part is transverse to $\mathfrak{F}$ and the short part is contained in A_{τ_n}. The transverse torus is obtained by pushing C_1 slightly to the inside of V_{n+1}, as shown in figure 15. It follows from the construction that T_n can be gotten arbitrarily near T_n', which is transverse to the normal field, so, we can assume that T_n is transverse to $\mathfrak{F}$ and to the normal field.

Now consider $p \in A_0$. Let X_t be the flow of X. The orbit of X through p enters the region V bounded by A_0 and never again leaves. Further, since $V = \overline{\cup V_n}$, $V_n \subseteq V_{n+1}$ and A_0 is compact, given $\delta > 0$ there exists $n_0 \in \mathbb{N}$ such that for $n \geq n_0$, $X_\delta(p) \in \operatorname{int}(V_n)$ for every $p \in A_0$. This implies that for n sufficiently large and for every $p \in A_0$, the segment of the orbit $\{X_t(p) \mid t \in [0,\delta]\}$ cuts T_n in only one point $P(p)$. This defines a map $P: A_0 \longrightarrow T_n$ that evidently is a diffeomorphism. So A_0 is diffeomorphic to T^2. In a similar manner it follows that V is diffeomorphic to V_n. So V is diffeomorphic to $D^2 \times S^1$. We leave it as an exercise for the reader to prove that $\mathfrak{F} \mid V$ is equivalent to the Reeb foliation of $D^2 \times S^1$.

Remark. Using Proposition 3 and Theorem 3 one can show that every leaf A_0 of $\mathfrak{F}$ which contains a vanishing cycle is the boundary of a Reeb component.

Indeed, let $f_0 : S^1 \longrightarrow M$ be a positive normal extension of f_0. By Proposition 3, there exists a sequence $\tau_n \longmapsto 0$ such that for every $m \geq 1$, the leaf A_{τ_m} contains a positive vanishing cycle $g_m : S^1 \longrightarrow A_{\tau_m}$, which has a simple, positive, normal, coherent extension. By Theorem 3, A_{τ_m} is compact and is the boundary of a Reeb component of $\mathfrak{F}$, which we denote by V^m. We claim that $A_{\tau_m} = A_{\tau_\ell}$ and $V^m = V^\ell$ for arbitrary $m, \ell \in \mathbb{N}$.

In fact, since for every $m \geq 1$, g_m is a positive vanishing cycle which has a simple coherent extension, it follows that for $p \in A_{\tau_m}$, V^m entirely contains the positive orbit $\mathcal{O}_+(p)$ of the normal field to $\mathfrak{F}$, passing through p at $t = 0$. On the other hand, if $\ell > m$, there exists a point $p = f_{\tau_\ell}(x_0) \in A_{\tau_\ell}$ such that $\mathcal{O}_+(p) \cap V^m \neq \varnothing$, so $V^\ell \cap V^m \neq \varnothing$. Since the boundaries of V^m and V^ℓ are leaves of $\mathfrak{F}$, $V^\ell \supseteq V^m$. By Theorem 2, the leaves of $\mathfrak{F}$ contained in $\operatorname{int}(V^\ell)$ are diffeomorphic to planes and therefore $A_{\tau_m} = \partial V^m$, which is compact, cannot be contained in the interior of V^ℓ, so $A_{\tau_m} = A_{\tau_\ell}$ and $V^m = V^\ell$. Hence we see that $A_{\tau_m} = A_0$ and $V^m = V^1$ for every $m \geq 1$, so A_0 is the boundary of a Reeb component of $\mathfrak{F}$.

§6. Other results of Novikov

A natural problem to propose is that of the characterization of three-dimensional manifolds which only admit foliations with Reeb components. By Theorem 2 this problem is equivalent to characterizing compact three-dimensional manifolds such that every foliation has a vanishing cycle. In this direction we have the following theorem due to Novikov [40].

Theorem 4. *Let $\mathfrak{F}$ be a transversely orientable, C^2, codimension-one foliation of a compact, orientable, three-dimensional manifold M such that $\pi_2(M) \neq \neq 0$. Then either $\mathfrak{F}$ has a Reeb component or the leaves of $\mathfrak{F}$ are compact with finite fundamental group.*

This applies for example to the manifold $S^2 \times S^1$.

This theorem is a consequence of Theorem 3 and of the following lemma:

Lemma 2. *Let $\mathfrak{F}$ be a transversely orientable, C^2, codimension-one foliation of a compact, orientable, three-dimensional manifold M such that $\pi_2(M) \neq \neq 0$. If $\mathfrak{F}$ has no vanishing cycles then all the leaves of $\mathfrak{F}$ are compact with finite fundamental group.*

Proof. Saying that $\pi_2(M) = 0$ is equivalent to saying that every continuous map $f : S^2 \longrightarrow M$ is homotopic to a constant.

Let $g : S^2 \longrightarrow M$ be a map not homotopic to a constant. By methods we used in Haefliger's construction, g can be approximated by another map which we will again denote by g such that $g^*(\mathfrak{F})$ is a foliation of S^2 defined by a vector field Y whose singularities are centers or saddles and such that Y admits no saddle connections, other than self-connections.

We will next see that the hypothesis that $\mathfrak{F}$ does not have vanishing cycles implies that if γ is a nonsingular trajectory of Y then γ is closed or γ is contained in a graph of Y. Moreover, if Γ is a closed trajectory or graph of Y then $g(\Gamma)$ (with the orientation given by Y) is homotopic to a constant in the leaf of $\mathfrak{F}$ that contains it.

Indeed, suppose that Y has a non-closed orbit γ that is not a self-connection of saddles. In this case by the Poincaré-Bendixson Theorem, the α or ω-limit set of γ is a closed orbit or a graph formed by a saddle and one or two separatrices. Call it Γ. The curve $g(\Gamma)$ is not homotopic to a constant in the leaf of $\mathfrak{F}$ which contains it. We can apply then the same arguments as in §2 in order to obtain a vanishing cycle of $\mathfrak{F}$. Since $\mathfrak{F}$ does not have vanishing cycles, an orbit γ as above cannot exist. So, all the nonsingular orbits of Y are closed or are separatrices of saddles forming self-connections. If there exists a closed path or a graph Γ such that $g(\Gamma)$ is not homotopic to a point in $A_{g(\Gamma)}$, again by the arguments of §2, we can obtain a vanishing cycle. So, if Γ is a closed orbit or graph of Y, $g(\Gamma)$ is homotopic to a constant in the leaf which contains it.

We will prove ahead that there is a map $g_1 : S^2 \longrightarrow M$, homotopic to g and such that $g_1(S^2) \subset A$, for some leaf A of $\mathfrak{F}$. We claim that this implies that $A \simeq P^2$ or S^2, therefore it is compact with finite fundamental group and so by Reeb's global stability theorem (Chapter IV) all the leaves of $\mathfrak{F}$ are compact with finite fundamental group. Let us prove the claim.

Let $\pi : \hat{A} \longrightarrow A$ be the universal covering of A. Then $\hat{A} \simeq \mathbb{R}^2$ or $\hat{A} \simeq S^2$. It suffices to prove that $\hat{A} \simeq S^2$. Suppose by contradiction that $\hat{A} \simeq \mathbb{R}^2$. Since S^2 is simply connected we can lift g_1 to a map $\hat{g}_1 : S^2 \longrightarrow \hat{A}$ such that

$\pi \circ \hat{g}_1 = g_1$. Since $\hat{A} \simeq \mathbb{R}^2$, $\hat{g}_1$ is homotopic to a constant. Let $\hat{F} : I \times S^2 \longrightarrow \hat{A}$ be a holonomy taking $\hat{g}_1$ to a constant. Then $F = \pi \circ \hat{F}$ is a homotopy taking g_1 to a constant, which is a contradiction, since g_1 is homotopic to g which is not homotopic to a constant, by hypothesis. Therefore, all the leaves of $\mathfrak{F}$ are compact with finite fundamental group.

We must still show that there exists $g_1 : S^2 \longrightarrow M$ homotopic to g and such that $g_1(S^2) \subset A$ for some leaf A of $\mathfrak{F}$. The idea, basically, is the following: using that for every closed orbit or graph Γ of Y, $g(\Gamma)$ is homotopic to a constant in $A_{g(\Gamma)}$, starting from the centers we are going to deform g progressively, obtaining a continuous family $g_t : S^2 \longrightarrow M$, $t \in [0,1]$, satisfying the following properties:

(1) For every $t \in [0,1]$ there exists a compact region $R_t \subset S^2$ such that if $s > t$ then $R_s \supseteq R_t$.
(2) The connected components of ∂R_t are centers, closed orbits or graphs of Y.
(3) If $R_t^1, \dots, R_t^{k_t}$ are the connected components of R_t then $g_t(R_t^j) \subset A_t^j$ for some leaf A_t^j of $\mathfrak{F}$, where $g_t : R_t^j \longrightarrow A_t^j$ is homotopic to a constant in A_t^j.
(4) $g_t \mid (S^2 - R_t) = g \mid (S^2 - R_t)$.
(5) R_0 is a center of Y and $R_1 = S^2$.

In the figure below we see two stages of the deformation in a neighborhood of a center.

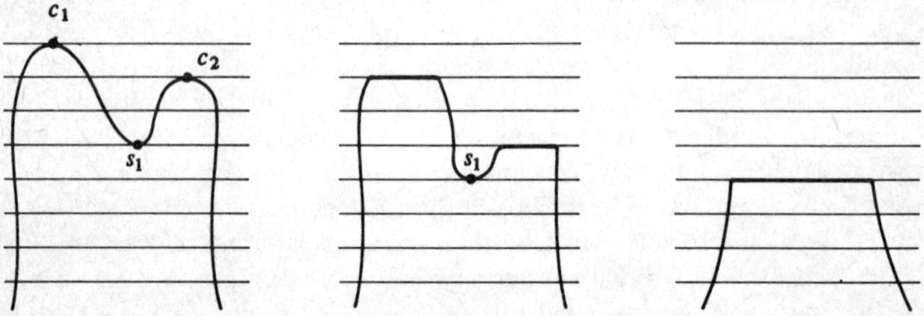

Figure 16

The following lemma implies Lemma 2.

Lemma 3. *Let $G : [0,\alpha] \times S^2 \longrightarrow M$ be continuous and set $g_t(p) = G(t,p)$. Suppose that the family g_t, $t \in [0,\alpha]$, satisfies properties (1) to (4) above, where $R_\alpha \neq S^2$. Then we can extend G to $[0,\alpha + \epsilon] \times S^2$, obtaining an extension of the family g_t, which satisfies properties (1) to (4).*

Proof. The technique of proof is similar to that used in Lemma 4 of Chapter IV and in Proposition 5 of this chapter. The only difference is that we cannot use the normal field to $\mathfrak{F}$ to obtain the extension of g_t to a neighborhood of $g_\alpha(\partial R_\alpha)$. We consider three cases:

(1) R_α has a connected component which is a center.
(2) R_α has a connected component S_α such that ∂S_α consists of closed orbits of Y.
(3) For every connected component S of R_α, ∂S contains a graph of Y.

In the first case the center of Y has a neighborhood V saturated by Y and such that all the orbits of Y in $V - \{c\}$ are closed, constituting a continuous family $t \longrightarrow \Gamma_t$, $t \in [\alpha,\alpha + \epsilon]$, where Γ_t is the boundary of a small disk V_t such that $V_t \cap R_\alpha = \{c\}$. In this case, for $t \in [\alpha,\alpha + \epsilon]$, $g(\Gamma_t)$ is the boundary of a small disk in A_t and we can extend the deformation as shown in figure 17. Here, the image $g_t(S^2)$ is changed only in a foliation box of $\mathfrak{F}$.

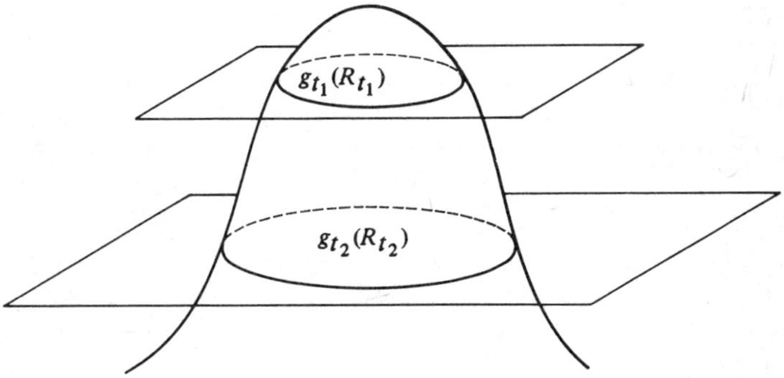

Figure 17

In the second case, let $\Gamma^1, \ldots, \Gamma^k$ be the connected components of ∂S_α. For each $i = 1, \ldots, k$, there exists a small tubular neighborhood V^i of Γ^i saturated by Y and such that all the orbits of Y in V^i are closed. Observe that $g_\alpha(\Gamma^1), \ldots, g_\alpha(\Gamma^k)$ are contained in the same leaf A_α of $\mathfrak{F}$ and are homotopic to a constant in this leaf. Since $g_\alpha \mid S_\alpha$ is homotopic to a constant in A_α, by Lemma 4 of Chapter IV, we can extend the family $g_\alpha \mid S_\alpha$ to a continuous family $\tilde{g}_t : S_\alpha \longrightarrow M$, $t \in [\alpha,\alpha + \delta]$, such that for every $t \in [\alpha,\alpha + \delta]$, $\tilde{g}_t(S_\alpha) \subset A_t$ for some leaf A_t of $\mathfrak{F}$. This implies that it is possible to define continuous families $t \longrightarrow \Gamma_t^i$, $t \in (\alpha,\alpha + \epsilon]$, $i = 1, \ldots, k$, such that for every $i \in \{1, \ldots, k\}$ and every $t \in (\alpha,\alpha + \epsilon]$, $\Gamma_t^i \subset V^i - R_\alpha$ and $g(\Gamma_t^i) = g_\alpha(\Gamma_t^i) \subset A_t$ where $0 < \epsilon \leq \delta$ is sufficiently small. Let S_t be the connected region of S^2 which contains S_α and whose boundary is $\Gamma_t^1 \cup \ldots \cup \Gamma_t^k$ and set $R_t = S_t \cup R_\alpha$. Then the family R_t satisfies properties (1) and (2) of the lemma. The construction of the family g_t that extends g_α and satisfies (3) and (4) is obtained using arguments analogous to Lemma 4 of Chapter IV, with the only difference being that it is not possible here to use the normal field of $\mathfrak{F}$. We leave the details to the reader.

We consider now the third and last case. This is divided into two subcases:

(a) Some connected component of $S^2 - R_\alpha$ contains a center c of Y.
(b) $S^2 - R_\alpha$ contains no centers of Y.

Subcase (a). In this case it is sufficient to add the center c to R_α, reducing the problem to the first case.

Subcase (b). In this case we will prove that R_α contains a connected component S_α with properties analogous to those of the second case, that is, if Γ^1 ,..., Γ^k are the connected components of ∂S_α then there exists a neighborhood V of ∂S_α such that $V - S_\alpha$ is saturated by Y and the orbits of Y in $V - S_\alpha$ are closed.

Observe that from the construction of g (Chapter VI), we see that if s and s' are distinct saddles of Y then $g(s)$ and $g(s')$ are in distinct leaves of $\mathfrak{F}$. It follows therefore that if S is a connected component of R_α then ∂S contains at most one graph of Y. On the other hand, given a graph Γ of Y which contains two separatrices, $S^2 - \Gamma$ has three connected components, say $B_j(\Gamma)$, $j = 1,2,3$. Two of these components, say $B_1(\Gamma)$ and $B_2(\Gamma)$, are such that $\overline{B_j(\Gamma)}$ is homeomorphic to a closed disk, while $\overline{B_3(\Gamma)}$ is not homeomorphic to a disk. Observe that $B_i(\Gamma)$ must contain a center for each $i = 1,2,3$.
It suffices to prove the following fact:
(*) There exists a component S_α of R_α such that if Γ_α is the graph contained in ∂S_α then Γ_α contains the two separatrices and $B_j(\Gamma_\alpha) \cap S_\alpha \neq \varnothing$ for $j = 1,2$ or $B_3(\Gamma_\alpha) \cap S_\alpha \neq \varnothing$.

If S_α satisfies (*) we can define a family of regions S_t, $t \in [\alpha, \alpha + \epsilon]$, satisfying (1) and (2) of Lemma 3. In the case that $B_3(\Gamma_\alpha) \cap S_\alpha \neq \varnothing$, the extension S_t of S_α is obtained inside the regions $B_1(\Gamma_\alpha)$ and $B_2(\Gamma_\alpha)$ and in the other case, inside $B_3(\Gamma_\alpha)$.

Let S_1 ,..., S_k be connected components of R_α. For each $j = 1$,..., k, ∂S_j contains a saddle point s_j. Let Γ_j be the union of the two separatrices of s_j, so that Γ_j is a graph of Y and at least one of its separatrices is contained in ∂S_j. Suppose S_1 does not satisfy (*). In this case it is easy to see that $B_{j_1}(\Gamma_1)$ contains a center $c \notin S_1$ for some $j_1 = 1,2,3$. So $B_{j_1}(\Gamma_1)$ contains a component of R_α distinct from S_1 which we can assume to be S_2. If S_2 does not satisfy (*), by the same argument, for some $j_2 = 1,2,3$, $B_{j_2}(\Gamma_2)$ contains a component of R_α distinct from S_1 and S_2, say S_3, with $S_3 \subset B_{j_1}(\Gamma_1)$. Proceeding inductively we obtain finally a component S_j of R_α satisfying (*).

We can now extend the family R_t, $t \in [0,\alpha]$, obtaining a new family, satisfying (1) and (2) of Lemma 3. In order to extend the family $t \longrightarrow g_t$, the argument is similar to the one used in the second case. ∎

End of the proof of Lemma 2. As was already said above we begin the deformation of g at a center of Y. Using Lemma 3 successively we obtain a homotopy $G : [0,\alpha) \times S^2 \longrightarrow M$ satisfying properties (1) to (4). For each $t \in [0,\alpha)$ there exists a region R_t such that $g_t(R_t) \subset A_t$ for some leaf A_t of $\mathfrak{F}$ and $g_t \mid (S^2 - R_t) = g \mid (S^2 - R_t)$. Observe that $\bigcup_{t\in[0,\alpha)} R_t = R_\alpha$ is a region of S^2 whose boundary ∂R_α consists of closed orbits or graphs of Y. Let S be a connected component of R_α. Let $\Gamma^1, \ldots, \Gamma^k$ be the closed orbits or graphs with a separatrix contained in ∂S. Since $S \subset S^2$ the closed curves $\Gamma^1, \ldots, \Gamma^k$ generate the fundamental group of S. Since $g \mid \Gamma^i : \Gamma^i \longrightarrow A$ is homotopic to a constant in A, we can extend $g \mid \partial S$ to a continuous map $h : S \longrightarrow A$. By Lemma 4 of Chapter IV, we can assume that there is a continuous family of maps $h_t : S \longrightarrow M$, $t \in [\alpha - \delta, \alpha + \delta]$ such that $h_\alpha = h$ and $h_t(S) \subset A_t$, $t \in [\alpha - \delta, \alpha + \delta]$. Set $S_t = S \cap R_t$. For every $t \in [\alpha - \delta, \alpha)$, $h_t \mid S_t$ is homotopic to a constant in A_t, so it is homotopic to $g_t \mid S_t$ by a homotopy $F_t : I \times (R_t \cap S) \longrightarrow A_t$. Using the (continuous) family of homotopies $t \longrightarrow F_t$, we can modify the family h_t obtaining a new family $\tilde{g}_t$ such that $\tilde{g}_{\alpha-\delta} \mid S_{\alpha-\delta} = g_{\alpha-\delta} \mid S_{\alpha-\delta}$ and $\tilde{g}_{\alpha-\delta} \mid S_{\alpha-\delta/2} = h_{\alpha-\delta/2} \mid S_{\alpha-\delta/2}$.

Observe that we can obtain the family $t \longrightarrow h_t$ such that $h_t \mid \partial S_\alpha = g \mid \partial S_\alpha$. The family $\tilde{g}_t$ thus defined extends $g_t \mid S_t$, $t \in [0,\alpha - \delta]$ to a family $\tilde{g}_t \mid S_t$, $t \in [0,\alpha]$ which satisfies properties (1) to (3). We leave the details to the reader. Proceeding in a similar manner in the other components of R_α, we can finally obtain a family $\tilde{g}_t : S^2 \longrightarrow A_t$, $t \in [0,\alpha]$, which satisfies property (4) also.

By Lemma 3 and by the above argument, we stop the process only when $R_\alpha = S^2$. This concludes the proof of Lemma 2. ∎

§7. The non-orientable case

Suppose that $\mathfrak{F}$ is a *transversely orientable* foliation of a compact-three dimensional manifold M. Suppose further that $\mathfrak{F}$ has a vanishing cycle which admits a simple, positive, coherent extension. Theorem 2 guarantees the existence of an open set $V \subset M$ saturated by leaves homeomorphic to $\mathbb{R}^2$. Further, V is homeomorphic to the set K_n obtained from $[\tau_{n+1}, \tau_n] \times D^2$ by identifying $(\tau_n, x) \in \tau_n \times D^2$ with $(\tau_{n+1}, g_n(x)) \in \tau_{n+1} \times D^2$ where $g_n : D^2 \longrightarrow D^2$ is a diffeomorphism onto $g_n(D^2)$.

In the proof of Theorem 3 we used the orientability of M only to guarantee the orientability of V, which has as a consequence that g_n preserves orientation.

When M is not orientable we must consider two cases:

(1) g_n preserves orientation; then K_n is homeomorphic to $D^2 \times S^1$ and $\mathfrak{F} \mid V$ is equivalent to the orientable Reeb foliation of $D^2 \times S^1$.

(2) g_n reverses orientation; then K_n is homeomorphic to the solid Klein bottle K and $\mathfrak{F} \mid V$ is equivalent to the nonorientable Reeb foliation of K.

Since the existence of vanishing cycles, with the hypothesis $\pi_1(M)$ finite or $\pi_2(M) \neq 0$, is independent of orientation of M, we have the following theorem:

Theorem 5. *Let $\mathfrak{F}$ be a transversely orientable C^2 codimension-one foliation of a compact three-dimensional manifold M.*

If $\pi_1(M)$ is finite then $\mathfrak{F}$ has a Reeb component (*orientable or not*).

If $\pi_2(M) \neq 0$ and $\mathfrak{F}$ has no Reeb components, then all the leaves of $\mathfrak{F}$ are compact with finite fundamental group.

VIII. TOPOLOGICAL ASPECTS OF THE THEORY OF GROUP ACTIONS

In this chapter M denotes a C^∞ differentiable manifold and G a simply connected Lie group.

Our objective here is to show how the theory of foliations can be applied to obtain global informations about the orbit structure of a locally free action of G on M. The central theorem of this chapter is the theorem on the rank of S^3 due to E. Lima, which states that two commutative vector fields on S^3 are necessarily linearly dependent at some point. In other words, there do not exist locally free actions of the group $\mathbb{R}^2$ on S^3. We generalize this theorem to higher dimensions showing that no locally free action of G on a compact simply connected manifold M has vanishing cycles.

§1. Elementary properties

Recall that a C^k action of G on M is a C^k map $\varphi : G \times M \longrightarrow M$ satisfying the properties below:

(1) $\varphi(e,x) = x$ for every $x \in M$, where e is the identity of G.
(2) $\varphi(g_1, \varphi(g_2,x)) = \varphi(g_1 \cdot g_2, x)$ for any $g_1, g_2 \in G$ and $x \in M$.

Denote by $\varphi_g : M \longrightarrow M$ the map $\varphi_g(x) = \varphi(g,x)$. From the definition it follows that $\varphi_{g^{-1}} = (\varphi_g)^{-1}$ for every $g \in G$, which proves that $\varphi_g : M \longrightarrow M$ is a C^k diffeomorphism.

The *orbit of a point* $x \in M$ (by the action φ) is the subset $\mathcal{O}_x = \mathcal{O}_x(\varphi) = \{\varphi(g,x) \mid g \in G\}$. The *isotropy group* of $x \in M$ (for the action φ) is the subgroup of G defined by $G_x = G_x(\varphi) = \{g \in G \mid \varphi(g,x) = x\}$. From the definition it is evident that G_x is a closed subgroup of G. Consider the equivalence relation $\sim$ on G such that $g_1 \sim g_2$ if and only if $g_1^{-1} \cdot g_2 \in G_x$. Let G/G_x be the quotient space of G by $\sim$ and $\pi : G \longrightarrow G/G_x$ the projection which takes $g \in G$ to its equivalence class $\pi(g)$.

Theorem 1. *There is on G/G_x a unique differentiable structure such that $\pi : G \longrightarrow G/G_x$ defines a fibered space with fiber G_x. In particular if we consider G/G_x with this structure, π is a submersion and if G_x is discrete then π is a covering map.*

Proof. Letting $\bar{g} = \pi(g)$ we have $\pi^{-1}(\bar{g}) = g \cdot G_x = \{g \cdot g' \mid g' \in G_x\}$, a set which is diffeomorphic to G_x by left translation $L_g : G \longrightarrow G$, $L_g(g') = g \cdot g'$. Consider now a disk D, imbedded in G, of class C^∞, transverse to G_x, with $e \in D$ and such that $\dim(D) = \operatorname{cod}(G_x)$. Shrinking D if necessary, we can assume that for every $g \in D$, D is transverse to $g \cdot G_x$ in g and moreover that D does not have distinct equivalent points. We have then that $\pi^{-1}(\pi(D)) = D \cdot G_x = \{g \cdot h \mid g \in D, h \in G_x\}$, so the map

$$\psi : D \times G_x \longrightarrow \pi^{-1}(\pi(D))$$

given by $\psi(g,h) = g \cdot h$ is a bijection, since it is surjective and if $g \cdot h = g' \cdot h'$ with $g,g' \in D$ and $h,h' \in G_x$ then $g \sim g'$ so $g = g'$ and $h = h'$. Further, since $g \cdot G_x$ is transverse to D at g, for every $g \in D$, and $\dim(D) = \operatorname{cod}(G_x)$, it follows that ψ is a C^∞ diffeomorphism. Therefore $\pi^{-1}(\pi(D))$ is open in G, so, by definition of the quotient topology, $\pi(D)$ is open in G/G_x and $\pi \mid D : D \longrightarrow \pi(D)$ is a homeomorphism, since it is a continuous, open bijection.

For each $g \in G$, consider the disk $D_g = g \cdot D = L_g(D)$.

Then $g \in D_g$, $L_g \mid D : D \longrightarrow D_g$ is a diffeomorphism and for every $g' \in D_g$, D_g is transverse to $g' \cdot G_x$ at g'. By an argument analogous to the above, we have that $\pi(D_g)$ is open in G/G_x and $\pi \mid D_g : D_g \longrightarrow \pi(D_g)$ is a homeomorphism. Moreover the map $\psi_g : D_g \times G_x \longrightarrow \pi^{-1}(\pi(D_g))$ defined by $\psi_g(g',h) = g' \cdot h$ is a C^∞ diffeomorphism. This proves that $\pi : G \longrightarrow G/G_x$ defines a fibered space structure. The collection $A = \{(\pi \mid D_g)^{-1} : \pi(D_g) \longrightarrow D_g \mid g \in G\}$ is a C^∞ atlas on G/G_x, which defines a C^∞ manifold structure on G/G_x. ■

Theorem 2. *Let $\varphi : G \times M \longrightarrow M$ be a C^k, $k \geq 1$, action and let $x \in M$. There exists a unique one-to-one C^k immersion, $\bar{\varphi}_x : G/G_x \longrightarrow \mathcal{O}_x$ such that $\bar{\varphi}_x \circ \pi = \varphi_x$. If G_x is discrete, the map $\varphi_x : G \longrightarrow \mathcal{O}_x$ is a covering map.*

Proof. Let $\bar{g} = \pi(g)$. Define $\bar{\varphi}_x : G/G_x \longrightarrow \mathcal{O}_x$ by $\bar{\varphi}_x(\bar{g}) = \varphi(g,x)$. It is immediate that $\bar{\varphi}_x$ is well-defined and that it is a bijection. We will show now that $\bar{\varphi}_x$ is an immersion. Given $D_g \subset G$ as before, we have that $\bar{\varphi}_x \circ (\pi \mid D_g) : D_g \longrightarrow \mathcal{O}_x$ is given by $\bar{\varphi}_x \circ (\pi \mid D_g)(g') = \varphi(g',x) = \varphi_x(g')$; so it is C^k, and if u is a tangent vector to D_g at g', $D(\overline{\varphi}_x \circ \pi \mid D_g) \cdot u = D\varphi_x(g') \cdot u$. On the other hand $g' \cdot G_x = \{h \in G \mid \varphi(h,x) = \varphi(g',x)\} = \varphi_x^{-1}(\varphi_x(g'))$. One sees therefore that $T_{g'}(g' \cdot G_x) = \{x \in T_{g'}(G) \mid D\varphi_x(g') \cdot u = 0]$; so, if $D\varphi_x(g') \cdot u = 0$ and $u \in T_{g'}(D_g)$ then $u = 0$, since $g' \cdot G_x$ and D_g have complementary dimensions. This shows that $\mathcal{O}_x$ is a manifold immersed injectively in M. If G_x is a discrete subgroup of G, $\pi : G \longrightarrow G/G_x$ is a covering, so $\varphi_x : G \longrightarrow \mathcal{O}_x$ is also a covering, since $\varphi_x = \overline{\varphi}_x \circ \pi$. In particular $\dim(G) = \dim(G/G_x) = \dim(\mathcal{O}_x)$. Conversely if $\dim(G) = \dim(\mathcal{O}_x)$, we have $\dim(G) = \dim(G/G_x)$, which implies that G_x is discrete. ■

When for every $x \in M$, $\dim(\mathcal{O}_x) = \dim G$, we say that the action is *locally free*. The orbits of a locally free action on M are leaves of a foliation in M (see Proposition 1 in Chapter II).

Let us see what the orbits of an action of $\mathbb{R}^2$ on a manifold M can be.

Proposition 1. *Let G be a closed subgroup of $\mathbb{R}^2$. Then G is isomorphic to one of the groups $\mathbb{R}^k \times \mathbb{Z}^\ell$ where k,ℓ are integers with $0 \leq k + \ell \leq 2$. In particular, the orbits of an action of $\mathbb{R}^2$ are immersions of some of the following manifolds: a point, $S^1, \mathbb{R}, S^1 \times S^1, S^1 \times \mathbb{R}, \mathbb{R}^2$.*

Proof. Let $G \subset \mathbb{R}^2$ be a closed subgroup. We distinguish two cases:

(1) G is not discrete
(2) G is discrete.

Consider case (1). In this case there exists a sequence (x_n), $x_n \in G - \{0\}$, $x_n \longrightarrow 0$. Let us write $x_n = (r_n,\theta_n)$ in polar coordinates, where $0 \leq \theta_n \leq 2\pi$. Let $\theta_0 \in [0,2\pi]$ be an accumulation point of the sequence (θ_n). We assert that the line with direction θ_0 is contained in G. Indeed, fix $z = (r,\theta_0)$, $r \neq 0$, and a subsequence $(\theta_{n_k} = \varphi_k)$ such that $\varphi_k \longrightarrow \theta_0$. Put $\rho_k = r_{n_k}$. Now, the set $\rho_k \cdot \mathbb{Z}$ divides $\mathbb{R}$ in intervals of length ρ_k and this implies that there exists $m_k \in \mathbb{Z}$ such that $|m_k\rho_k - r| < \rho_k$. This implies that $\lim_{k\to\infty}(m_k\rho_k,\varphi_k) = (r,\theta_0)$ and so $\lim_{k\to\infty} m_k x_{n_k} = z$. Since G is closed it follows that $z \in G$ and so the line $\mathbb{R}_{\theta_0} = \{(r,\theta_0) \mid r \in \mathbb{R}\} \subset G$. Now, we have two subcases to consider:

(a) There exists a sequence $x_n = (r_n,\theta_n) \in G - \{0\}$ such that $x_n \longrightarrow 0$ and the sequence (θ_n) has two accumulation points θ_1,θ_2, which define different directions in $\mathbb{R}^2$.
(b) There exists $\theta_0 \in [0,\pi]$ such that for any sequence $x_n = (r_n,\theta_n)$ such that $x_n \longrightarrow 0$, the set of accumulation points of (θ_n) is contained in $\{\theta_0,\theta_0 + \pi\}$.

In subcase (a), G contains two different vector subspaces of dimension 1 of $\mathbb{R}^2$ and so $G = \mathbb{R}^2$. In subcase (b), G contains the vector subspace $\mathbb{R}_{\theta_0} = \{(r,\theta_0) \mid r \in \mathbb{R}\}$, but does not contain any other line which passes through the origin. We have two possibilities: (i) $G = \mathbb{R}_{\theta_0}$ or (ii) $G \neq \mathbb{R}_{\theta_0}$. Let us consider possibility (ii). In this case for any $x \in G - \mathbb{R}_{\theta_0}$, the line $x + \mathbb{R}_{\theta_0} = \{x + y \mid y \in \mathbb{R}_{\theta_0}\} \subset G$. Let $\ell = \mathbb{R}_{\theta_0}^{\perp}$ be the subspace of $\mathbb{R}^2$ which is perpendicular to R_{θ_0}. Then $\ell \cap G$ is a subgroup of $\mathbb{R}^2$. Moreover $\ell \cap G$ is discrete. In fact, if $\ell \cap G$ was not discrete then $\ell \subset G$ and we were in case (a). Therefore $\ell \cap G$ is discrete and this implies that $\ell \cap G = \bar{x} \cdot \mathbb{Z}$, where $\bar{x} \in \ell \cap G$ is such that $|\bar{x}| = \delta$, where $\delta = \min\{|x|; x \in \ell \cap G - \{0\}\}$. It follows that $G = \mathbb{R}_{\theta_0} \oplus \bar{x} \cdot \mathbb{Z} \approx \mathbb{R} \times \mathbb{Z}$.

Suppose now that $G \neq \{0\}$ is discrete. If there exists $x \in \mathbb{R}^2$ such that $G = x\mathbb{Z}$ then clearly $G \approx \mathbb{Z}$. If this is not the case, let $x_1, x_2 \in G$ be such that $|x_1| = \inf\{|x|; x \in G - \{0\}\}$ and $|x_1| = \inf\{|x| \mid x \in G - x_1\mathbb{R}\}$. We will prove that $G = x_1\mathbb{Z} + x_2\mathbb{Z} \approx \mathbb{Z} \oplus \mathbb{Z}$. For this it suffices to prove that in the parallelogram with vertices $0, x_1, x_1 + x_2, x_2$ there do not exist other elements of G. In fact, in the triangle with vertices $0, x_1, x_2$ there do not exist other elements of G by the definition of x_1 and x_2, and if there were a $g \in G$ in the triangle $x_1, x_1 + x_2, x_2$ then $x_1 + x_2 - g$ would be in the triangle $0, x_1, x_2$.

The rest of the proposition follows from Theorem 2.

Example. Let us see now an example of an action of $\mathbb{R}^2$ on S^3.

Let $D^2 \times S^1 = \{(x,y) \in \mathbb{R}^2 \times \mathbb{R}^2 \mid x_1^2 + x_2^2 \leq 1, y_1^2 + y_2^2 = 1\}$. Consider the vector fields

$$X(x_1,x_2,y) = (\rho(r)x_1 - \beta x_2)\frac{\partial}{\partial x_1} + (\beta x_1 + \rho(r)x_2)\frac{\partial}{\partial x_2}$$

$$Y(x_1,x_2,y_1,y_2) = -\alpha x_2\frac{\partial}{\partial x_1} + \alpha x_1\frac{\partial}{\partial x_2} - y_2\frac{\partial}{\partial y_1} + y_1\frac{\partial}{\partial y_2},$$

where $r = \sqrt{x_1^2 + x_2^2}$ and $\rho(r)$ is a C^∞ nonnegative function such that $\rho(r) = 0$ if and only if $r = 1$, $d^n\rho/dr^n(1) = 0$ for all $n \geq 1$ and $\rho(r) = 1$ for r in a neighborhood of zero.

It is easy to see that the flows X_t and Y_t commute, i.e., $X_s \circ Y_t = Y_t \circ X_s$ $(s,t \in \mathbb{R})$ on $D^2 \times S^1$, so they define a C^∞ action of $\mathbb{R}^2$ on $D^2 \times S^1$, having two compact orbits which are $\{0\} \times S^1$ and $\partial(D^2 \times S^1)$. In the case that α is an irrational number, all the remaining orbits are planes dense in $D^2 \times S^1$. When α is rational they are imbedded cylinders.

Taking two copies of $D^2 \times S^1$ and identifying them along the boundary $\partial(D^2 \times S^1) = S^1 \times S^1$ by means of a diffeomorphism which takes parallels of one to meridians of the other and vice-versa, we obtain examples of actions of $\mathbb{R}^2$ on S^3 in which the orbits of dimension < 2 are two linked circles

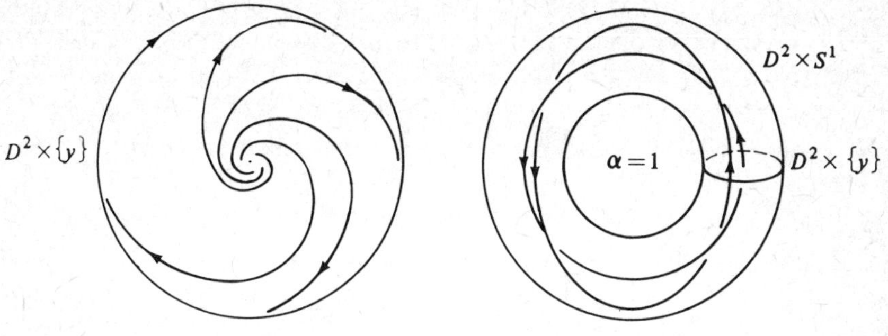

(orbits of X) (orbits of Y)

Figure 1

Other examples of actions can be obtained by taking the suspension of an action $\varphi : \mathbb{Z} \oplus \mathbb{Z} \longrightarrow \mathrm{Diff}(M)$. As we saw in Chapter V, these are locally free actions of $\mathbb{R}^2$, that is, the dimension of every orbit is two.

§2. The theorem on the rank of S^3

We will show in this section that it is impossible to define locally free actions of $\mathbb{R}^2$ on S^3. All the actions considered will be C^r, $r \geq 2$.

An important result in the direction of this theorem is the following proposition.

Proposition 2 ([28]). *Let X and Y be commuting vector fields on $D^2 \times S^1$ such that on points of $\partial(D^2 \times S^1)$ they are linearly independent and tangent to $\partial(D^2 \times S^1)$. There exists a point of $D^2 \times S^1$ where X and Y are linearly dependent.*

Proof. Let $\varphi : \mathbb{R}^2 \times (D^2 \times S^1) \longrightarrow D^2 \times S^1$ be the action induced by X and Y. By hypothesis $\partial(D^2 \times S^1)$ is an orbit of φ, so there exist $r_1, r_2 \in \mathbb{R}^2 - \{0\}$, $r_1 \notin r_2\mathbb{R}$, such that the isotropy group of $\partial D^2 \times S^1$, is $\mathbb{Z} \cdot r_1 + \mathbb{Z} \cdot r_2$. If we take $r = m_1 r_1 + m_2 r_2$ where $m_1 \neq 0$ or $m_2 \neq 0$, then the flow $\varphi_r(t,x) = \varphi(tr,x), x \in \partial D^2 \times S^1$, has all orbits closed with period 1. Now, it is easy to see that there exists $(m_1, m_2) \in \mathbb{Z}^2 - \{0\}$, such that the orbits of φ_r are tangent to the meridians of $S^1 \times S^1$, so that any of such orbits is the boundary of a disk in $D^2 \times S^1$. After a change of basis in $\mathbb{Z} \cdot r_1 + \mathbb{Z} \cdot r_2$ we may assume that $r = r_1$. Let $e_1(x) = \frac{\partial}{\partial t} \varphi(t \cdot r_1, x) \Big|_{t=0}$ and $e_2(x) =$

$= \frac{\partial}{\partial t} \varphi(t \cdot r_2, x) \,\Big|_{t=0}$. Since φ is an $\mathbb{R}^2$-action, e_1 and e_2 are linear combinations of X and Y. Since the orbits of e_1 are closed and bound a disk in $D^2 \times S^1$, we can suppose (after a diffeomorphism) that e_1 is tangent to the boundary of $D^2 \times \{\theta\} \subset D^2 \times S^1$ and e_2 is transverse to $D^2 \times \{\theta\}$ in the points of the boundary $S^1 \times \{\theta\}$. Consequently the proposition will be proved with the following lemma.

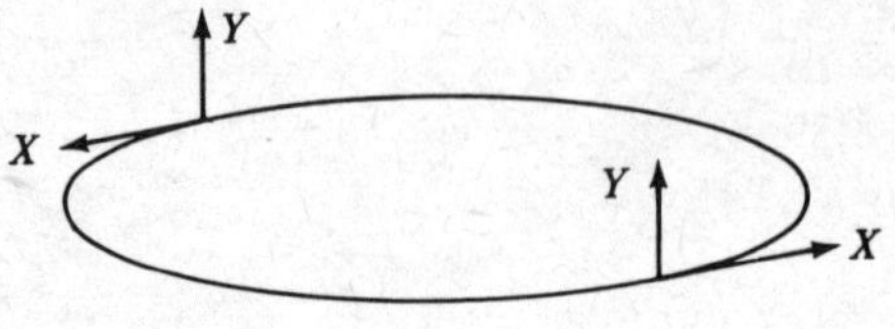

Figure 2

Lemma 1. *Consider the disk $D^2 \subset \mathbb{R}^3 : x_1^2 + x_2^2 \leq 1, x_3 = 0$ and the vector fields $e_1(x) = -x_2\, \partial/\partial x_1 + x_1\, \partial/\partial x_2$, $e_2(x) = \partial/\partial x_3$, $x = (x_1, x_2, 0) \in \partial D^2$. Then for any continuous extension $(e_1(x), e_2(x))_{x \in D^2}$ of $(e_1(x), e_2(x))_{x \in \partial D^2}$ there exists $x_0 \in D^2$ where $e_1(x_0)$ and $e_2(x_0)$ are linearly dependent.*

Proof. First we observe that if there exists an extension $(e_1(x), e_2(x))_{x \in D^2}$ of linearly independent vector fields, then we have an extension by orthonormal fields. In fact if $f_2(x) = e_2(x) - \langle \bar{e}_1(x), e_2(x) \rangle \bar{e}_1(x)$, $\bar{e}_1(x) = \dfrac{e_1(x)}{|e_1(x)|}$ and $\bar{e}_2(x) = \dfrac{f_2(x)}{|f_2(x)|}$, the pair $(\bar{e}_1(x), \bar{e}_2(x))$ would be an orthonormal system.

To prove the lemma, it suffices to show then that there do not exist orthonormal extensions of the pair $(e_1(x), e_2(x))_{x \in \partial D^2}$.

Let $V_{2,3}$ be the space of ordered pairs of orthonormal vectors of $\mathbb{R}^3$. The set $(e_1(x), e_2(x))_{x \in \partial D^2}$ can be understood as a path $\alpha : \partial D^2 \longrightarrow V_{2,3}$, $\alpha(x) = (e_1(x), e_2(x))$. It is sufficient then to show that α is not homotopic to a point.

To prove this we identify $V_{2,3}$ with the real projective space $\mathbb{P}^3$, of dimension 3, in the following manner. To each $(f_1, f_2) \in V_{2,3}$ we associate a vector $w = w(f_1, f_2) \in B(0, \pi) \subset \mathbb{R}^3$ of the ball with center $0 \in \mathbb{R}^3$ and radius π defined as follows: the matrix F, whose columns are f_1, f_2, $f_1 \times f_2$ is orthogonal and has as axis of rotation the subspace ℓ of dimension 1 in $\mathbb{R}^3$. Let P be the plane orthogonal to ℓ. In P the matrix F acts as a rotation of angle $\Theta \in [0, \pi]$. For $\Theta = 0$ one verifies that $F = I$ and we set $w = 0$. For $0 < \Theta < \pi$, we associate the vector $w \in \ell$ with magnitude Θ and the same direction as $v \times F(v)$, where $v \in P - \{0\}$. For $\Theta = \pi$, $v \times F(v) = 0$, and in this case the direction of w is not well-defined. This corresponds to identifying antipodal vectors of magnitude π on ℓ. We obtain in this way a homeomorphism on the quotient space of $B(0,\pi)$ by the equivalence relation which identifies antipodal points of the boundary, which, as is known, is homeomorphic to $\mathbb{P}^3$. The path α, in this model, is represented by the diameter of $B(0,\pi)$, parallel to the x_3-axis and traversed once from the bottom to the top. This path, after identification of its endpoints, is nontrivial in $\mathbb{P}^3$.

Definition. The rank of a manifold is the maximum number of linearly independent commuting vector fields that the manifold admits. This concept was introduced by J. Milnor.

Theorem 3 ([28]). *Every compact three-dimensional manifold with finite fundamental group has rank one. For example, S^3 has rank one.*

Proof. As is well-known every manifold of odd dimension has Euler characteristic zero. So it admits a non-singular vector field and its rank is ≥ 1. Taking a double cover, we can assume that the manifold is orientable. Proceeding by contradiction, suppose that there exists a locally free action of $\mathbb{R}^2$ on some compact three-dimensional manifold with finite fundamental group. By Novikov's theorem, there exists a compact orbit bounding an imbedded solid torus, where there are defined two linearly independent commuting vector fields, which is impossible by Proposition 2. ■

§3. Generalization of the rank theorem

In this section we will see that the theorem on the rank of S^3 generalizes to actions of simply connected groups on compact manifolds with finite fundamental group.

Proposition 3. *Let $\varphi : G \times M \longrightarrow M$ be a locally free action of a Lie group G on a manifold M and $f : I \times I \longrightarrow M$ be a C^0 map such that for every $t \in I$ the curve $s \longrightarrow f(s,t)$ is contained in an orbit F_t of φ. Let $x(t) = f(0,t)$ and suppose that, for every $t \in I$, $x'(t) \notin T_{x(t)}(F_t)$. Then there exists a unique continuous function $\hat{f} : I \times I \longrightarrow G$ such that $\hat{f}(0,t) = e$, and $\varphi(\hat{f}(s,t), x(t)) = f(s,t)$ for any $(s,t) \in I \times I$.*

Proof. By Theorem 2, for every t the map $\varphi_{x(t)} : G \longrightarrow F_t$ given by $\varphi_{x(t)}(g) = \varphi(g, x(t))$ is a covering of F_t. By the lifting theorem for paths, there exists a unique continuous curve $\hat{f}_t : I \longrightarrow G$ such that $\hat{f}_t(0) = e$ and for every $s \in I$ one has $\varphi_{x(t)}(\hat{f}_t(s)) = f(s,t)$. Define $\hat{f}(s,t) = \hat{f}_t(s)$. It is evident that $\hat{f}(0,t) = e$ and $\varphi(\hat{f}(s,t), x(t)) = f(s,t)$ for any $(s,t) \in I \times I$. Moreover, if $h : I \times I \longrightarrow G$ satisfies the conditions $h(0,t) = e$ and

$\varphi(h(s,t),x(t)) = f(s,t)$ for every $(s,t) \in I \times I$, we must have $h \equiv f$, by the uniqueness of the lift of the curve $s \mapsto f(s,t)$ to the covering $\varphi_{x(t)} : G \longrightarrow F_t$. It remains to be proved that $\hat{f}$ is continuous.

Let $x_0 \in M$, $y_0 = \varphi(g_0,x_0)$. Let Σ, $x_0 \in \Sigma$, be a transverse section to the orbits of φ. Since $\varphi_{g_0} : M \longrightarrow M$ is a diffeomorphism which preserves the orbits of φ it follows that $T_{y_0}M = D\varphi(g_0,x_0)(T_{g_0}G \times T_{x_0}\Sigma)$. So $D\varphi(g_0,x_0) : T_{g_0}G \times T_{x_0}\Sigma \longrightarrow T_{y_0}M$ is an isomorphism. By the inverse function theorem there exist neighborhoods $g_0 \in V \subset G$ and $x_0 \in D \subset \Sigma$ such that $U = \varphi(V \times D)$ is a neighborhood of y_0 and $\psi = \varphi \mid V \times D$ is a diffeomorphism of $V \times D$ onto U.

Let us prove that $\hat{f}$ is continuous at all the points of the type $(0,t_0)$, $t_0 \in I$. Fix $t_0 \in I$ and let $x_0 = x(t_0)$, $y_0 = f(0,t_0)$, $g_0 = \hat{f}(0,t_0) = e$ and consider D, V, U and $\psi = \varphi \mid V \times D : V \times D \longrightarrow U$ as above. Let $I_0 = [0,\delta)$ and $J_0 = (t_0 - \delta, t_0 + \delta)$ be intervals such that $f(I_0 \times J_0) \subset U$ and $x(J_0) \subset D$. Define $g : I_0 \times J_0 \longrightarrow V \subset G$ by $(g(s,t),x(t)) = \psi^{-1}(f(s,t))$. Then g is continuous, $g(0,t) = e$ and $\varphi(g(s,t),x(t)) = f(s,t)$ for every $(s,t) \in I_0 \times J_0$. By uniqueness of the lift of the curve $s \longrightarrow f(s,t)$ across $\varphi_{x(t)} : G \longrightarrow F_t$, one concludes that $g(s,t) = \hat{f}(s,t)$ for $(s,t) \in I_0 \times J_0$, so $\hat{f}$ is continuous on $(0,t_0)$.

We prove now that the set $A = \{(s,t) \in I \times I \mid f \text{ is continuous at } (s,t)\}$ is open in $I \times I$. Let $(s_0,t_0) \in A$, $x_0 = x(t_0)$, $y_0 = f(s_0,t_0)$, $g_0 = \hat{f}(s_0,t_0)$ and consider D, V, U, and $\psi = \varphi \mid V \times D$ as above. Since $\hat{f}$ is continuous at (s_0,t_0), there exist open sets $I_0, J_0 \subset I$ such that $s_0 \in I_0$, $t_0 \in J_0$, $x(J_0) \subset D$, $\hat{f}(I_0 \times J_0) \subset V$ and $f(I_0 \times J_0) \subset U$. For every $(s,t) \in I_0 \times J_0$ we have $(\hat{f}(s,t),x(t)) = \psi^{-1}(f(s,t))$, which proves that f is continuous on $I_0 \times J_0$, so A is open in $I \times I$.

For each $t \in I$, let $\alpha(t) = \sup\{s \in I \mid \hat{f} \text{ is continuous at } (\mu,t) \text{ for every } \mu \in [0,s)\}$. It suffices to prove now that $(\alpha(t),t) \in A$ for every $t \in I$. Fix $t_0 \in I$, $s_0 = \alpha(t_0)$, $x_0 = x(t_0)$, $y_0 = f(s_0,t_0)$ and $g_0 = \hat{f}(s_0,t_0)$. Consider D, V, U and $\psi = \varphi \mid V \times D$ as above. Since $s \mapsto \hat{f}(s,t_0)$ is continuous (by construction), there exists $\delta > 0$ such that if $s \in [s_0 - \delta, s_0]$ then $\hat{f}(s,t_0) \in V$ and $\hat{f}$ is continuous at (s_1,t_0), $s_1 = s_0 - \delta$. Since A is open there exists $\delta_1 > 0$ such that the segment $\ell = \{s_1\} \times [(t_0 - \delta_1, t_0 + \delta_1) \cap I] \subset A$. Set $I_0 = (s_0 - \delta, s_0 + \delta) \cap I$ and $J_0 = (t_0 - \delta_1, t_0 + \delta_1) \cap I$. If δ and δ_1 are sufficiently small, we have $f(I_0 \times J_0) \subset U$, $x(J_0) \subset D$ and $\hat{f}(\ell) \subset V$. Now define $g : I_0 \times J_0 \longrightarrow V \subset G$ by $(g(s,t),x(t)) = \psi^{-1}(f(s,t))$. However it is clear that $g(s_1,t) = \hat{f}(s_1,t)$ for $(s_1,t) \in \ell$ and that $\varphi(g(s,t),x(t)) = f(s,t)$ for $(s,t) \in I_0 \times J_0$. By uniqueness of the lift of the curve $s \mapsto f(s,t)$, it follows that $g = \hat{f} \mid I_0 \times J_0$ and therefore $\hat{f}$ is continuous on (s_0,t_0) as desired. ∎

Theorem 4. *Let $\varphi : G \times M \longrightarrow M$ be a locally free action of a simply connected Lie group G on a manifold M where* $\dim M \geq \dim G + 1$. *Then the folia-*

tion of M by orbits of φ does not have vanishing cycles. In particular, if $\dim G =$ $= 2$ *and* $\dim M = 3$, $\mathfrak{F}$ *has no Reeb components.*

Proof. Suppose by contradiction that there exists a leaf F_0 of $\mathfrak{F}$ which has a vanishing cycle $\gamma : I \longrightarrow F_0$. Then γ is a closed curve not homotopic to a constant in F_0 and there is an extension $f : I \times I \longrightarrow M$ such that for every $s \in I, f(s,0) = \gamma(s)$ and for every $t > 0$, $t \in I$, the curve $f_t(s) = f(s,t)$ is a closed curve contained in a leaf F_t of $\mathfrak{F}$, homotopic to a constant in F_t. Let $\hat{f} : I \times I \longrightarrow G$ be as in Proposition 3. Now, for $t > 0$, the curve $f_t(s)$ is homotopic to a constant in F_t, so the curve $\hat{f}_t(s) = \hat{f}(s,t)$, which is a lift of f_t to G, is necessarily closed, that is, $\hat{f}_t(0) = \hat{f}_t(1)$. By continuity we have $\hat{f}_0(0) = \hat{f}_0(1)$, or, $\hat{f}_0$ is a closed curve. But $\hat{f}_0$ is a lift of γ to G, so it cannot be closed since G is simply connected and γ is not homotopic to a constant in F_0, which is a contradiction. We conclude therefore that $\mathfrak{F}$ has no vanishing cycles. ■

Corollary 1. *Let G be a simply connected Lie group and M a compact manifold with finite fundamental group such that* $\dim(M) = \dim(G) + 1 \geq 3$. *Then there is no locally free action of G on M.*

Proof. By Proposition 1 of Chapter VII we know that a foliation $\mathfrak{F}$ of codimension 1 on M necessarily has a vanishing cycle, so it cannot have a locally free action $\varphi : G \times M \longrightarrow M$. ■

Corollary 2. *Let G be a simply connected Lie group and M a manifold such that* $\dim(M) = \dim(G) + 1 \geq 3$. *Let $\varphi : G \times M \longrightarrow M$ be a locally free action. Then the following properties are true:*

(a) *if F is an orbit of φ and $i : F \longrightarrow M$ is the* canonical immersion *then $i_* : \pi_1(F) \longrightarrow \pi_1(M)$ is injective, that is, if γ is a closed curve in F homotopic to a constant in M then γ is homotopic to a constant in F,*

(b) *if γ is a curve transverse to the orbits of φ, then the homotopy class of γ in $\pi_1(M)$ has infinite order.*

Proof. (b) is immediate from Theorem 4 and from the proof of Proposition 1 of Chapter VII. Let us consider (a). Suppose γ is a closed curve in F, homotopic to a constant in M. Let $f : D^2 \longrightarrow M$ be a C^r $(r \geq 2)$ map such that $f \mid S^1 = \gamma$ and f is in general position with respect to the foliation $\mathfrak{F}$ defined by orbits of φ. Let X be a vector field on D^2 tangent to the foliation with singularities $f^*(\mathfrak{F})$ (see Proposition 2 of Chapter VI). $S^1 = \partial D^2$ is a closed orbit of X. We claim that all the nonclosed orbits of X are separatrices of saddles. Indeed, if X has a nonclosed orbit which is not a separatrix of a saddle, let us say δ, the ω-limit set of δ is a closed orbit or a graph and this implies that there exists a closed curve μ in D^2 transverse to the orbits of X.

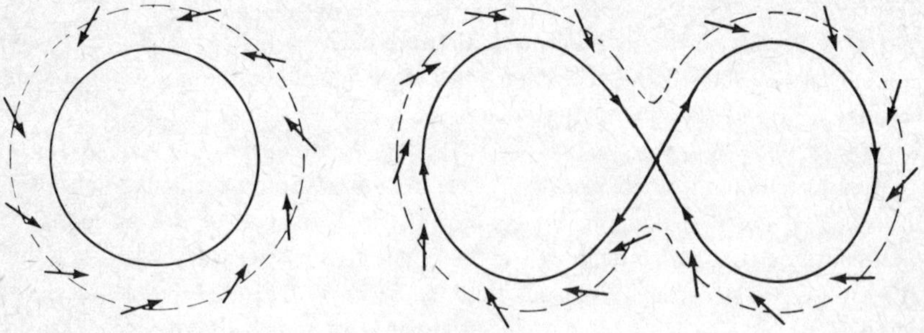

Figure 3

Repeating the argument of Proposition 1 of Chapter VII on the disk $D_1^2 \subset D^2$ bounded by μ, one concludes that $\mathfrak{F}$ has a vanishing cycle, which is not possible. Therefore all the regular orbits are separatrices of saddles. Supposing by contradiction that $S^1 = \partial D^2$ is not homotopic to a constant in its leaf, repeating the argument of Proposition 1 of Chapter VII, it is possible to prove that $\mathfrak{F}$ has a vanishing cycle. We leave the details to the reader.

Corollary 3. *Let G be a simply connected group and M a (noncompact) manifold with finite fundamental group such that* $\dim(M) = \dim(G) + 1 \geq 3$. *Let $\varphi : G \times M \longrightarrow M$ be a locally free action and $\mathfrak{F}$ the foliation whose leaves are the orbits of φ. If F is a leaf of $\mathfrak{F}$ then $\pi_1(F)$ is finite and F is a closed subset of M. In particular every minimal set of $\mathfrak{F}$ is a leaf and if M is simply connected then all the leaves are diffeomorphic to G.*

Proof. Let F be a leaf of $\mathfrak{F}$. By (a) of Corollary 2, $i^* : \pi_1(F) \longrightarrow \pi_1(M)$ is injective so $\pi_1(F)$ is finite and $\#\pi_1(F) \leq \#\pi_1(M)$. In particular if M is simply connected, F is also and therefore if $x \in F$ the map $\varphi_x : G \longrightarrow F$ given by $\varphi_x(g) = \varphi(g,x)$ is a diffeomorphism, since it is a covering.

Suppose by contradiction that $\mathfrak{F}$ has a leaf not closed in M. Then there is a sequence $x_n \in F$, $x_n \longrightarrow x_0 \notin F$. Let Q be a foliation box of x_0. The leaf F cuts Q in an infinite number of plaques, therefore we can construct a closed curve transverse to $\mathfrak{F}$. Since $\pi_1(M)$ is finite this curve has finite order, which contradicts (b) of Corollary 2. ■

§4. The Poincaré-Bendixson Theorem for actions of $\mathbb{R}^2$

One of the versions of the Poincaré-Bendixson theorem for a flow ξ on the sphere S^2 or in the plane says that every minimal set of ξ is an orbit.

In this section we generalize this theorem to locally free actions of $\mathbb{R}^2$ on manifolds of dimension three. Other results in this direction can be found in [6], [38] and [45].

Let $\varphi : G \times M \longrightarrow M$ be an action. A subset of M is *invariant* (under φ) if it is the union of orbits of φ.

A *minimal set* of φ is a subset $\mu \subset M$ which is closed, invariant, non-empty and such that no subset properly contained in μ has these three properties.

The theorem below is a special case of Corollary 3 of Theorem 4. We think it is interesting to include this new proof because it is more geometric than the first and also because it contains the original argument of [28] in the proof of the fact that an action of $\mathbb{R}^2$ on S^3 (if it exists) must have an invariant torus.

Theorem 5. *Let $\varphi : \mathbb{R}^2 \times M \longrightarrow M$ be a locally free action on a simply connected three-dimensional manifold. Then any minimal set of φ is an orbit (in fact by Corollary 1 of Theorem 4, the manifold is not compact).*

Proof. Given $p \in M$ we define $\partial\mathcal{O}_p(\varphi) = \bigcap_{n=1}^{\infty} \overline{\mathcal{O}_p(\varphi) - K_n}$ where $K_n \subset K_{n+1} \subset \mathcal{O}_p(\varphi)$ are compact neighborhoods of p and $\bigcup_{n=1}^{\infty} K_n = \mathcal{O}_p(\varphi)$. It is clear that $\overline{\mathcal{O}_p(\varphi)} = \mathcal{O}_p(\varphi) \cup \partial\mathcal{O}_p(\varphi)$. Now let $\mu \subset M$ be a minimal set of φ and $p \in \mu$.

Proceeding by contradiction we suppose that $\mu \neq \mathcal{O}_p(\varphi)$. Since μ is minimal we have that $\mu = \overline{\mathcal{O}_p(\varphi)}$. So $\partial\mathcal{O}_p(\varphi)$ is nonempty and, since it is closed and invariant, $\partial\mathcal{O}_p(\varphi) = \mu$. This means that if U is a coordinate neighborhood of the foliation $\mathcal{F}(\varphi)$ induced by φ, with $p \in U$, the orbit $\mathcal{O}_p(\varphi)$ has as intersection with U an infinite number of plaques, each being an accumulation point of plaques in $\mathcal{O}_p(\varphi)$.

Thus it follows (Proposition 4, Chapter VII) that there is a closed path $\alpha : S^1 \longrightarrow M$ passing through p and tranverse to the leaves of $\mathcal{F}(\varphi)$. Since M is simply connected, α extends to a map $A : D^2 \longrightarrow M$ in general position with respect to the leaves of $\mathcal{F}(\varphi)$. The foliation $A^*\mathcal{F}(\varphi)$ of D^2 has singularities which are saddles and centers. Further it is transverse to the boundary. Let $q \in \partial D^2$ be such that $A(q) = p$.

Modifying A slightly if necessary, we can assume that the point q is not contained in a separatrix of a saddle. There will exist, then, in the limit set of the point q a *limit cycle* $\gamma : S^1 \longrightarrow D^2$, inducing a curve $\tilde{\gamma} : S^1 \longrightarrow M$, $\tilde{\gamma} = A \circ \gamma$, $\tilde{\gamma}(S^1) \subset \mathcal{O}_{x_0}(\varphi) \subset \mu$ with nontrivial holonomy. So $\mathcal{O}_{x_0}(\varphi)$ is homeomorphic to $\mathbb{R} \times S^1$ or $S^1 \times S^1$. Since $\mathcal{O}_{x_0}(\varphi) \neq \mu$, $\mathcal{O}_{x_0}(\varphi)$ cannot be a torus. Consequently $\mathcal{O}_{x_0}(\varphi)$ is an immersed cylinder with $\partial\mathcal{O}_{x_0}(\varphi) \supset \mathcal{O}_{x_0}(\varphi)$. We next show that this is impossible.

Since $G_{x_0}(\varphi) \neq 0$ there exist $r_1, r_2 \in \mathbb{R}^2$, $r_1 \notin \mathbb{R}r_2$, such that the orbit Γ through x_0 of the flow φ_1, $\varphi_1(t,x) = \varphi(tr_1,x)$, is periodic of period t_0 and the orbits of φ_2, $\varphi_2(t,x) = \varphi(tr_2,x)$, for $x \in \Gamma$ are transverse to Γ. Since φ_1 and φ_2 are commutative, for every $x \in \mathcal{O}_{x_0}(\varphi)$ the orbit of φ_1 through x is periodic of period t_0.

Let $\mathcal{C}$ be a two-dimensional cylinder imbedded in M, transverse to the orbits of the action φ with $\Gamma \subset \mathcal{C}$. We next show that arbitrarily near Γ there exists a closed curve $\Gamma' \subset \mathcal{O}_{x_0} \cap \mathcal{C}$ such that $\Gamma' \cap \Gamma = \emptyset$ and $\Gamma' \cup \Gamma$ is the boundary of a closed cylinder $B \subset \mathcal{O}_{x_0}(\varphi)$. Let V be a cylindrical neighborhood of Γ in $\mathcal{C}$, such that the orbits of φ_2 are transverse to V. If $\epsilon > 0$ is small, the set $V_\epsilon = \{\varphi_2(t,x) \mid -\epsilon < t < \epsilon, x \in V\}$ is a neighborhood of Γ in M and the orbits of $\varphi_2 \mid V_\epsilon$ define a projection $\pi : V_\epsilon \longrightarrow V$. Since Γ is a periodic orbit of φ_1 with period t_0 and $\Gamma \subset V_\epsilon$, there exists a neighborhood $U \subset V_\epsilon$ of Γ such that, for every $x \in U$, $\varphi_1(t,x) \in V_\epsilon$ for $|t| < 2t_0$. Since $\partial\mathcal{O}_{x_0}(\varphi) \supset \mathcal{O}_{x_0}(\varphi)$,

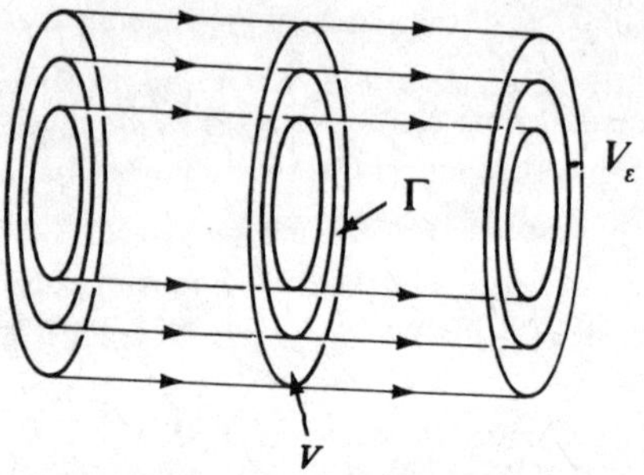

Figure 4

there exist sequences $x_n \in \Gamma$ and $s_n \in \mathbb{R}$ where $|s_n| \longrightarrow \infty$, such that $\varphi_2(s_n,x_n) \longrightarrow x_0 \in U$. If n is sufficiently large, $y_n = \varphi_2(s_n,x_n) \in U \cap \mathcal{O}_{x_0}(\varphi)$, so $\varphi_1(t,y_n) \in V_\epsilon$ for $|t| < 2t_0$ and since the orbit $\mathcal{O}_n$ of y_n under φ_1 has period t_0, we have that $\mathcal{O}_n \subset V_\epsilon$. If $\Gamma_n = \pi(\mathcal{O}_n)$ it is clear that $\Gamma_n \subset \mathcal{O}_{x_0}(\varphi) \cap \mathcal{C}$ and $\Gamma_n \cap \Gamma = \emptyset$. On the other hand, if d is the intrinsic metric of $\mathcal{O}_{x_0}(\varphi)$, $d(\Gamma,\mathcal{O}_n) \longrightarrow \infty$; so, $\Gamma \cup \Gamma_n$ is the boundary of a cylinder $B_n \subset \mathcal{O}_{x_0}(\varphi)$ since $d(\Gamma_n,\mathcal{O}_n)$ is bounded because $\mathcal{O}_n \subset \{\varphi_2(t,x) \mid x \in \Gamma_n, |t| < \epsilon\}$. This proves what we wanted.

Let $A_n \subset \mathcal{C}$ be the annulus bounded by Γ and Γ_n. Since $\lim_{n \to \infty} \Gamma_n = \Gamma$ there exists n_0 such that $A_{n_0} \cap B_{n_0} = \Gamma \cup \Gamma_{n_0}$. We have then that $T = A_{n_0} \cup B_{n_0}$ is a topological torus imbedded in M.

Since M is simply connected, T is the boundary of an open bounded set $W \subset M$ (see [28]). Let X be the vector field given by $X(x) = \dfrac{\partial}{\partial t}\varphi_2(t,x)\Big|_{t=0}$. We have that X is tangent to T on B_{n_0} and X is transverse to T on A_{n_0}. We can assume that for every $x \in A_{n_0}$, $X(x)$ points inward toward W. This implies that every orbit of φ_2 which enters W does not leave in positive time, so $\mathcal{O}_{x_0} \cap A_{n_0} = \Gamma_{n_0} \cup \Gamma$, a contradiction. ■

Remark. The above theorem can be used to prove, by contradiction, the rank theorem of S^3.

Indeed, since S^3 is compact, every action $\varphi : \mathbb{R}^2 \times S^3 \longrightarrow S^3$ has a minimal set μ. If φ were locally free, μ would have to be a compact orbit. So, μ would be diffeomorphic to a torus $S^1 \times S^1$. Using Alexander's theorem (Proc. Nat. Acad. Sci. (1924) pp. 6-8) according to which for any imbedded torus μ, one of the connected components of $S^3 - \mu$ is a solid torus, it follows again from Proposition 2 that there is a contradiction. This proof of the rank theorem was given in ([28]), when Novikov's theorem had not yet been proved.

One problem still open about the rank of manifolds is to find out the rank of S^{2n+1}.

It is well-known that there do not exist 2-dimensional plane fields on S^5 ([53]). On the other hand, the existence of locally free actions of R^2 on S^7 is unknown.

In [48], H. Rosenberg, R. Roussarie, and D. Weil prove the following result: a compact orientable three-dimensional manifold has rank 2 if and only if it is a fibration over S^1 with fiber T^2.

§5. Actions of the group of affine transformations of the line

The Lie group of affine transformations of the line is the set G_2 of all affine diffeomorphisms $f: \mathbb{R} \longrightarrow \mathbb{R}$ with the operation of composition $(f,g) \mapsto f \circ g$. The affine transformations are those of the form $f(x) = ax + b$ where $a,b \in \mathbb{R}$, $a \neq 0$, are fixed. Given $f,g \in G_2$, $f(x) = ax + b$, $g(x) = a'x + b'$, the composition $f \circ g(x) = (aa')x + (ab' + b)$ is also affine so G_2 is a two-dimensional Lie group which has two connected components. We consider only the connected component $G_2^+ = \{f(x) = ax + b \mid a > 0\}$ of affine transformations that preserve the orientation of $\mathbb{R}$.

The following facts can be easily verified:

(a) G_2^+ is not abelian, so is not isomorphic to $\mathbb{R}^2$,
(b) G_2^+ is diffeomorphic to $\mathbb{R}^2$, so is simply connected,
(c) G_2^+ is isomorphic to the subgroup G^+ of $GL(2,\mathbb{R})$, defined by the set of matrices $\begin{bmatrix} a & b \\ 0 & 1 \end{bmatrix}$ such that $b \in \mathbb{R}$ and $a > 0$.

In this section we will prove that the two-dimensional orbits of an action of G_2^+ on a manifold M^n are diffeomorphic to planes or cylinders. Finally, we will see an example of a locally free action of G_2^+ on a compact manifold M^3 which has a countably infinite set of cylindrical orbits whose complement consists of orbits diffeomorphic to the plane. As we will see, all the orbits of this action are dense in M^3.

Proposition 4. *Let φ be an action of G^+ on a manifold M^n. Then the two-dimensional orbits of φ are diffeomorphic to planes or cylinders.*

Proof. Let $p \in M$ and $G_p = \{g \in G^+ \mid \varphi(g,p) = p\}$ be the isotropy group of p. As we know, the orbit $\mathcal{O}(p)$ is diffeomorphic to the manifold $G^+ \mid G_p$, obtained as the quotient of G^+ by the equivalence relation $\sim$ on G^+ which identifies g_1 and g_2 if $g_1 g_2^{-1} \in G_p$. If $\mathcal{O}(p)$ has dimension 2, it is clear that G_p is a discrete subgroup of G^+. It suffices then to prove that every discrete subgroup of G^+ is of the form $H = \{g^n \mid n \in \mathbb{Z}\}$ where $g = \begin{bmatrix} a & b \\ 0 & 1 \end{bmatrix}$ is fixed. Let us identify G^+ with the half plane $P^+ = \{(x,y) \in \mathbb{R}^2 \mid x > 0\}$.

With this identification, the product on P^+ is given by $(x,y) \circ (x',y') \mapsto (xx', xy' + y)$ and $g = (a,b)$. First suppose that $a \neq 1$. In this case the line $r = \{(1,y) \mid y \in \mathbb{R}\}$ is identified by $\sim$ with the line $r \circ g = \{(a, b+y) = (1,y) \circ (a,b) \mid y \in \mathbb{R}\}$.

The lines r and $r \circ g$ bound a strip $A = \{(x,y) \mid 1 \leq x \leq \max(a, a^{-1})\}$. It is easy to see that the two distinct points contained in the interior of A are not identified by $\sim$; so $\mathcal{O}(p)$ is diffeomorphic to the quotient of A by $\sim$; that is, the manifold obtained from A be identifying the boundaries r and $r \circ g$ by the diffeomorphism $(1,y) \in r \mapsto (a, b+y) \in r \circ g$, which is clearly a cylinder.

In the case that $a = 1$ and $b \neq 0$ we can make the same argument, substituting the line r by the line $s = \{(x,0) \mid x \in \mathbb{R}\}$ and the strip A by the strip $B = \{(x,y) \mid 0 \leq y \leq |b|x\}$ bounded by the lines s and $s \circ g = \{(x,y) \mid y = |b|x\}$.

In the case that $a = 1$ and $b = 0$, clearly $\mathcal{O}(p)$ is diffeomorphic to $\mathbb{R}^2$. Consider then a discrete subgroup $H \subset G^+$. First we are going to prove that H is contained in a line ℓ which passes through $(1,0)$. Suppose that (a,b), $(c,d) \in H$ where $a \neq 1$. In this case it is easy to see that $(a,b)^n = (a^n, b(a^n - 1)(a-1)^{-1})$ for every $n \in \mathbb{Z}$ Therefore

$$h_n = (a,b)^{-n} \circ (c,d) \cdot (a,b)^n = (c, \frac{1}{a-1}[bc(1 - a^{-n}) + b(a^{-n} - 1)] + da^{-n}).$$

If $a > 1$ one sees that $\lim_{n \to \infty} h_n = (c, (bc - b)/(a-1)) = h \in H$, since $h_n \in H$ for every $n \in \mathbb{Z}$ and H is closed. Since H is discrete one concludes that $h_n = h$ for every $n \geq n_0$, which implies that (a,b) and (c,d) commute, as can easily be seen using the relation $h_{n+1} \circ h_n^{-1} = (1,0) = 1$. On the other hand, this implies the relation $b(c-1) = d(a-1)$ and so $(c,d) \in \ell = \{(x,g) \mid b(x-1) = (a-1)y\}$. In the case that $a = 1$ and $c \neq 1$ the same argument applies, so we can assume that $a = c = 1$ and in this case $H \subset \{(1,y) \mid y \in \mathbb{R}\}$. Suppose now that $H \subset \ell \cap P^+ = \{(x,y) \mid x > 0$ and $y = k(x-1)\}$. In this case $\ell \cap P^+$ is a subgroup of G^+ isomorphic to the additive group $\mathbb{R}$ by the isomorphism $f(t) = (e^t, k(e^t - 1))$, as can easily be seen. Therefore $f^{-1}(H)$ is a discrete subgroup of $\mathbb{R}$, so $H = \{g^n \mid n \in \mathbb{Z}\}$ for some $g \in H$. In the case that $H \subset \ell = \{(1,y) \mid y \in \mathbb{R}\}$, the argument is similar. ■

Example. First we are going to define a locally free action $\tilde{\varphi}$ of G^+ on $\mathbb{R}^3$. The manifold M^3 will be constructed as the quotient of $\mathbb{R}^3$ by an equivalence relation $\sim$ defined in such a way that $\tilde{\varphi}$ induces an action φ of G^+ on M.

(A) Construction of $\tilde{\varphi}$.

Let $A = \begin{bmatrix} 2 & 1 \\ 1 & 1 \end{bmatrix}$. As can be directly verified the eigenvalues of A are λ and λ^{-1} where $0 < \lambda = (3 - \sqrt{5})/2 < 1$. Let $v = (v_2, v_3)$ be an eigenvector of A relative to λ. Define $\tilde{\varphi} : G^+ \times \mathbb{R}^3 \longrightarrow \mathbb{R}^3$ by

$$\tilde{\varphi}\,(g;(x,z)) = (x + \frac{1}{a}\log\alpha,\ z + \frac{\beta}{\alpha}\ e^{-ax}\cdot v)$$

where in the above expression $g = \begin{bmatrix} \alpha, & \beta \\ 0 & 1 \end{bmatrix}$, $z \in \mathbb{R}^2$ and $a = ln(\lambda) < 0$.

As can be verified directly $\tilde{\varphi}(g' \circ g; p) = \tilde{\varphi}(g'; \tilde{\varphi}(g;p))$ and $\tilde{\varphi}(I,p) = p$ for any $g, g' \in G^+$ and $p \in \mathbb{R}^3$. Therefore $\tilde{\varphi}$ is an action of G^+ on $\mathbb{R}^3$.

(B) Construction of M.

Consider on $\mathbb{R}^3 = \mathbb{R} \times \mathbb{R}^2$ the equivalence relation $\sim$ which identifies two points (x,z) and (x',z') if one of the following conditions are true: (a) $x = x'$ and $z - z' \in \mathbb{Z}^2$ or (b) $x - x' = k \in \mathbb{Z}$ and $z' = A^k z$.

Making the identification (a) we get the manifold $\mathbb{R} \times T^2$. Note that A and A^{-1} are matrices with integer entries. This implies that if $z - z' \in \mathbb{Z}^2$ then $A^k z - A^k z' \in \mathbb{Z}^2$ so A induces a diffeomorphism $\overline{A} : T^2 \longrightarrow T^2$. The identification (b) is equivalent to identifying two points (x,p) and $(x',p') \in \mathbb{R} \times T^2$ when $x - x' = k \in \mathbb{Z}$ and $p' = \overline{A}^k(p)$. Note that the equivalence relation $\sim$ is generated by the diffeomorphisms f_1, f_2 and $f_3 : \mathbb{R}^3 \longrightarrow \mathbb{R}^3$ defined by $f_1(x,z) = (x + 1, A^{-1}z)$, $f_2(x,z) = (x, z + e_1)$ and $f_3(x,z) = (x, z + e_2)$ where $e_1 = \binom{1}{0}$ and $e_2 = \binom{0}{1}$. This means that two points q and q' of $\mathbb{R}^3$ are equivalent if and only if $q = h(q')$ where $h = f_1^{m_1} \circ f_2^{n_1} \circ f_3^{\ell_1} \circ \ldots \circ f_1^{m_k} \circ f_2^{n_k} \circ f_3^{\ell_k}$, $k \in \mathbb{N}$ and $m_i, n_i, \ell_i \in \mathbb{Z}$ for $i = 1, \ldots, k$. Denote by H the set of all diffeomorphisms of this type.

Let $M = \mathbb{R}^3/\sim$ with the quotient topology and $\pi : \mathbb{R}^3 \longrightarrow M$ the natural projection of $\sim$. It is easily verified that π is a covering map and therefore we can induce on M a C^∞ differentiable manifold structure as was done in Chapter I. An atlas of M is the set of all local charts $(\pi(V), (\pi \mid V)^{-1})$ where $V = \{(x, z_2, z_3) \mid |x - \alpha| < 1/2,\ |z_2 - \beta_2| < 1/2,\ |z_3 - \beta_3| < 1/2\}$, where $(\alpha, \beta_2, \beta_3) \in \mathbb{R}^3$. The changes of coordinates of this atlas are restrictions of diffeomorphisms of H.

(C) Construction of φ.

Observe that $\tilde{\varphi}$ commutes with the diffeomorphisms f_1, f_2 and f_3, that is, for every $(g,p) \in G^+ \times \mathbb{R}^3$ and $i = 1,2,3$ one has $\tilde{\varphi}(g, f_i(p)) = f_i(\tilde{\varphi}(g,p))$. We can then define a C^∞ action on M setting

$$\varphi(g, \pi(p)) = \pi(\tilde{\varphi}(g,p)) .$$

It is easy to verify that φ is well-defined by this expression, that is, if $\pi(p) = \pi(p')$ then $\pi(\tilde{\varphi}(g,p)) = \pi(\tilde{\varphi}(g,p'))$.

(D) Geometric description of the orbits of φ.

First let us see what happens to the orbits of $\tilde{\varphi}$ as we make the identification (a). Observe that the orbit $\tilde{\mathcal{O}}(p)$ of a point $p \in \mathbb{R}^3$ by $\tilde{\varphi}$ is the plane that passes through p and is parallel to the subspace of $\mathbb{R}^3$ generated by the vectors $(1,0,0)$ and $(0, v_2, v_3)$. In making the identification (a) we do not change the first coordinate and each plane $\{x\} \times \mathbb{R}^2$ is transformed in the quotient to the surface $\{x\} \times T^2$ where $T^2 = \mathbb{R}^2/\mathbb{Z}^2$. On the other hand $\tilde{\mathcal{O}}(p)$ cuts the plane $\{x\} \times \mathbb{R}^2$ along the line $t \longrightarrow p' + tv$, $p' = (p_2, p_3)$, which has slope $\Theta = v_3/v_2 = -(1 + \sqrt{5})/2$ in this plane. Therefore after identification (a), the orbit $\tilde{\mathcal{O}}(p)$ is transformed into an immersed submanifold $\mathbb{R} \times \xi$, where ξ is an orbit of the irrational flow of slope Θ in T^2. As we know ξ is dense in T^2 (see exercise 13 of Chapter II), so $\mathbb{R} \times \xi$ is dense in $\mathbb{R} \times T^2$. This implies that all the orbits of φ are dense in M.

We will now see what occurs after identification (b). The immersed submanifold $\mathbb{R} \times \xi$ meets $\{k\} \times T^2$ in the line $\{k\} \times \xi$ ($k \in \mathbb{Z}$) and this line will be identified by (b) with the line $\{0\} \times \overline{A}^k(\xi)$. Now, the irrational flow of slope Θ is invariant under $\overline{A}$, since v is an eigenvector for A, so $\overline{A}(\xi)$ is an orbit of this flow. In the case that ξ contains a periodic point q of period k of $\overline{A}$ ($\overline{A}^k(q) = q$), the endpoints $\{0\} \times \xi$ and $\{k\} \times \xi$ of the strip $\{x\} \times \xi$, $0 \le x \le k$, will be identified and we hence have a cylindrical orbit of φ.

Conversely, if the endpoints $\{0\} \times \xi$ and $\{k\} \times \xi$ ($k > 0$) are identified, we have $\overline{A}^k(\xi) = \xi$ and this implies that ξ contains a unique fixed point q_0 of $\overline{A}^k$, since $\overline{A}^k \mid \xi$ is a contraction, that is, if q and $q' \in \xi$ then $d(\overline{A}^k(q), \overline{A}^k(q')) \le \lambda^k d(q,q')$. With this argument we establish a one-to-one correspondence between cylindrical orbits of φ and periodic orbits of $\overline{A}$. As is known, $\overline{A}$ has countably infinite periodic orbits (see [34]), so φ has countably infinite cylindrical orbits. The remaining orbits of φ are diffeomorphic to the plane, by the previous proposition.

APPENDIX
FROBENIUS' THEOREM

Let $\mathfrak{F}$ be a k-dimensional, C^r $(r \geq 1)$ foliation defined on a manifold M of dimension n. The foliation $\mathfrak{F}$ induces on M a C^{r-1} k-dimensional plane field, denoted by $T\mathfrak{F}$ (see Chapter II). Here $T\mathfrak{F}(p) = T_p\mathfrak{F}$ is the subspace of T_pM, tangent to the leaf of $\mathfrak{F}$ which passes through p. A natural question is the following. Let P be a C^r $(r \geq 1)$ k-plane field on M. Under what conditions does there exist a foliation $\mathfrak{F}$ such that $T\mathfrak{F} = P$?

As we saw earlier this does not happen in general for an arbitrary k-plane field $(k \geq 2)$. Frobenius' Theorem provides necessary and sufficient conditions for such a foliation to exist. With the purpose of proving this theorem, we introduce in §1 of this appendix the notion of Lie bracket between two vector fields, proving its main properties. In §2 we prove Frobenius' Theorem. In §3 we state and proof, without going into details the dual version of Frobenius' Theorem in terms of differential forms.

§1. Vector fields and the Lie bracket

Recall that a C^r $(r \geq 0)$ vector field on M is a C^r map $X : M \longrightarrow TM$ such that $\pi \circ X =$ identity map of M, where $\pi : TM \longrightarrow M$ is the natural projection of the tangent fibration. Denote by $\mathfrak{X}^r(M)$ the set of all C^r vector fields on M.

Definition. Let $X \in \mathfrak{X}^0(M)$. A C^1 curve $\gamma : (a,b) \longrightarrow M$ is called an *integral curve of* X if $\gamma'(t) = X(\gamma(t))$ for every $t \in (a,b)$.

On a local chart $x : U \subset M \longrightarrow \mathbb{R}^n$ a vector field X is represented by the vector field $x_*(X)$ defined on $x(U) \subset \mathbb{R}^n$ by the expression

$$X_*(p) = x_*(X)(p) = dx_{x^{-1}(p)} \cdot X(x^{-1}(p)) .$$

Since $x : U \longrightarrow \mathbb{R}^n$ is a C^∞ map, the vector field X_* has the same class of differentiability as X. A fact that can be easily verified is the following. Let $\gamma : (a,b) \longrightarrow U \subset M$ be a C^1 curve. Then γ is an integral curve of X if and only if $x \circ \gamma : (a,b) \longrightarrow x(U)$ is an integral curve of X_*. Putting the above fact together with the theorem on the existence and uniqueness of solutions of ordinary differential equations on $\mathbb{R}^n$, we can state the following result.

Theorem (*Existence and uniqueness*). *Let X be a C^r, $r \geq 1$, vector field on M. Given $p_0 \in M$ and $t_0 \in \mathbb{R}$ there exist $\epsilon > 0$ and a neighborhood U_0 of p_0 in M such that for every $p \in U_0$ there exists an integral curve γ of X defined on $(t_0 - \epsilon, t_0 + \epsilon)$ with $\gamma(t_0) = p$. If $\gamma_1 : I_1 \longrightarrow M$ and $\gamma_2 : I_2 \longrightarrow M$ are two integral curves of X such that $t_0 \in I_1 \cap I_2$ and $\gamma_1(t_0) = \gamma_2(t_0)$ then $\gamma_1 = \gamma_2$ on $I_1 \cap I_2$.*

From the above theorem we can conclude that for each $p \in M$ there exist $a < 0 < b$ and an integral curve $\gamma : (a,b) \longrightarrow M$ with $\gamma(0) = p$ such that for any other integral curve $\gamma_1 : I_1 \longrightarrow M$ such that $0 \in I_1$ and $\gamma_1(0) = p$ then $I_1 \subset (a,b)$ and $\gamma_1 = \gamma \mid I_1$. The interval (a,b) is called the *maximal interval of definition* of the solution which passes through p; we will denote it by I_p.

Let $V = \{(t,p) \in \mathbb{R} \times M \mid t \in I_p\}$. Define $\varphi : V \longrightarrow M$ such that for each fixed $p \in M$, the curve $t \longrightarrow \varphi(t,p)$ is the integral curve of X with $\varphi(0,p) = p$.

Definition. The map φ defined above is called the *flow of X*. Denote the map $p \in M \longrightarrow \varphi(t,p)$, $t \in I_p$, by X_t. For convenience, we omit the domain of such a map.

From the existence and uniqueness theorem, it follows that V is open in $\mathbb{R} \times M$ and that φ has the following property:

$$\varphi(s + t,p) = \varphi(s,\varphi(t,p)) = \varphi(t,\varphi(s,p))$$

whenever the expressions make sense, that is, are compatible with the domain V. If $V = \mathbb{R} \times M$, the above identities can also be written as $X_{s+t} = X_s \circ X_t = X_t \circ X_s$.

The theorem on the differentiable dependence of the solutions of an ordinary differential equation on $\mathbb{R}^n$ with respect to initial conditions implies the following:

Theorem. *Let X be a $C^r (r \geq 1)$ vector field defined on M. Then $\varphi : V \longrightarrow M$ is C^r.*

Suppose now that $t_0 \in \mathbb{R}$ is such that $\varphi(t_0,p)$ is defined for all $p \in M$. In this case $X_{t_0} : M \longrightarrow M$ is a C^r diffeomorphism. In fact, if $\varphi(t_0,p)$ is defined for all $p \in M$, it is easy to see that $\varphi(-t_0,p)$ is also, therefore $X_{t_0} \circ X_{-t_0} =$

$= X_{-t_0} \circ X_{t_0}$ = the identity on M. Since X_{t_0} and X_{-t_0} are C^r, by the above theorem we conclude that X_{t_0} is a C^r diffeomorphism. More generally, if V_t is the open set in M, defined by $V_t = \{p \in M \mid t \in I_p\}$ then V_t is the domain of X_t, where $X_t(V_t) = V_{-t}$. Therefore $X_t : V_t \longrightarrow V_{-t}$ is a C^r diffeomorphism whose inverse is $X_{-t} : V_{-t} \longrightarrow V_t$. If M is compact it is possible to prove that $V = \mathbb{R} \times M$ and hence $V_t = M$ for every $t \in \mathbb{R}$ (see [34]).

Change of variables

Let $f : N \longrightarrow P$ be a local $C^r (r \geq 1)$ diffeomorphism. If X is a C^s vector field on P we can define a vector field $f^*(X)$ on N by

$$f^*(X)(q) = (Df(q))^{-1} \cdot X(f(q)) .$$

The vector field $f^*(X)$ is C^ℓ where $\ell = \min(s, r-1)$. If f is a diffeomorphism we can write

$$f^*(X)(q) = (Df(q))^{-1} \cdot X(f(q)) = Df^{-1}(f(q)) \cdot X(f(q)) .$$

In this case the operator f^* has an inverse f_* defined by

$$f_*(Y)(p) = Df(f^{-1}(p)) \cdot Y(f^{-1}(p)) .$$

The following proposition is a direct consequence of these definitions.

Proposition 1. *Let $f : N \longrightarrow P$ and $g : P \longrightarrow M$ be $C^r (r \geq 2)$ diffeomorphisms. The following properties hold:*

(a) $(g \circ f)^* = f^* \circ g^*$ *and* $(g \circ f)_* = g_* \circ f_*$.
(b) *Let X be a $C^s (s \geq 1)$ vector field defined on P.* Let $\varphi : V \longrightarrow P$ *and $\varphi^* : V^* \longrightarrow N$ be the flows of X and $f^*(X)$ respectively. Then* $V^* = \{(t,p) \in \mathbb{R} \times N \mid (t, f(p)) \in V\}$ *and further* $f(\varphi^*(t,p)) = \varphi(t, f(p))$ *for every* $(t,p) \in V^*$.

Property (b) above tells us that if we know the integral curves of $f^*(X)$ and the diffeomorphism f, then we know the integral curves of X, and conversely. In the notation $X_t(p) = \varphi(t,p)$, (b) is written as $f \circ X_t^* = X_t \circ f$.

Lie Bracket

Let X and Y be two vector fields on M. For simplicity we are going to assume that X is C^2, Y is C^1 and the flow of X is defined on $\mathbb{R} \times M$.

Fix $p \in M$ and $t \in \mathbb{R}$; the vector

$$v(t) = X_t^*(Y)(p) = DX_{-t}(X_t(p)) \cdot Y(X_t(p))$$

is tangent to M at p. So $t \mapsto v(t)$ is a C^1 curve in T_pM.

Definition. The Lie bracket of X and Y is the vector field $[X, Y]$ defined by by

$$[X, Y](p) = \frac{d}{dt}(X_t^*(Y))_{t=0} .$$

Note that the definition of $[X, Y](p)$ is local, that is, depends only on values of X and Y in a neighborhood of p. Consequently the hypothesis of the flow of X being defined on $\mathbb{R} \times M$ is unnecessary.

We will next see that the bracket $[X, Y]$ is defined even when X is C^1.

Lemma 1. (Expression of $[X, Y]$ on a local chart).

Let $x : U \subset M \longrightarrow \mathbb{R}^n$ be a local chart. Denote by $\partial/\partial x_i$ the vector field on U defined by $\partial/\partial x_i(q) = (dx(q))^{-1} \cdot e_i$ where $\{e_1 ,..., e_n\}$ is the canonical basis of $\mathbb{R}^n$. If *X and Y are vector fields on M, we can write $X = \sum_{i=1}^n a_i\, \partial/\partial x_i$, $Y = \sum_{i=1}^n b_i\, \partial/\partial x_i$. Then $[X, Y] = \sum_{i=1}^n c_i\, \partial/\partial x_i$, where*

$$c_i = \sum_{j=1}^{n} (a_j \frac{\partial b_i}{\partial x_j} - b_j \frac{\partial a_i}{\partial x_j}) .$$

In particular we can define $[X, Y]$ even when X is C^1. Further, if X and Y are C^r, $[X, Y]$ is C^{r-1}.

Proof. We can write $x_*(X) = \sum_{i=1}^n \alpha_i e_i = \overline{X}$ and $x_*(Y) = \sum_{i=1}^n \beta_i e_i = \overline{Y}$, where $\alpha_i = a_i \circ x^{-1}$ and $\beta_i = b_i \circ x^{-1}$, $i = 1 ,..., n$. With this notation, taking into account Proposition 1, it is easy to see that $(\overline{X}_t)^*(\overline{Y}) = x_*((X_t)^*(Y))$ where $\overline{X}_t$ and X_t are the flows of $\overline{X}$ and X respectively. At a point $q \in x(u)$, this identity can be written as below

$$(\overline{X}_t)^*(\overline{Y})(q) = Dx(p) \cdot ((X_t)^*(Y)(p)),\ p = x^{-1}(q) .$$

Since p is fixed, $Dx(p) : T_pM \longrightarrow \mathbb{R}^n$ is a linear transformation and therefore

$$[\overline{X}, \overline{Y}](q) = \frac{d}{dt}\{Dx(p) \cdot ((X_t^*)(Y)(p))\}_{t=0} =$$

$$Dx(p) \cdot \frac{d}{dt}\{X_t^*(Y)(p)\}_{t=0} = Dx(p) \cdot [X, Y](p) = x_*([X, Y])(q) .$$

We can say then that $x_*[X, Y] = [x_*(X), x_*(Y)]$. It now suffices to calculate $[\overline{X}, \overline{Y}](q)$. For this we are going to look at $\overline{X}$ and $\overline{Y}$ as functions $\overline{X}, \overline{Y} : x(U) \longrightarrow \mathbb{R}^n$. We will also use the notation $W_t = D\overline{X}_{-t}(\overline{X}_t(q))$. We will look at $t \longrightarrow W_t$ as a curve (C^1) in $\mathcal{L}(\mathbb{R}^n)$, the set of linear transformations from $\mathbb{R}^n$ to $\mathbb{R}^n$. With this notation we have that $W_0 = I$ and that

$$[\overline{X},\overline{Y}](q) = \frac{d}{dt}\{W_t \cdot \overline{Y}(\overline{X}_t(q))\}_{t=0} =$$

$$\frac{d}{dt} W_t \Big|_{t=0} \cdot \overline{Y}(q) + D\overline{Y}(q) \cdot \frac{d}{dt}(\overline{X}_t(q))_{t=0}.$$

On the other hand, it can be verified that $W_t \cdot D\overline{X}_t(q) = I$, therefore

$$\frac{d}{dt} W_t \Big|_{t=0} = -\frac{d}{dt}(D\overline{X}_t(q))_{t=0} = -D\left(\frac{d}{dt}\overline{X}_t(q)\right)_{t=0} = -D\overline{X}(q) .$$

Thus we conclude that

$$[\overline{X},\overline{Y}](q) = D\overline{Y}(q) \cdot \overline{X}(q) - D\overline{X}(q) \cdot \overline{Y}(q) .$$

Therefore the ith component of $[\overline{X},\overline{Y}](q)$ is

$$\sum_{j=1}^{n} \left(\alpha_j \frac{\partial \beta_i}{\partial x_j} - \beta_j \frac{\partial \alpha_i}{\partial x_j}\right) .$$

This concludes the proof. ■

An important consequence of this lemma is the following.

Proposition 2. *The following properties hold:*

(a) *If $f : N \longrightarrow P$ is a $C^r (r \geq 2)$ diffeomorphism and $X, Y \in \mathfrak{X}^s(P)$, $s \geq 1$, then*

$$f^*([X,Y]) = [f^*(X), f^*(Y)] .$$

(b) *If X and Y are two vector fields defined on an open set $U \subset \mathbb{R}^n$, then*

$$[X,Y](q) = DY(q) \cdot X(q) - DX(q) \cdot Y(q), \; q \in V .$$

(c) $[X,Y] = -[Y,X]$

(d) $[\ ,\ ]$ *is bilinear*: $[aX_1 + bX_2, Y] = a[X_1,Y] + b[X_2,Y]$ *and* $[X, aY_1 + bY_2] = a[X,Y_1] + b[X,Y_2]$.

The proof is immediate from the lemma.

Commuting flows and actions of $\mathbb{R}^k$.

Recall that a C^r action of a Lie group G on M is a C^r map $\varphi : G \times M \longrightarrow M$ such that

(a) $\varphi(e,p) = p$ for every $p \in M$, where e is the identity of G,

(b) $\varphi(g_1 \cdot g_2, p) = \varphi(g_1, \varphi(g_2,p))$ for any $g_1, g_2 \in G$ and $p \in M$.

It follows immediately from the definition that for every $g \in G$, the map $p \in M$

$\longrightarrow \varphi_g(p) = \varphi(g,p)$ is a C^r diffeomorphism, whose inverse is $\varphi_{g^{-1}}(p) = \varphi(g^{-1},p)$.

If $G = \mathbb{R}^k$, the additive group, an action $\varphi : \mathbb{R}^k \times M \longrightarrow M$ is generated by k vector fields on M. Indeed, let $\{v_1 ,\ldots, v_k\}$ be a basis for $\mathbb{R}^k$ and for each $j \in \{1 ,\ldots, k\}$ set $\varphi_j(t,p) = \varphi(tv_j,p)$. As is easy to verify, for every $j \in \{1 ,\ldots, k\}$, φ_j is a C^r flow, to which is associated the vector field X^j, of class C^{r-1}, defined by $X^j(p) = \frac{d}{dt}\varphi_j(t,p)\big|_{t=0}$. It is easy to see that the flow of X^j is φ_j, so we can write

$$\varphi\Big(\sum_{i=1}^{k} t_i v_i, p\Big) = X^1_{t_1} \circ X^2_{t_2} \circ \ldots \circ X^k_{t_k}(p) .$$

We say that the vector fields $X^1 ,\ldots, X^k$ are generators of the action φ. Since addition in $\mathbb{R}^k$ is commutative, we have that

$$X^i_t \circ X^j_s = X^j_s \circ X^i_t ,$$

for any $i,j \in \{1 ,\ldots, k\}$ and $s,t \in \mathbb{R}$.

Proposition 3. *Let $X^1 ,\ldots, X^k$ be $C^r\,(r \geq 1)$ vector fields defined on M. Suppose that for every $i \in \{1 ,\ldots, k\}$, the flow X^i_t is defined on $\mathbb{R} \times M$. The following statements are equivalent:*

(a) *The fields $X^1 ,\ldots, X^k$ are generators of a C^r action $\varphi : \mathbb{R}^k \times M \longrightarrow M$.*
(b) *For any $i,j \in \{1 ,\ldots, k\}$ and $s,t \in \mathbb{R}$, we have $X^i_t \circ X^j_s = X^j_s \circ X^i_t$.*
(c) *For any $i,j \in \{1 ,\ldots, k\}$ we have $[X^i,X^j] = 0$.*

Proof. (a) $\Rightarrow$ (b) has already been done.
(b) $\Rightarrow$ (c): The condition $X^i_t = X^j_{-s} \circ X^i_t \circ X^j_s$ implies that

$$X^i(p) = \frac{d}{dt}X^i_t(p)\Big|_{t=0} = \frac{d}{dt}(X^j_{-s} \circ X^i_t \circ X^j_s(p))_{t=0} = (X^j_s)^*(X^i)(p)$$

as is easy to see. Differentiating the above expression with respect to s at $s = 0$, we get $[X^j,X^i](p) = \frac{d}{ds}(X^j_s)^*(X^i)(p)\big|_{s=0} = 0$, since $X^i(p)$ does not depend on s.

(c) $\Rightarrow$ (b). The chain rule implies that

$$\frac{d}{dt}(X^i_t)^*(X^j)(p)) = \frac{d}{ds}((X^i_{t+s})^*(X^j)(p))_{s=0} =$$

$$\frac{d}{ds}(X^i_s)^* \circ ((X^i_t)^* \circ (X^j))(p)\Big|_{s=0} = [X^i,(X^i_t)^*(X^j)](p) .$$

On the other hand $(X^i_t)^*(X^i) = X^i$ so

$$(X_t^i)^* [X^i, X^j] = [(X_t^i)^* (X^i), (X_t^i)^* (X^j)] = [X^i, (X_t^i)^* (X^j)]$$

and therefore $\frac{d}{dt}((X_t^i)^* (X^j)(p)) = 0$, or, $(X_t^i)^* (X^j)(p)$ is independent of t, so $(X_t^i)^* (X^j)(p) = X^j(p)$.

By Proposition 1, the flow of $(X_t^i)^* (X^j)$ is given by

$$s \longrightarrow (X_t^i)^{-1} \circ X_s^j \circ X_t^i = X_{-t}^i \circ X_s^j \circ X_t^i .$$

Thus one sees immediately that $X_t^i \circ X_s^j = X_s^j \circ X_t^i$ for any $s, t \in \mathbb{R}$.

(b) $\Rightarrow$ (a): Let $\{e_1, \ldots, e_k\}$ be the canonical basis of $\mathbb{R}^k$. It suffices to define

$$\varphi\left(\sum_{i=1}^{k} t_i e_i, p\right) = X_{t_1}^1 \circ X_{t_2}^1 \circ \ldots \circ X_{t_k}^k(p) . \quad ■$$

Definition. A local action of $\mathbb{R}^k$ on M is a map $\varphi : V \longrightarrow M$, where V is an open set of $\mathbb{R}^k \times M$ containing $\{0\} \times M$, and φ satisfies the properties $\varphi(0,p) = p$ for every $p \in M$, and $\varphi(u + v, p) = \varphi(u, \varphi(v,p)) = \varphi(v, \varphi(u,p))$, whenever (u,p), (v,p) and $(u + v, p) \in V$.

Corollary 1. *Let $X^1, \ldots, X^k$ be $C^r (r \geq 1)$ vector fields defined on M. If $[X^i, X^j] = 0$ for any $i, j \in \{1, \ldots, k\}$ then there exists a basis $\{v_1, \ldots, v_k\}$ of $\mathbb{R}^k$ and a local action $\varphi : V \subset \mathbb{R}^k \times M \longrightarrow M$ such that $\varphi(tv_i, p) = X_t^i(p)$ for every $(t,p) \in \mathbb{R} \times M$ such that both sides are defined.*

The proof is similar to that of Proposition 3.

Corollary 2. *Let $X^1, \ldots, X^k$ be $C^r (r \geq 1)$ vector fields defined on M such that $[X^i, X^j] = 0$ for any $i, j \in \{1, \ldots, k\}$. Suppose that for every $q \in M$ the subspace $P(q)$ of $T_q M$, generated by $X^1(q), \ldots, X^k(q)$ has dimension k. Then the k-plane field P is tangent to a unique C^r foliation $\mathfrak{F}$ of dimension k.*

Proof. We will see how the local charts of $\mathfrak{F}$ are defined. Let $\varphi : V \longrightarrow M$ be the local action given by Corollary 1. Given $p \in M$, fix an imbedded disk $D^{n-k} = D \subset M$ such that $\overline{D}$ is compact, $p \in D$ and for every $q \in D$ we have

$$(*) \qquad T_q M = P(q) \oplus T_q D .$$

Since $\overline{D}$ is compact and V is open, it is easy to see that there exists a ball B^k such that $0 \in B^k \subset \mathbb{R}^k$ and $B^k \times D \subset V$. Set $\psi = \varphi \mid B^k \times D$. Given $(u,v) \in \mathbb{R}^k \times T_p D$, $u = \sum_{i=1}^{k} u_i e_i$, it is easy to see that

$$\partial_1 \psi(0,p) \cdot u = \sum_{i=1}^{k} u_i X^i(p) \quad \text{and} \quad \partial_2 \psi(0,p) \cdot v = v .$$

So from the condition $(*)$ we have that $D\psi(0,p) : \mathbb{R}^k \times T_p D \longrightarrow T_p M$ is an isomorphism. By the inverse function theorem there exist small disks $B' \subset B^k$ and $D' \subset D$ such that $0 \in B'$, $p \in D'$ and $\psi' = \psi \mid B' \times D'$: $B' \times D' \longrightarrow \psi(B' \times D') \subset M$ is a C^r diffeomorphism onto the open set $\psi(B' \times D')$ of M. Set $U = \psi(B' \times D')$ and $\xi = (\psi')^{-1}$. Then (U, ξ) is a local C^r chart on M and the submanifolds of the form $\xi^{-1}(B' \times \{q\})$, $q \in D'$, are the plaques of a C^r foliation $\mathfrak{F}_U$ defined on U and tangent to P.

We claim that $\mathfrak{F}_U$ is the unique foliation on U that is tangent to P. Indeed, the leaves of $\mathfrak{F}_U$ are mapped by ξ to disks of the form $B' \times \{q\}$, $q \in D'$. On the other hand, the plane field P is mapped by $D\xi$ to the horizontal plane field

$$P^*(q) = D\xi(\xi^{-1}(q)) \cdot P(\xi^{-1}(q)) = \mathbb{R}^k \times \{0\},\ 0 \in T_q D' .$$

As is easy to see, the foliation of $B' \times D'$ whose leaves are $B' \times \{q\}. q \in D'$, is the only one on $B' \times D'$ that is tangent to the plane field $q \mapsto \mathbb{R}^k \times \{0\}$. This proves the claim.

The set of all charts (U, ξ) constructed as above defines a C^r foliation $\mathfrak{F}$ on M. Indeed, if $(\tilde{U}, \tilde{\xi})$ is another local chart as above with $U \cap \tilde{U} \neq \varnothing$ then, by the preceding claim, $\mathfrak{F}_U$ coincides with $\mathfrak{F}_{\tilde{U}}$ on $U \cap \tilde{U}$ and this implies that the change of coordinate map $\tilde{\xi} \circ \xi^{-1} : \xi(U \cap \tilde{U}) \subset B \times D \longrightarrow \tilde{\xi}(U \cap \tilde{U}) \subset \tilde{B} \times \tilde{D}$ is of the form $(x, y) \longrightarrow (h_1(x, y), h_2(y))$. ■

§2. Frobenius' theorem

Before stating the theorem we are going to fix some notations. Given a k-plane field P on M we say that the vector field X is *tangent* to P if $X(q) \in P(q)$ for every q in the domain of X.

A $C^r (r \geq 1)$ k-plane field P is called *involutive* if given X and Y, C^1 fields tangent to P, then $[X, Y]$ is tangent to P. We say that P is *completely integrable* if there exists a C^r foliation of dimension k on M such that $T\mathfrak{F} = P$, where $T\mathfrak{F}$ is the plane field tangent to $\mathfrak{F}$.

Theorem 1 (Frobenius). *Let P be a $C^r (r \geq 1)$ k-plane field defined on M. Then P is completely integrable if and only if it is involutive. Further if either of these conditions hold, the foliation tangent to P is unique.*

Proof. Suppose P is involutive. The idea is to use Corollary 2 of Proposition 3. We prove that for every $p \in M$ there exist C^r vector fields, $X^1, \ldots, X^k$, defined on a neighborhood U of p, such that $[X^i, X^j] = 0$ for any $i, j \in \{1, \ldots, k\}$ and $P(q)$ is the subspace of $T_q M$ generated by $\{X^1(q), \ldots, X^k(q)\}$ for every $q \in U$. By Corollary 2, there will exist a unique C^r k-dimensional foliation on U, $\mathfrak{F}_U$, such that $T\mathfrak{F}_U = P \mid U$. In this manner, it will be possible to define a cover $\{U_i\}_{i \in I}$ of M be open sets such that for each $i \in I$ there will be defined a foliation $\mathfrak{F}_i$ on U_i. By uniqueness of such foliations it follows that if $U_i \cap U_j \neq \emptyset$ then $\mathfrak{F}_i \mid U_i \cap U_j = \mathfrak{F}_j \mid U_i \cap U_j$. In this way we will have defined a unique foliation $\mathfrak{F}$ on M such that $T\mathfrak{F} = P$.

Fix then $p \in M$. From the definition of C^r k-plane fields, there exist k C^r vector fields, $Y^1, \ldots, Y^k$ defined on a neighborhood V of P such that $P(q)$ is the subspace generated by $\{Y^1(q), \ldots, Y^k(q)\}$ for every $q \in V$. We can assume that V is the domain of a local (C^∞) chart $x : V \longrightarrow \mathbb{R}^n$. For each $j \in \{1, \ldots, k\}$ we can write $Y_j = \sum_{i=1}^n a_{ij}\, \partial/\partial x_i$ where $\{\partial/\partial x_1, \ldots, \partial/\partial x_n\}$ is the basis associated to the local chart x and $a_{ij} : V \longrightarrow \mathbb{R}$ is of class C^r. Consider the matrix $A = (a_{ij})_{1 \le i \le n}^{1 \le j \le k}$. The fact that $Y^1(p), \ldots, Y^k(p)$ are linearly independent implies that the matrix $A(p)$ has rank k, that is, there exists a $k \times k$ submatrix B of A such that $\det(B(p)) \neq 0$. Since the determinant is a continuous function we have that $\det(B(q)) \neq 0$ for every $q \in U$, a neighborhood of p. Permuting the variables $(x_1, \ldots, x_n)$ if necessary, we can assume that $B = (a_{ij})_{1 \le i,j \le k}$. In this way the matrix $B^{-1} = (b_{ij})_{1 \le i,j \le k}$ is defined and has C^r coefficients on U.

Set $X^j = \sum_{i=1}^k b_{ij} Y^i$, $j = 1, \ldots, k$.

In the first place, it is clear that $\{X^1(q), \ldots, X^k(q)\}$ is a basis of $P(q)$ for every $q \in U$. On the other hand,

$$X^j = \sum_{i=1}^{k} b_{ij} \sum_{\ell=1}^{n} a_{\ell i}\, \partial/\partial x_\ell = \sum_{\ell=1}^{n} \Big(\sum_{i=1}^{k} a_{\ell i} b_{ij} \Big) \partial/\partial x_\ell \, .$$

For $\ell = j$ we have that $\sum_{i=1}^k a_{ji} b_{ij} = 1$ and for $\ell \neq j$, with $1 \le \ell \le k$, we have that $\sum_{i=1}^k a_{\ell i} b_{ij} = 0$; so

$$X^j = \partial/\partial x_j + \sum_{\ell=k+1}^{n} c_\ell^j\, \partial/\partial x_\ell, \quad \text{for} \quad j = 1, \ldots, k \, .$$

We claim that $[X^i, X^j] = 0$ for any $i, j \in \{1, \ldots, k\}$. In fact, from Lemma 1 one has that $[\partial/\partial x_r, \partial/\partial x_s] = 0$ for any $r, s \in \{1, \ldots, n\}$ so

$$[X^i, X^j] = \Big[\sum_{\ell=k+1}^{n} c_\ell^i\, \partial/\partial x_\ell, \partial/\partial x_j \Big] + \Big[\partial/\partial x_i, \sum_{m=k+1}^{n} c_m^j\, \partial/\partial x_m \Big]$$

$$+ \sum_{\ell, m=k+1}^{n} [c_\ell^i\, \partial/\partial x_\ell, c_m^j\, \partial/\partial x_m] = \sum_{r=k+1}^{n} d_r\, \partial/\partial x_r \, .$$

Since P is involutive and $\{X^1(q), \ldots, X^k(q)\}$ is a basis of $P(q)$ for every $q \in U$, we must have that

$$\sum_{r=k+1}^{n} d_r\, \partial/\partial x_r = \sum_{s=1}^{k} f_s X^s = \sum_{s=1}^{k} f_s\, \partial/\partial x_s + \sum_{s=k+1}^{n} g_s\, \partial/\partial x_s \, .$$

From the above relation we can conclude that $f_1 = \ldots = f_s = 0$, so $[X^i, X^j] =$ $= 0$, as desired.

Suppose now that P is completely integrable. Let $\mathfrak{F}$ be a foliation such that $T\mathfrak{F} = P$. Consider two cases:

1st case. $\mathfrak{F}$ is $C^r (r \geq 2)$: Fix $p \in M$ and a foliation chart for $\mathfrak{F}$, $\varphi : U \longrightarrow \mathbb{R}^k \times \mathbb{R}^{n-k}$. Let X and Y be C^1 fields tangent to $\mathfrak{F}$ and defined on U. Since the leaves of $\mathfrak{F} \mid U$ are mapped to plaques $\mathbb{R}^k \times \{y\}$, $y \in \mathbb{R}^{n-k}$, the fields $X_* = \varphi_*(X)$ and $Y_* = \varphi_*(Y)$ are of the form $X_*(x,y) = (f(x,y),0)$ and $Y_*(x,y) = (g(x,y),0)$ where $f,g : \mathbb{R}^k \times \mathbb{R}^{n-k} \longrightarrow \mathbb{R}^k$. Using (b) of Proposition 2, one sees that $[X_*,Y_*](x,y) = (h(x,y),0)$ where $h(x,y) = = f(x,y) \cdot \partial g/\partial x - g(x,y) \cdot \partial f/\partial x$. So $[X, Y] = \varphi^*[X_*,Y_*]$ is tangent to P.

2nd case. $\mathfrak{F}$ is C^1 : Although $\mathfrak{F}$ is C^1, the fact that P is C^1 (by hypothesis) implies that the leaves of $\mathfrak{F}$ are C^2. Indeed, let Q be a plaque of $\mathfrak{F}$ contained in a coordinate chart of M, $x : U \longrightarrow \mathbb{R}^n$. Then $x(Q)$ is a submanifold of $\mathbb{R}^n = \mathbb{R}^k \times \mathbb{R}^{n-k}$, which we can assume is the graph of a function $\psi : D^k \subset \mathbb{R}^k \longrightarrow \mathbb{R}^{n-\ell}$. In this way Q is mapped by x onto the submanifold $x(Q)$ defined by the equation $y = \psi(x)$, $x \in D^k$ and $P \mid Q$ is mapped onto the k-plane field $\tilde{P}$ along $x(Q)$ defined by $\tilde{P}(x,\psi(x)) = = \{(u, D\psi(x) \cdot u) \mid u \in \mathbb{R}^k\}$. However $\tilde{P}$ is C^1, so $D\psi$ is C^1, as is easy to see, therefore ψ and Q are C^2. Since Q is C^2, given $p \in Q$, there exists a C^2 diffeomorphism $\varphi : U' \longrightarrow \mathbb{R}^n = \mathbb{R}^k \times \mathbb{R}^{n-k}$, $p \in U$, such that $\varphi(U' \cap Q) = \mathbb{R}^k \times \{0\}$. We can now apply the same argument as in the first case.

Remark. The argument in this theorem proves that if P is a C^r involutive plane field on M then the leaves of the foliation associated to P are of class C^{r+1}, although $\mathfrak{F}$ is in general C^r.

Examples.

(1) Every 1-plane field is completely integrable.

(2) Define a 2-plane field P on $\mathbb{R}^3$, setting $P(x_1,x_2,x_3) =$ the plane generated by $(1,0,0)$ and $(0,e^{-x_1},e^{x_1})$. Then P has no integral surfaces. Indeed, if $X^1(x_1,x_2,x_3) = (1,0,0)$ and $X^2(x_1,x_2,x_3) = = (0,e^{-x_1},e^{x_1}) \in P(x_1,x_2,x_3)$ then $[X^1,X^2](x_1,x_2,x_3) = (0,-e^{-x_1},e^{x_1}) \notin P(x_1,x_2,x_3)$. Therefore P is not involutive at any point $(x_1,x_2,x_3) \in \mathbb{R}^3$.

§3. Plane fields defined by differential forms

In this section we will see a version of Frobenius' theorem in terms of differential forms. We assume here that the reader has some basic knowledge of the theory of differential forms, such as the exterior product, exterior derivative, co-induced forms, etc. For more details about the subject we refer the reader to [52].

Let $\omega^1, \ldots, \omega^k$ be C^r $(r \geq 0)$ 1-forms defined and linearly independent on an open set $U \subset M^n$. For each $q \in U$ define

$$P(q) = \{v \in T_q(M) \mid \omega_q^1(v) = \ldots = \omega_q^k(v) = 0\} .$$

Since $\omega_q^1, \ldots, \omega_q^k$ are linearly independent elements of the dual $(T_qM)^*$, it is clear that $P(q)$ is a subspace of codimension k of T_qM. We claim that the k-plane field $q \mapsto P(q)$ is C^r. Indeed, given $p \in U$, let $x : V \longrightarrow \mathbb{R}^n$ be a local chart with $p \in V \subset U$, where $x(q) = (x_1(q), \ldots, x_n(q))$, $x_j : V \longrightarrow \mathbb{R}$ for $j = 1, \ldots, n$. Consider also for each $j \in \{1, \ldots, n\}$ the C^∞ 1-form dx_j. Then $\{dx_1(p), \ldots, dx_n(p)\}$ is a basis of $(T_pM)^*$, so it is possible to determine indices $i_1, \ldots, i_{n-k}$ such that $\{\omega_p^1, \ldots, \omega_p^k, dx_{i_1}(p), \ldots, dx_{i_{n-k}}(p)\}$ is a basis of $(T_pM)^*$, this because $\omega_p^1, \ldots, \omega_p^k$ are linearly independent. Set, for convenience, $dx_{i_1} = \omega^{k+1}, \ldots, dx_{i_{n-k}} = \omega^n$ so that $\{\omega_p^1, \ldots, \omega_p^n\}$ is a basis of $(T_pM)^*$. By continuity there exists a neighborhood W of p with $W \subset V$, such that $\{\omega_p^1, \ldots, \omega_q^n\}$ is a basis of $(T_qM)^*$ for every $q \in W$. Consider now vector fields $X^1, \ldots, X^n$ defined on W, such that for every $q \in W$ the set $\{X^1(q), \ldots, X^n(q)\}$ is the dual basis to $\{\omega_q^1, \ldots, \omega_q^n\}$ that is, $\omega_q^i(X^j(q)) = \delta_{ij}$ ($\delta_{ij} = 0$ if $i \neq j$ and $\delta_{ii} = 1$). It is easy to see now that the vector fields $X^1, \ldots, X^n$ are C^r and further that $P(q)$ is generated by the vectors $X^{k+1}(q), \ldots, X^n(q)$, $q \in W$. Conversely, given a C^r plane field P of codimension k defined on $U \subset M$, it is possible to locally define P as the kernel of k C^r 1-forms. Indeed, given $p \in U$, consider $n - k$ C^r vector fields $X^{k+1}, \ldots, X^n$ such that $P(q)$ is the subspace of T_qM generated by $X^{k+1}(q), \ldots, X^n(q)$ for every $q \in V \subset U$. By a similar argument to that made in the case of forms, it is possible to define k C^∞ vector fields $X^1, \ldots, X^k$ on V such that $X^1, \ldots, X^n$ are linearly independent on a neighborhood W with $p \in W \subset V$. It suffices now to consider the dual basis $\{\omega^1, \ldots, \omega^n\}$ which consists of C^r forms. We have then

$$P(q) = \{v \in T_qM \mid \omega_q^1(v) = \ldots = \omega_q^k(v) = 0\} .$$

We sum up what was done above in the following:

Proposition 4. *A C^r $(r \geq 0)$ codimension k plane field can be defined locally as the kernel of k C^r linearly independent 1-forms. Conversely if $\omega^1, \ldots, \omega^k$ are C^r linearly independent 1-forms then $P(q) = \{v \in T_qM \mid \omega_q^1(v) = \ldots = \omega_q^k(v) = 0\}$ define a C^r codimension k plane field.*

Proposition 4 assures that there must be a version of Frobenius' theorem in terms of differential forms. Indeed, this version is given by the following result.

Theorem 2. *Let P be a C^r $(r \geq 1)$ codimension k plane field defined on an open set $U \subset M$ by k linearly independent C^r 1-forms $\omega^1, \ldots, \omega^k$. Then P is completely integrable if and only if for every $j \in \{1, \ldots, k\}$ we have*

$(*)_j$ $$d\omega^j \wedge \omega^1 \wedge \ldots \wedge \omega^k = 0 \ .$$

Proof. Suppose that the equalities $(*)_j$ are true. It suffices to prove that P is involutive. For this we will need two lemmas.

Lemma 2. *Let η be a 2-form on U such that $\eta \wedge \omega^1 \wedge \ldots \wedge \omega^k = 0$. Given $p \in U$ there exist a neighborhood V of p with $V \subset U$ and k 1-forms $\alpha_1, \ldots, \alpha_k$ such that $\eta = \alpha_1 \wedge \omega^1 + \ldots + \alpha_k \wedge \omega^k$.*

Proof. Since $\omega^1, \ldots, \omega^k$ are linearly independent, it is possible to define $n-k$ C^∞ 1-forms $\omega^{k+1}, \ldots, \omega^n$ and a neighborhood V of p with $V \subset U$, such that for every $q \in V$, $\{\omega_q^1, \ldots, \omega_q^n\}$ is a basis of $(T_qM)^*$. We can write $\eta = \sum_{i<j} a_{ij}\omega^i \wedge \omega^j$. Since $\eta \wedge \omega^1 \wedge \ldots \wedge \omega^k = 0$ one sees that

$$\sum_{i<j} a_{ij}\omega^i \wedge \omega^j \wedge \omega^1 \wedge \ldots \wedge \omega^k = 0, \quad \text{or,} \quad a_{ij} = 0 \ \text{ if } \ k < i < j \ .$$

So $\eta = \sum_{\substack{i \le k \\ i<j}} a_{ij}\omega^i \wedge \omega^j = \sum_{i=1}^k \alpha_i \wedge \omega^i$ where $\alpha_i = -\sum_{j>i} a_{ij}\omega^j$. ■

Lemma 3. *Let ω be a $C^r\,(r \ge 1)$ 1-form and X and Y two C^1 vector fields defined on an open set $U \subset \mathbb{R}^n$. Then*

$$d\omega(X,Y) = d(\omega(X)) \cdot Y - d(\omega(Y)) \cdot X + \omega([X,Y]).$$

We denote the function $q \longrightarrow \omega_q(X(q))$ by $\omega(X)$.

Proof. Suppose $\omega = \sum_{i=1}^n a_i dx_i$ where $a_i : U \longrightarrow R$ is of class C^r for $i = 1, \ldots, n$ and $\{dx_1, \ldots, dx_n\}$ is the dual basis of the canonical basis $\{e_1, \ldots, e_n\}$. Also set $X = \sum_{i=1}^n b_i e_i$ and $Y = \sum_{i=1}^n c_i e_i$. We have then that

(i) $\omega(X) = \sum_{i=1}^n a_i b_i, \quad \omega(Y) = \sum_{i=1}^n a_i c_i$ and

$$\omega([X,Y]) = \sum_{i=1}^n a_i \sum_{j=1}^n \left(b_j \frac{\partial c_i}{\partial x_j} - c_j \frac{\partial b_i}{\partial x_j}\right) = \sum_{i,j} a_i \left(b_j \frac{\partial c_i}{\partial x_j} - c_j \frac{\partial b_i}{\partial x_j}\right),$$

(ii) $d\omega = \sum_{i<j} a_{ij} dx_i \wedge dx_j$, $a_{ij} = \dfrac{\partial a_j}{\partial x_i} - \dfrac{\partial a_i}{\partial x_j}$ so $d\omega(X,Y) = \sum_{i<j} a_{ij}(b_i c_j - b_j c_i)$.

By (i) we have that

$$d\omega(X) \cdot Y = \sum_{i,j} \frac{\partial(a_i b_i)}{\partial x_j} \cdot c_j = \sum_{i,j} c_j \left(a_i \frac{\partial b_i}{\partial x_j} + b_i \frac{\partial a_i}{\partial x_j}\right)$$

and

$$d\omega(Y) \cdot X = \sum_{i,j} \frac{\partial(a_i c_i)}{\partial x_j} b_j = \sum_{i,j} b_j \left(a_i \frac{\partial c_i}{\partial x_j} + c_i \frac{\partial a_i}{\partial x_j}\right) .$$

So

$$\text{(iii)} \quad d\omega(X) \cdot Y - d\omega(Y) \cdot X = \sum_{i,j} a_i \left(c_j \frac{\partial b_i}{\partial x_j} - b_j \frac{\partial c_i}{\partial x_j}\right) + \\ + \sum_{i,j} \frac{\partial a_i}{\partial x_j} (b_i c_j - b_j c_i) .$$

Summing the expression for $\omega([X, Y])$ from (i) with (iii), we obtain

$$d\omega(X) \cdot Y - d\omega(Y) \cdot X + \omega([X, Y]) = \sum_{i,j} \frac{\partial a_i}{\partial x_j} (b_i c_j - b_j c_i) .$$

Thus,

$$\sum_{i,j} \frac{\partial a_i}{\partial x_j} (b_i c_j - b_j c_i) = \sum_{i<j} \left(\frac{\partial a_i}{\partial x_j} - \frac{\partial a_j}{\partial x_i}\right)(b_i c_j - b_j c_i) = d\omega(X, Y) . \blacksquare$$

Suppose then that X and Y are two C^1 vector fields tangent to P. Since $d\omega^j \wedge \omega^1 \wedge \ldots \wedge \omega^k = 0$, from Lemma 2 we obtain

$$d\omega^j = \sum_{i=1}^{k} \alpha_i^j \wedge \omega^i$$

and therefore

$$d\omega^j(X, Y) = \sum_{i=1}^{k} (\alpha_i^j(X)\omega^i(Y) - \alpha_i^j(Y)\omega^i(X)) = 0$$

since for every $i = 1, \ldots, k$ we have $\omega^i(X) = \omega^i(Y) = 0$. On the other hand from Lemma 3, we obtain

$$0 = d\omega^j(X, Y) = d(\omega^j(X)) \cdot Y - d(\omega^j(Y)) \cdot X + \omega^j([X, Y]) .$$

Since $\omega^j(X) = \omega^j(Y) = 0$, one sees that $\omega^j([X, Y]) = 0$ for every $j = 1, \ldots, k$, i.e., $[X, Y]$ is tangent to P.

We leave the converse to the reader. ■

Exercises

Chapter I

1. Let $f : M \longrightarrow N$ be a submersion, where N is connected.
   (a) Prove that if A is an open subset of M then $f(A)$ is open in N.
   (b) Prove that if M is compact then N is also.

2. Let $\alpha : [a,b] \longrightarrow M$ be a continuous simple (i.e., injective) curve. Suppose that d is a metric on M. Prove that, given $\epsilon > 0$, there exists a cover of $\alpha([a,b])$, by domains of local charts $U_1, \ldots, U_k$ such that, if $U_i \cap U_j \neq \varnothing$, then $j \in \{i-1, i, i+1\}$ and the diameter of U_i in the metric d is less than ϵ for all $i \in \{1, \ldots, k\}$.

3. Let M be a connected manifold. Prove that, given $p, q \in M$, there exists a simple C^∞ curve $\alpha : [0,1] \longrightarrow M$ such that $\alpha(0) = p$, $\alpha(1) = q$ and $(d\alpha)(t)/dt \neq 0$ for every $t \in [0,1]$.

4. Let M be a connected 1-dimensional manifold. Prove that M is diffeomorphic to $\mathbb{R}$ or to S^1.
   Hint. Use Exercise 3.

5. Let $\langle\ ,\ \rangle$ be a Riemannian metric on M. If $f : M \longrightarrow \mathbb{R}$ is a $C^r\,(r \geq 1)$ function, one defines the gradient of f in the metric $\langle\ ,\ \rangle$ as the unique vector field ∇f such that $df_p(u) = \langle \nabla f(p), u \rangle_p$, for any $p \in M$ and $u \in T_pM$.
   (a) Prove that ∇f is well-defined and is of class C^{r-1}.
   (b) Prove that $p \in M$ is a singularity of ∇f if and only if it is a critical point of f.
   (c) Prove that ∇f is normal to the level surfaces of f and therefore has no periodic orbits.

Chapter II

6. Let $\mathcal{F}$ be a foliation on M and F a leaf of $\mathcal{F}$. Consider on F the intrinsic manifold structure. Prove that F has a countable basis.
Hint. Show that F meets a foliation box of $\mathcal{F}$ in a countable number of plaques.

7. Let $\mathcal{F}$ be the Reeb foliation of $D^2 \times S^1$ and N an n-dimensional manifold. Prove that if $f : D^2 \times S^1 \longrightarrow N$ is continuous and constant on each leaf of $\mathcal{F}$ then f is constant.

8. Let $S^3 = \{(z_1,z_2) \in \mathbf{C}^2 \mid |z_1|^2 + |z_2|^2 = 1\}$. Prove that the map $\pi : S^3 \longrightarrow S^2$ defined by

$$\pi(z_1,z_2) = (|z_1|^2 - |z_2|^2, 2\,\mathrm{Re}(z_1\bar{z}_2), 2\,\mathrm{Im}(z_1\bar{z}_2)) ,$$

is a submersion and that $\pi^{-1}(p) \simeq S^1$ for all $p \in S^2$. Show that there does not exist a submersion $f : S^3 \longrightarrow \mathbb{R}^2$.

9. Let $f : M \longrightarrow N$ be a $C^r (r \geq 1)$ submersion and $\mathcal{F}$ the foliation on M given by level surfaces of f. Prove that if N is orientable then $\mathcal{F}$ is transversely orientable.

10. Let $\mathcal{F}$ be a transversely orientable foliation on N and $f : M \longrightarrow N$ a map transverse to $\mathcal{F}$. Prove that $f^*(\mathcal{F})$ is transversely orientable.

11. Let $\mathcal{F}$ and $\mathcal{G}$ be two C^r foliations on M. Suppose that for every $p \in M$ one has $T_pM = T_p\mathcal{F} + T_p\mathcal{G}$. Prove that there exists a foliation $\mathcal{H}$ on M, of class C^r, such that if F and G are leaves of $\mathcal{F}$ and $\mathcal{G}$ respectively, then the connected components of $F \cap G$ are leaves of $\mathcal{H}$. Show that $\dim(\mathcal{H}) =$ $= \dim(\mathcal{F}) + \dim(\mathcal{G}) - \dim M$.

12. Let M be a compact manifold of dimension n. Prove that if there exists a foliation of codimension one on M then the Euler characteristic of M is zero.
Hint. If there exists a continuous nonsingular vector field on M then the Euler characteristic of M is zero [26].

13. Let $\alpha \in \mathbb{R} - \mathbb{Q}$.
(a) Prove that the set $\{m + n\alpha \mid m,n \in \mathbf{Z}\}$ is dense in $\mathbb{R}$.
(b) Define $f_\alpha : S^1 \longrightarrow S^1$ by $f_\alpha(z) = e^{2\pi i\alpha} \cdot z$. Prove that for every $z \in S^1$ the orbit of z by f_α, $\mathcal{O}(z) = \{f_\alpha^n(z) \mid n \in \mathbf{Z}\}$ is dense in S^1.
(c) Let X^α be the vector field obtained by the suspension of f_α to a compact manifold M^2 (see note 1 of Chapter II). Prove that M^2 is diffeomorphic to T^2 and that all the orbits of X^α are dense.

14. Let P be a C^r k-plane field and $\langle\ ,\ \rangle$ a (C^∞) Riemannian metric on M. Define $P^\perp$ by $P^\perp(q) = \{v \in T_qM \mid \langle u,v\rangle_q = 0 \text{ for every } u \in P(q)\}$.

(a) Prove $P^{\perp}$ is of class C^r.

(b) If $A_q : T_qM \longrightarrow P^{\perp}(q)$ is the orthogonal projection on $P^{\perp}(q)$ and X is a C^s $(s \leq r)$ vector field, prove that the vector field Y, defined by $Y(q) = A_q(X(q))$ is of class C^s.

15. Let $\sim$ be the equivalence relation on $\mathbb{C}^{n+1} - \{0\}$ defined by $z \sim w$ if and only if there exists $\lambda \in \mathbb{C} - \{0\}$ such that $z = \lambda w$. Let $\mathbb{C}P(n)$ be the quotient space of $\mathbb{C}^{n+1} - \{0\}$ by $\sim$. Denote by $[z]$ the equivalence class of $z \in \mathbb{C}^n - \{0\}$.

(a) Let $U_i = \{[z_1, \ldots, z_{n+1}] \in \mathbb{C}P(n) \mid z_i \neq 0\}$, $1 \leq i \leq n+1$, and $\varphi_i : U_i \longrightarrow \mathbb{C}^n$ defined by $\varphi_i[z_1, \ldots, z_{n+1}] = z_i^{-1}(z_1, \ldots, z_{i-1}, z_{i+1}, \ldots, z_{n+1})$. Prove that $\mathcal{A} = \{(U_i, \varphi_i) \mid i = 1, \ldots, n+1\}$ is a (holomorphic) atlas for $\mathbb{C}P(n)$.

(b) Let $\mathfrak{F}$ be the foliation of S^{2n-1} defined by the Hopf fibration. If M is the quotient space of S^{2n-1} by the equivalence relation which identifies the points which are in the same leaf of $\mathfrak{F}$, prove that M is homeomorphic to $\mathbb{C}P(n-1)$.

16. Let $f : \mathbb{R} \longrightarrow \mathbb{R}$ be of class C^r $(r \geq 2)$ and $G \subset \mathbb{R}^2$ the graph of f. For each $p \in G$, let ℓ_p be the normal line to G at p with $p \in \ell_p$.

(a) Prove that there exists an open neighborhood U of G such that, if F_p is the connected segment of $\ell_p \cap U$ that contains p, then $F_p \cap F_{p'} \neq \emptyset$ implies that $p = p'$.

(b) Prove that the segments F_p, $p \in G$, are the leaves of a C^{r-1} foliation on U.

(c) Prove that if f is not C^{r+1} then $\mathfrak{F}$ is not C^r.

(d) Give an example of a foliation on $\mathbb{R}^2$, of class C^r, but not C^{r+1}, whose leaves are C^{∞}.

Chapter III

17. Prove that all the leaves of the Reeb foliation of S^3 are imbedded.

18. Let $f : \mathbb{R}^2 \longrightarrow \mathbb{R}$ be defined by $f(x, y) = (1 - x^2)e^y$. Prove that f is a submersion. Let $\mathfrak{F}$ be the foliation of $\mathbb{R}^2$ given by level surfaces of f and $\pi : \mathbb{R}^2 \longrightarrow \mathbb{R}^2/\mathfrak{F}$ the projection onto the leaf space of $\mathfrak{F}$. Prove that π is not closed.

19. Let $\mathfrak{F}$ be a C^r $(r \geq 1)$ 1-dimensional foliation on $\mathbb{R}^2$ and $\pi : \mathbb{R}^2 \longrightarrow \mathbb{R}^2/\mathfrak{F}$ be the projection onto the leaf space of $\mathfrak{F}$.

(a) Prove that all the leaves of $\mathfrak{F}$ are closed, but that $\mathfrak{F}$ has no compact leaves.

(b) Prove that $\mathbb{R}^2/\mathfrak{F}$ has a C^r 1-dimensional manifold structure (in general non-Hausdorff). Show that π is a C^r submersion.

(c) Prove that $\mathbb{R}^2/\mathfrak{F}$ is simply connected.

Hint. Prove that if $\alpha : [0,1] \longrightarrow \mathbb{R}^2/\mathfrak{F}$ is a closed curve then there exists a closed curve $\tilde{\alpha} : [0,1] \longrightarrow \mathbb{R}^2$ such that $\alpha = \pi \circ \tilde{\alpha}$.

(d) Give an example of a foliation $\mathfrak{F}$ on $\mathbb{R}^2$ such that $\mathbb{R}^2/\mathfrak{F}$ has three points which are pairwise nonseparable. We say that p and q are nonseparable if, given arbitrary neighborhoods U and V of p and q then $U \cap V \neq \varnothing$.

20. Given an example of a foliation $\mathfrak{F}$ of codimension 1 on T^3 which has every leaf dense ($T^3 = S^1 \times S^1 \times S^1$). Using the method of turbilization (note 4 of Chapter II) modify $\mathfrak{F}$, obtaining a foliation $\mathfrak{F}'$ on T^3 with the following properties:
(a) There exists a solid torus $D^2 \times S^1$ imbedded in T^3 such that $D^2 \times S^1$ is saturated by $\mathfrak{F}'$ and $\mathfrak{F}'/D^2 \times S^1$ is the Reeb foliation.
(b) All the leaves of $\mathfrak{F}'$ are dense in $T^3 - (D^2 \times S^1)$.

21. Let $\mathfrak{F}$ be a 1-dimensional C^1 foliation on T^2. Prove that $\mathfrak{F}$ has at most one exceptional minimal set.
Hint. Suppose that μ is an exceptional minimal set of $\mathfrak{F}$. Show that there exists a circle C imbedded in T^2, such that C is transverse to $\mathfrak{F}$, $C \cap \mu \neq \varnothing$ and $C \cap \mu' = \varnothing$ for every minimal set μ' such that $\mu' \cap \mu = \varnothing$ (see Prop. 4 of Chapter VII). Using the fact that $T^2 - C \simeq \mathbb{R} \times S^1$ and the Poincaré-Bendixson theorem [34], prove that $\mathfrak{F}$ has no exceptional minimal sets in $T^2 - C$.

Chapter IV

22. Show with an example that the global stability theorem is not true in $\mathbb{R}^3 - \{0\}$.

23. Let $\mathfrak{F}$ be a transversely orientable, C^r ($r \geq 1$) foliation, F a leaf of $\mathfrak{F}$ and $f : \Sigma' \subset \Sigma \longrightarrow \Sigma$ an element of holonomy of F. Prove that $Df(p) : T_p\Sigma \longrightarrow T_p\Sigma$ preserves the orientation of $T_p\Sigma$, that is, takes a basis of $T_p\Sigma$ to another basis with the same orientation.

24. Let $\mathfrak{F}$ be a C^1 codimension-one foliation on M. If F is a nonclosed leaf of $\mathfrak{F}$, prove that, given $p \in F$, there exists a closed curve transverse to $\mathfrak{F}$ passing through p.
Hint. Use Lemma 3.

25. Let $f : M \longrightarrow N$ be a C^r ($r \geq 1$) submersion and $\mathfrak{F}$ the foliation on M given by level surfaces of f. Suppose M and N are connected. Let $F \subset f^{-1}(q)$ be a leaf of $\mathfrak{F}$.
(a) Prove that, given $p \in F$, there exists a neighborhood V of q and a section Σ transverse to $\mathfrak{F}$ through p, such that $f \mid \Sigma : \Sigma \longrightarrow V$ is a C^r diffeomorphism.
(b) Show that the holonomy of F is trivial.
(c) Suppose that F is a compact leaf of $\mathfrak{F}$. Let $\pi : V \longrightarrow F$ be a C^r tubular

neighborhood of F. Show that there exists a neighborhood $U \subset V$ of F, saturated by $\mathfrak{F}$ and such that the intersection of each leaf of $\mathfrak{F} \mid U$ with a fiber $\pi^{-1}(q) \cap U$ contains only one point.

(d) With the notation of (c), define $h : U \longrightarrow F \times f(U)$ in the following way: $h(p) = (\pi(p), f(p))$. Prove that h is a C^r diffeomorphism which takes the leaves of $\mathfrak{F}$ contained in U to the sets $F \times \{y\}$, $y \in f(U)$.

(e) If all the leaves of $\mathfrak{F}$ are compact, prove that they are diffeomorphic to F.

26. (a) Let $f : \mathbb{R} \longrightarrow \mathbb{R}$ be a C^1 diffeomorphism. Suppose that there exists $0 < \lambda < 1$ such that $f(\lambda x) = \lambda f(x)$ for every $x \in \mathbb{R}$. Prove that f is linear.

(b) Let $\mathfrak{F}$ be a $C^r(r \geq 1)$ codimension-one foliation on $T^2 \times \mathbb{R}$ such that $F = T^2 \times \{0\}$ is a leaf of $\mathfrak{F}$. Fix a section $\Sigma = \{p\} \times (-\epsilon, \epsilon)$ transverse to $\mathfrak{F}$ which we identify to $(-\epsilon, \epsilon)$. Suppose that one of the elements of the holonomy of F, $g : \Sigma \longrightarrow \Sigma$, is given by $g(x) = \lambda x$ where $0 < \lambda < 1$. Prove that all the elements of holonomy of $\mathfrak{F}$ are linear.

(c) If $\mathfrak{F}$ is as in (b), describe $\mathfrak{F} \mid V$ where V is a neighborhood of F.

27. Let $\omega_1, \ldots, \omega_k$ be linearly independent, $C^r(r \geq 1)$ closed 1-forms defined on M. Let P be the codimension k plane field defined by $P(q) = \{v \in T_qM \mid \omega_i(q) \cdot v = 0,\ i = 1, \ldots, k\}$.

(a) Prove that there exists a unique C^r foliation on M such that $T_q\mathfrak{F} = P(q)$ for every $q \in M$. Show that $\mathfrak{F}$ is transversely orientable.

(b) Prove that the holonomy of any leaf of $\mathfrak{F}$ is trivial.

28. Let $f : M \longrightarrow \mathbb{R}$ be a $C^r(r \geq 1)$ submersion and $\mathfrak{F}$ the foliation on M given by level surfaces of f. Suppose that M is connected and that all the leaves of $\mathfrak{F}$ are compact.

(a) Show that there does not exist a closed curve transverse to $\mathfrak{F}$.

(b) Let F be a leaf of $\mathfrak{F}$. Prove that there exists a C^r diffeomorphism $h : F \times \mathbb{R} \longrightarrow M$ such that $h^*(\mathfrak{F})$ is the foliation whose leaves are of the form $F \times \{t\}$, $t \in \mathbb{R}$.

Hint. For the construction of h use the trajectories of a vector field transverse to $\mathfrak{F}$.

29. Let $M^n (n \geq 3)$ be a compact connected manifold with boundary (see [39]) $\partial M = \cup_{i=1}^{k} F_i$ where $k \geq 1$ and F_1 is simply connected. Suppose that $\mathfrak{F}$ is a $C^r(r \geq 1)$ transversely orientable codimension 1 foliation on M such that $F_1, \ldots, F_k$ are leaves of $\mathfrak{F}$.

(a) Prove that $k = 2$.

(b) Prove that there exists a C^r diffeomorphism $h : F_1 \times [0,1] \longrightarrow M$

such that $h^*(\mathfrak{F})$ is the foliation whose leaves are $F_1 \times \{t\}$, $t \in [0,1]$.

(c) Give an example of a codimension 1 foliation $\mathfrak{F}$ defined on a compact manifold M^3 with boundary $\partial M \simeq S^2$ and such that ∂M is a leaf of $\mathfrak{F}$.

30. Let $\mathfrak{F}$ be a C^r transversely orientable codimension-one foliation defined on a connected manifold of dimension $n \geq 2$. Suppose that all the leaves of $\mathfrak{F}$ are compact.
(a) Show that any two leaves of $\mathfrak{F}$ are diffeomorphic.
(b) Suppose that M is compact. Prove that there exists a submersion $f: M \longrightarrow S^1$ such that the leaves of $\mathfrak{F}$ are the level surfaces $f^{-1}(\theta)$, $\theta \in S^1$.
(c) Prove that there does not exist a foliation of codimension 1 on $\mathbb{R}^3$ such that all its leaves are compact.

Chapter V

31. Let $\mathfrak{F}$ be a $C^r (r \geq 1)$ foliation transverse to the fibers of the fibration (E, π, B, F). Prove that any element of holonomy of $\mathfrak{F}$ is C^r.

32. Let (E, π, B, F) and (E', π', B, F) be two fibered spaces where B and F are connected. Let $\mathfrak{F}$ and $\mathfrak{F}'$ be $C^r (r \geq 1)$ foliations transverse to the fibers of E and E' respectively. Let $H: E \longrightarrow E'$ be a continuous one-to-one map which takes fibers of E to fibers of E' and leaves of $\mathfrak{F}$ to leaves of $\mathfrak{F}'$. Prove that H is a homeomorphism and that there exists a homeomorphism $h: B \longrightarrow B$ such that $\pi' \circ H = h \circ \pi$. Suppose now that h and the restriction of H to some fiber F_0 of E are C^r diffeomorphisms. Prove that H is a C^r diffeomorphism.

33. Let $f: M \longrightarrow N$ be a submersion where M is compact and connected. Let $p \in N$ be fixed.
(a) Prove that $(M, f, N, f^{-1}(p))$ is a fibered space.
(b) Show by an example that (a) is not true if M is not compact, N being compact or not.

34. Let (E, π, B, F) be a fibered space where E is connected and F compact. Suppose $\mathfrak{F}$ is a foliation on E such that for every $p \in E$ one has $T_pE = T_p\mathfrak{F} + T_p(F_p)$ where $F_p = \pi^{-1}(\pi(p))$. Prove that if L is a leaf of $\mathfrak{F}$ then L cuts all the fibers of E.

35. Let $f: S^3 \longrightarrow S^2$ be a submersion. Prove that there does not exist a foliation $\mathfrak{F}$ of codimension 1 on S^3 whose leaves are transverse to the fibers $f^{-1}(p)$, $p \in S^2$. Conclude that there does not exist a fibered space structure with discrete structure group, whose total space is S^3 and base is S^2.

36. Let E be a fibered space with base S^1 and compact fiber. Prove that E is equivalent to a fibration with discrete structure group.

Hint. Prove that there exists a vector field on E transverse to the fibers.

37. Let (E, π, B, F) and (E', π', B', F') be two fibered spaces and $\mathfrak{F}$ and $\mathfrak{F}'$ C^r foliations transverse to the fibers of E and E', with holonomies φ and φ' respectively. We say that φ and φ' are C^r-conjugate at b_0 and b_0' if there exist C^r diffeomorphisms $h : B \longrightarrow B'$ and $f : \pi^{-1}(b_0) \longrightarrow (\pi')^{-1}(b_0')$ such that $h(b_0) = b_0'$ and for every $[\alpha] \in \pi_1(B, b_0)$ one has that $f \circ \varphi([\alpha]) = \varphi'([h \circ \alpha]) \circ f$. Prove that, if φ and φ' are C^r-conjugate, then there exists a diffeomorphism $H : E \longrightarrow E'$ taking leaves of $\mathfrak{F}$ to leaves of $\mathfrak{F}'$ with $\pi' \circ H = h \circ \pi$.

38. Let (E, π, B, F) be a fibered space, $\mathfrak{F}$ a C^r foliation transverse to the fibers of E and φ the holonomy of $\mathfrak{F}$. Let $\mathfrak{F}(\varphi)$ be the suspension of φ to $E(\varphi)$. Prove that the holonomy of $\mathfrak{F}(\varphi)$ is φ.

39. Let $\mathfrak{F}$ be a C^2 transversely orientable foliation. Let F be a compact leaf of $\mathfrak{F}$. Prove that F has a tubular neighborhood diffeomorphic to a product.
Hint. Prove that the normal fibration of F is diffeomorphic to a product.

Chapter VI

40. Show that there does not exist a foliation of codimension one on S^3 whose leaves are all homeomorphic to $\mathbb{R}^2$.

41. Let $\mathfrak{F}$ be an analytic, codimension-one, transversely orientable foliation on M^n. Let F be a leaf of $\mathfrak{F}$. Suppose that $\gamma : S^1 \longrightarrow M$ is homotopic to a constant and that $\gamma(S^1) \subset F$. Show that the holonomy of γ is trivial.
Hint. If the holonomy of γ is not trivial, construct a curve $\tilde{\gamma} : S^1 \longrightarrow M$ transverse to $\mathfrak{F}$ and homotopic to γ.

42. Let $\mathfrak{F}$ be an analytic foliation of codimension one defined on a simply connected manifold.
    (a) Let $\gamma : \mathbb{R} \longrightarrow M$ be a curve transverse to $\mathfrak{F}$. If F is a leaf of $\mathfrak{F}$ such that $\gamma(\mathbb{R}) \cap F \neq \varnothing$ prove that $\gamma(\mathbb{R}) \cap F$ contains exactly one point.
    (b) Prove that all the leaves of $\mathfrak{F}$ are closed.
    (c) If $\gamma(\mathbb{R})$ cuts all the leaves of $\mathfrak{F}$ prove there exists a submersion $f : M \longrightarrow \mathbb{R}$ such that $\mathfrak{F}$ is the foliation whose leaves are the level surfaces of f.

43. Show that an analytic foliation of codimension one on $\mathbb{R}^n$ $(n \geq 2)$ has no compact leaf.

44. Let $\mathfrak{F}$ and $\mathcal{G}$ be two analytic foliations defined on a connected manifold M. Suppose there exists an open set $U \subset M$ such that $\mathfrak{F} \mid U = \mathcal{G} \mid U$. Prove that $\mathfrak{F} = \mathcal{G}$.
Hint. Use the fact that two analytic functions which coincide on an open set are equal on the intersection of their domains.

45. Let $\mathfrak{F}$ be the foliation by parallel planes in $\mathbb{R}^3$ and $f : \mathbb{R}^3 - \{0\} \longrightarrow S^2$

$\times$ $\mathbb{R}$ defined by $f(x) = (\|x\|^{-1} \cdot x, \log\|x\|)$. Set $\mathfrak{F}^* = (f^{-1})^*(\mathfrak{F} \mid \mathbb{R}^3 - \{0\})$.

(a) Prove that $\mathfrak{F}^*$ is invariant by translations of $S^2 \times \mathbb{R}$ of the type $(g,t) \mapsto (y, t + t_0)$, t_0 fixed.

(b) Prove that $\mathfrak{F}^*$ induces an analytic foliation on $S^2 \times S^1$ which has one leaf F diffeomorphic to T^2 and two Reeb components, R_1 and R_2, such that $\partial R_1 = \partial R_2 = F$.

(c) Generalize the above procedure, obtaining an analytic foliation of codimension one on $S^{n-1} \times S^1$ which has one leaf diffeomorphic to $S^{n-2} \times S^1$ and the others diffeomorphic to $\mathbb{R}^{n-1}$.

Chapter VII

46. Let M be a compact manifold of dimension 3, without boundary, with finite fundamental group. If X is a vector field on M, without singularities and transverse to a C^2 codimension-one foliation, then X has nonperiodic orbits and at least one periodic orbit.

47. Let F be a compact leaf of a C^2 codimension-one foliation $\mathfrak{F}$ on M^n $(n \geq 3)$. Suppose there exists a closed curve $\gamma : S^1 \longrightarrow F$ such that γ is homotopic to a constant in M but not in F. Prove that $\mathfrak{F}$ has a vanishing cycle.

48. Give an example of a foliation on S^3 which has a Reeb component R such that $S^3 - R$ is not homeomorphic to a solid torus.
Hint. Consider the Reeb foliation of S^3 and a closed transverse curve that is knotted. Turbulize.

49. Let $\mathfrak{F}$ be a codimension-one foliation on $\mathbb{R}^3$. Prove that
(a) if $\mathfrak{F}$ has a compact leaf then $\mathfrak{F}$ has a Reeb component,
(b) if F is a compact leaf of $\mathfrak{F}$ then F is diffeomorphic to T^2.
Hint (for b). Prove that the Euler characteristic of F is zero.

50. Let $\mathfrak{F}$ be a codimension-one foliation on $D^2 \times S^1$ such that $T^2 = \partial(D^2 \times S^1)$ is the only compact leaf of $\mathfrak{F}$. Show that $\mathfrak{F}$ is topologically equivalent to the Reeb foliation of $D^2 \times S^1$.
Hint. Prove that the curve $S^1 \times \{0\} \subset T^2$ is a vanishing cycle for $\mathfrak{F}$ and that the leaves of $\mathfrak{F}$, contained in the interior of $D^2 \times S^1$, are diffeomorphic to $\mathbb{R}^2$.

51. Let $\pi : \mathbb{R}^2 \longrightarrow M$ be a covering. Prove that if $M \neq \mathbb{R}^2$ then there exists a covering transformation $f: \mathbb{R}^2 \longrightarrow \mathbb{R}^2$ which has infinite order (i.e., $f^k = id \Rightarrow k = 0$).
Hint. Let α be a simple closed curve in M not homotopic to a constant. Let f be the transformation relative to $[\alpha]$. If $f^k = id$ then $[\alpha]^k = 1$ so, if $\tilde{\alpha}$ is a lift of $\alpha^k = \alpha * \cdots * \alpha$ (k times) then $\tilde{\alpha}$ is a simple closed curve in

$\mathbb{R}^2$. Let A be the region of $\mathbb{R}^2$ bounded by $\tilde{\alpha}$. Then $f(A) = A$ so by Brouwer's fixed point theorem ([35]), f has a fixed point in A. Conclude that $f = id$.

Chapter VIII

52. On $\mathbb{R}^n$ consider the additive group structure. Let $G \subset \mathbb{R}^n$ be a closed subgroup.
    (a) Suppose G is not discrete. Show that G contains a subgroup of dimension 1.
    (b) Given $v \in G - \{0\}$, let $L = \mathbb{R} \cdot v$, the subspace generated by v. If H is a subspace of dimension $n - 1$ complementary to L, prove that G is isomorphic to $(H \cap G) \oplus L$, if $L \subset G$, or $(H \cap G) \oplus \mathbb{Z} \cdot v$, if $L \not\subset G$.
    (c) Using (a) and (b), prove by induction on n that G is isomorphic to $\mathbb{R}^k \times \mathbb{Z}^\ell$ where $0 \leq k + \ell \leq n$.

53. Let φ be a C^2 action of $\mathbb{R}^2$ on $\mathbb{R}^3$.
    (a) Prove that all the orbits of φ are closed subsets of $\mathbb{R}^3$ homeomorphic to $\mathbb{R}^2$.
    (b) Suppose there exists a curve $\gamma : \mathbb{R} \longrightarrow \mathbb{R}^3$ which cuts all the orbits of φ transversely. Prove that φ is conjugate to the product action $\psi(s,t,(x,y,z)) = (x + s, y + t, z)$. We say that φ and ψ are conjugate if there exists $h : \mathbb{R}^3 \longrightarrow \mathbb{R}^3$ such that $h \circ \varphi(u,p) = \psi(u,h(p))$ for any $(u,p) \in \mathbb{R}^2 \times \mathbb{R}^3$.
    (c) Give an example of an action of $\mathbb{R}^2$ on $\mathbb{R}^3$ which is not conjugate to the product action ψ.

    Hint. Prove that a curve transverse to the orbits of φ cuts each orbit in at most one point. In (b) define $h : \mathbb{R}^2 \times \mathbb{R} \longrightarrow \mathbb{R}^3$ by $h(u,t) = \varphi(u,\gamma(t))$. Prove that h is a diffeomorphism and conjugates φ with ψ.

54. Actions of $\mathbb{R}^2$ linearly induced on S^n. Let A and B be commuting $(n + 1) \times (n + 1)$ matrices. Define $\varphi : \mathbb{R}^2 \times \mathbb{R}^{n+1} \longrightarrow \mathbb{R}^{n+1}$ by

$$\varphi(s,t;p) = \exp(sA + tB) \cdot p .$$

    (a) Prove that φ is an action of $\mathbb{R}^2$ on $\mathbb{R}^{n+1}$.
    (b) Define $\tilde{\varphi} : \mathbb{R}^2 \times S^n \longrightarrow S^n$ by

$$\tilde{\varphi}(s,t;p) = \frac{\varphi(s,t;p)}{\|\varphi(s,t;p)\|}, \quad p \in S^n .$$

    Show that $\tilde{\varphi}$ defines an action of $\mathbb{R}^2$ on S^n.
    (c) Prove that a necessary and sufficient condition for all the orbits of $\tilde{\varphi}$ to have dimension ≤ 1 is that A, B and I be linearly dependent, where

I is the $(n+1) \times (n+1)$ identity matrix.

(d) Suppose that A, B and I are linearly independent and that the eigenvalues of A and B are pairwise distinct. For $n = 2$ and 3 describe the orbits of $\tilde{\varphi}$ in terms of the Jordan canonical forms of A and B.

55. Let X and Y be C^1 vector fields on S^2 such that $[X, Y] = 0$. Prove that there exists $p \in S^2$ such that $X(p) = Y(p) = 0$.
Hint. If Y_t is the flow of Y and $X(p) = 0$, prove that $X(Y_t(p)) = 0$, for every $t \in \mathbb{R}$. Next show that the α and ω-limit sets of p for Y consist of singularities of X. Now use the Poincaré-Bendixson Theorem.

56. Let $f : S^2 \longrightarrow S^2$ be a diffeomorphism and X a vector field on S^2. Suppose that $f^*(X) = X$.
(a) Prove that $f \circ X_t = X_t \circ f$ for every $t \in \mathbb{R}$, where X_t is the flow of X.
(b) Show that there exists $p \in S^2$ that is simultaneously a periodic point of f and a singularity of X.
Hint. Every diffeomorphism of S^2 has a periodic point. If f is a diffeomorphism of $\mathbb{R}^2$ without periodic points, then the ω-limit set of any point is empty (Brouwer).

57. Prove that the rank of the following manifolds is one: $\mathbb{P}^3$, $S^2 \times S^1$, $\mathbb{P}^2 \times S^1$ and $S^2 \times \mathbb{R}$.

BIBLIOGRAPHY

[1] R. ABRAHAM e J. ROBBIN – *Transversal mappings and flows,* W. A. Benjamin, Inc., N.Y. 1967.

[2] D. V. ANOSOV – *Geodesic flows on closed riemannian manifolds of negative curvature,* Proc. Steklov Math. Inst. vol. 90, 1967, An. Math. Soc. Transl., 1969.

[3] V. I. ARNOLD e A. AVEZ – *Problèmes ergodiques de la mécanique classique,* Gauthier-Villars, Paris, 1967.

[4] P. BOHL – *Uber die hinsichtlich der unabhängigen und abhängigen variabeln periodisch differentialgleichung erster Ordnung,* Acta Math. 40 (1916), pg. 321-336.

[5] C. CAMACHO – *Structural stability theorems for integrable differential forms on 3-manifolds,* Topology vol. 17, 1978, pg. 143-155.

[6] C. CAMACHO – *Poincaré-Bendixson theorem for* $\mathbb{R}^2$*-actions.* Bol. da Soc. Bras. de Mat., vol. 5, n.° 1, 1974, pg. 11-16.

[7] A. DENJOY – *Sur les courbes définies par les équations différentielles à la surface* du tore, J. Math. Pures Appl., vol. 11, 1932, pg. 333-375.

[8] A. DENJOY – *Un demi-siècle de notes communiquées aux académies,* vol. II, champ réel, Gauthier-Villars, Paris, pg. 575-585.

[9] A. H. DURFEE – *Foliations of odd-dimensional spheres,* Ann. of Math., vol. 96, 1972. pg. 407-411.

[10] C. EHRESMANN e W. SHIH – *Sur les espaces feuilletées: théorème de stabilité,* C. R. Ac. Sc. Paris, vol. 243, 1956, pg. 344-346.

[11] C. EHRESMANN – *Les connexions infinitésimales dans un espace fibré différentiable,* Colloque de Topologie, Bruxelles, june 1950, pg. 29-55.

[12] W. FRANZ – *Topologia general y algebraica,* Seleciones Cientificas, Madrid., Translated from German: Topologie, I: Allgemeine topologie e II: Algebraische topologie.

[13] R. GERARD e A. SEC – *Feuilletages de Painlevé,* Bull. Soc. Math. France, vol. 100, 1972, pg. 47-72.

[14] GOTTSCHALK-HEDLUND – *Topological Dynamics,* Am. Math. Soc. Coloquium Publ., vol. XXXVI, 1955.

[15] M. L. GROMOV – *Stable mappings of foliations into manifolds,* Math. USSR Isv., vol. 3, 1969, pg. 671-694.

[16] A. HAEFLIGER – *Sur les feuilletages des variétés de dimension n par des feuilles fermées de dim. n*-1, Colloque de topologie de Strasbourg, 1955.

[17] A. HAEFLIGER – *Structures feuilletées e cohomologie à valeur dans un faisceau de groupoids,* Comment. Math. Helv., vol. 32, 1958, pg. 249-329.

[18] A. HAEFLIGER – *Variétés feuilletées,* Ann. Scuola Norm. Sup Pisa, série 3, vol. 16, 1962, pg. 367-397.
[19] G. HECTOR – *Quelques exemples de feuilletages espèces rares,* Ann. de L'Institut Fourier, vol. 26, fasc. 1, 1976, pg. 238-264.
[20] HIRSCH-SMALE – *Differential equations, dynamical systems and linear algebra,* Academic Press, 1974.
[21] E. KAMKE – *Über die partielle differentialgleichung*

$$f(x,y)\frac{\partial z}{\partial x} + g(x,y)\frac{\partial z}{\partial y} = h(x,y) ,$$

I. Math. Zeitschrift, vol. 41, 1936, pg. 56-66. II: Math. Zeitschrift, vol. 42, 1936, pg. 287-300.
[22] W. KAPLAN – *Regular curve – families filling the plane,* I: Duke Math J., vol. 7, 1940, pg. 154-155. II: Duke Math. J., vol. 8, 1941, pg. 11-46.
[23] H. KNESER – *Reguläre kurvenscharen auf den ringflächen,* Math. Ann. vol. 91, 1924, pg. 135-154.
[24] H. B. LAWSON – *Codimension-one foliations of spheres,* Ann. of Math., vol. 94, 1971, pg. 494-503.
[25] W. LICKORISH – *A foliation for 3-manifolds,* Ann. of Math., vol. 82, 1965, pg. 414-420.
[26] E. L. LIMA – *Introdução à topologia diferencial,* Notas de Matemática # 23, IMPA, 1961.
[27] M. HIRSCH – *Differential Topology,* Graduate Texts in Math., Springer Verlag, 1976.
[28] E. L. LIMA – *Commuting vector fields on S^3*, Ann. of Math., vol. 81, 1965, pg. 70-81.
[29] M. K. GREENBERG – *Lectures on Algebraic Topology,* Mathematics Lecture Notes, Benjamin, N.Y. 1967.
[30] A. LINS NETO – *Local structural stability of C^2 integrable 1-forms,* Ann. Inst. Fourier, Grenoble, vol. 27, fasc. 2, 1977, pg. 197-225.
[31] B. MALGRANGE – *Frobenius avec singularités, I. Codimension un,* Publ. Math. de IHES, vol. 46, 1976, pg. 163-173.
[32] W. S. MASSEY – *Algebraic topology: an introduction,* Harcourt, Brace & World, 1967.
[33] A. S. MEDEIROS – *Structural stability of integrable differential forms,* Proc. of III ELAM, Lect. Notes in Math., vol. 597, Springer, 1977, pg. 395-428.
[34] W. MELO e J. PALIS – *Geometric Theory of Dynamical Systems,* Springer Verlag, 1982.
[35] J. MILNOR – *Topology from the differentiable viewpoint,* University Press of Virginia, 1965.
[36] J. MILNOR – *Morse theory,* Princeton University Press, 1963.
[37] R. MOUSSU – *Sur l'existence d'intégrales premières pour un germe de forme de Pfaff,* Ann. Inst. Fourier, Grenoble, vol. 26, fasc. 2, 1976, pg. 171-220.
[38] R. MOUSSU e F. PELLETIER – *Sur le théorème de Poincaré-Bendixson,* Ann. Inst. Fourier, vol. 24, fasc. 1, 1974, pg. 131-148.
[39] J. MUNKRES – *Elementary differential topology,* Princeton University Press, 1966.
[40] S. P. NOVIKOV – *Topology of foliations,* Trans. Moscow Math. Soc., 1965, pg. 268-304.
[41] P. PAINLEVÉ – *Leçons sur la théorie analytique des équations différentielles professées à Stockholm,* Hermann, 1895.
[42] C. F. B. PALMEIRA – *Open manifolds foliated by planes,* Ann. of Math., vol. 107, # 1, 1978, pg. 109-131.

[43] M. M. PEIXOTO – *Teoria geométrica das equações diferenciais,* 7.° Col. Bras., de Mat., IMPA, 1969.

[44] A. PHILLIPS – *Foliations of open manifolds.*
I: Comment. Math. Helv., vol. 43, 1968, pg. 204-211.
II: Comment. Math. Helv., vol. 44, 1969, pg. 367-370.

[45] J. PLANTE – *A generalization of the Poincaré-Bendixson theorem for foliations of codimension one,* Topology, vol. 12, 1973, pg. 177-182.

[46] G. REEB – *Sur certains propriétés topologiques des variétés feuilletées,* Actual. Sci. Ind., # 1183, Hermann, Paris, 1952.

[47] G. REEB – *Les espaces localement numériques non separés et leurs applications à un problème classique,* Coll. de top. de Strasbourg, julho de 1955.

[48] H. ROSENBERG, R. ROUSSARIE e D. WEIL – *A classification of closed orientable 3-manifolds of rank two,* Ann. of Math., vol. 91 # 2, 1970, pg. 449-464.

[49] R. SACKSTEDER – *On the existence of exceptional leaves in foliations of codimension one,* Ann. Ins. Fourier, Grenoble, vol. 14, fasc. 2, 1964, pg. 221-225.

[50] R. SACKSTEDER – *Foliations and pseudo-groups,* Amer. J. Math., vol 87, 1965, pg. 79-102.

[51] H. SEIFERT e W. THREFALL – *Leciones de Topologia,* Inst. Jorge Juan de Mat., Madrid, 1951. Translated from German: *Lehrbuch der topolgie.*

[52] M. SPIVAK – *Calculus on manifolds,* W. A. Benjamin, N. Y., 1965.

[53] N. STEENROD – *The topology of fibre bundles,* Princeton Univ. Press, 1951.

[54] I. TAMURA – *Every odd dimensional homotopy sphere has a foliation of codimension one,* Comment. Math. Helv., vol. 47, 1972, pg. 73-79.

[55] W. THURSTON – *A local construction of foliations for three-manifolds,* Proc. Symp. in Pure Math., Stanford Univer. 1973, Diff. Geometry A. M. S., vol. 27, parte 1, 1975, pg. 315-319.

[56] W. THURSTON – *The theory of foliations of codimension greater than one,* Comm. Math. Helv., vol. 49 1974, pg. 214-231.

[57] W. THURSTON – *A generalization of the Reeb stability theorem,* Topology, vol. 3, # 4, 1974, pg. 347-352.

[58] A. H. WALLACE – *Modifications and cobounding manifolds,* Can. J. Math., vol. 12, 1960, pg. 503-528.

[59] T. WAZEWSKI – *Sur l'équation* $P(x,y)dx + Q(x,y)dy = 0$, Mathematica, vol. 8, 1934, pg. 103-116 e vol. 9, pg. 179-182.

[60] J. WOOD – *Foliations on 3-manifolds,* Ann. of Math., vol. 89 # 2, 1969, pg. 336-358.

[61] C. ZEEMAN – *Uma introdução informal à topologia das superficies,* Monografias de Matemática, vol. 20, IMPA, 1975.

[62] H. B. LAWSON – *Foliations,* Bull. of the Ann. Math. Soc., vol. 80 # 3, 1974, pg. 369-418.

[63] J. MILNOR – *Foliations and foliated vector bundles,* Preprint, Princeton, 1970.

[64] G. REEB – *Feuilletages, résultats anciens et nouveaux,* Univ. of Montreal, 1972.

[65] G. REEB – *Structures feuilletées,* Proc. of Diff. Top., Fol. and Gelfand-Fuks Cohom., Lecture Notes in Math., vol. 652, 1976, pg. 104-113.

INDEX